RED BLOOD CELL
AGING

ADVANCES IN EXPERIMENTAL MEDICINE AND BIOLOGY

Recent Volumes in this Series

RED BLOOD CELL AGING

Edited by

Mauro Magnani

University of Urbino
Urbino, Italy

and

Antonio De Flora

University of Genoa
Genoa, Italy

PLENUM PRESS • NEW YORK AND LONDON

Library of Congress Cataloging in Publication Data

Red blood cell aging / edited by Mauro Magnani and Antonio De Flora.
     p.   cm. — (Advances in experimental medicine and biology; v. 307)
   Proceedings of a symposium held Sept. 24–26, 1990, in Urbino, Italy.
   Includes bibliographical references and index.
   ISBN 0-306-44021-0
   1. Erythrocytes — Aging — Congresses. I. Magnani, Mauro. II. De Flora, Antonio.
III. Series.
   [DNLM: 1. Erythrocyte Aging — physiology — congresses. 2. Erythrocytes —
physiology — congresses. W1 AD559 v.307 / WH 150 R287 1990]
QP96.R27   1991
599′.0113 — dc20
DNLM/DLC   91-24227
for Library of Congress   CIP

Proceedings of an international symposium on Red Blood Cell Aging,
held September 24–26, 1990, in Urbino, Italy

ISBN 0-306-44021-0

© 1991 Plenum Press, New York
A Division of Plenum Publishing Corporation
233 Spring Street, New York, N.Y. 10013

Printed in the United States of America

PREFACE

The mammalian erythrocyte is a very suitable model for the study of aging at the cellular and molecular level. It is not only a matter of apparent simplicity in terms of biochemistry, biophysics and physiology but more likely this cell offers a great possibility for elucidating some basic problems in the process of aging. In fact, nowadays, it is possible to follow individual cells all along their life span in circulation, it is possible to obtain these cells when young, middle aged or old and it is possible to obtain cells from individuals of defined ages and transfuse them into compatible recipients to investigate the role of the environment where the cell lives, and finally it is possible to easily manipulate the red cell content in terms of enzymatic activities and/or metabolic properties to investigate the possible effect of these manipulations on cell survival.

This book, Red Blood Cell Aging, is based on a symposium held in Urbino, Italy, at the end of 1990 and examines the impact of age on the membrane, metabolism, structural and enzymatic proteins of mammalian erythrocytes. The various contributions to this symposium not only described those processes of aging which affect the cell but also provided a nearly complete picture of the event(s) and mechanism(s) that every day permits to recognize among 25 trillion circulating red cells (in an average adult) that 1 percent that have reached the end of their 120 day life span in circulation.

However, once again, several reports end with different conclusions suggesting that while significant progress in understanding red cell aging has been made we still do not know cell senescence completely. Even if we must admit that our knowledge of the field is still fragmentary, a number of reports in this book prove to be of value for the diagnosis and understanding of several diseases which directly involve the red cell as in cases of enzyme defects, malaria infection and sickle cell anemia. Furthermore, the majority of the reports in this book provide support to the concept mentioned above that the erythrocyte may be a useful model for elucidating some basic phenomena of aging in more complex cells and intact organisms.

We would like to dedicate this book to the late Professor Giorgio Fornaini who will be remembered as a distinguished pioneer in this field.

Mauro Magnani
Antonio De Flora

CONTENTS

RETICULOCYTE MATURATION

MEMBRANES

ENZYMOLOGY

THE SENESCENT RED CELL AND ITS REMOVAL

RETICULOCYTE MATURATION

SYNTHESIS OF THE TRANSFERRIN RECEPTOR IN PERIPHERAL SHEEP RETICULOCYTES:

EVIDENCE FOR INCOMPLETE OLIGOSACCHARIDE PROCESSING

Jinhi Ahn and Rose M. Johnstone

Department of Biochemistry
McGill University
Montreal, Canada

INTRODUCTION

Studies involving red cells have provided generations of investigators with new insights into the behaviour of more complex cells. This has been particularly true in the study of membrane proteins and their functions, as well as the organization of the cytoskeleton. The behaviour of the red cell is also striking in its development and maturation, undergoing vast changes in composition, appearance and variety of catalytic functions in its voyage from stem cell to circulating erythrocyte. In fact, it may almost be true to say that the red cell continuously changes until its senescent properties are recognized and it is eliminated from the circulation.

In the peripherial circulation there are two highly significant transitions for red cells, the cross-over between reticulocyte and erythrocyte and the cross-over between erythrocyte and frankly senescent cell. The presence or absence of a minimal amount of reticulum has traditionally been the mark of the reticulocyte to erythrocyte transition. But other distinctive changes also occur. As the reticulocyte develops into the mature erythrocyte, significant changes take place in the plasma membrane-associated activities which may serve to delineate the transition between the two stages.

The selective nature of the losses in the plasma membrane is very striking. A cursory examination of the SDS-PAGE protein pattern of sheep red cell membranes fails to indicate the specific nature of the losses, although it is evident that less protein is obtained in membranes of mature cells than in an equivalent number of reticulocytes. However, when specific functions are assayed in red cells from a variety of mammalian species, the selective nature of the maturation-associated loss of membrane functions is striking. For example, pig erythrocytes (1), but not their reticulocytes, are devoid of glucose transporters. Dog red cells are devoid of Na^+/K^+ ATP-ase (2), an enzyme which is present in the corresponding reticulocytes

(3). Sheep reticulocytes lose most of their nucleoside transporters (3) as they mature, but retain substantial numbers of glucose transporters, which are lost in the course of maturation (5). This latter function could also be used to measure the change between the reticulocyte and the erythrocyte.

RESULTS

To obtain some insight into the mechanisms of selective protein externalization, we (6) followed the sheep transferring receptor with gold-conjugated antibodies. We established that the transferrin receptor is segregated into small bodies of 50 nm contained inside larger sacs of 300-800 nms (Fig. 1). Upon fusion of the sac with the plasma membrane, the vasicular contents are released. Studies show similar results with rat reticulocytes (7,8). These 50 nm vesicles, or exosomes, may be retrieved either from the culture medium of a suspension of reticulocytes or from the circulation of phlebotomized animals. The data show (Fig. 2A) that in vesicles, the transferrin receptor (94kD) is the single major protein. There is no Band 3 in the exosomes. It is obvious that the peptide composition of the exosomes is quite distinct from that of either mature red cell or reticulocyte membranes. In contrast, the relative proportions of the phospholipids, including the high sphingomyelin (Sph) content (Fig. 2B) are identical in the plasma membranes and exosomes. In the sheep exosomes, ten different plasma membrane functions have been detected. No evidence was obtained for the presence of soluble enzymes or mitochondrial functions. Although lysosomal activities were present in the same exosomal pellet as the plasma membrane functions, further studies using antitransferrin receptor-coated magnetic beads (8) revealed that the lysosomal activities are not included in the exosome fraction which carries the transferrin receptor. Plasma membrane functions (e.g. nucleoside transporter and cholinesterase) whose activity in the membrane also descreased during maturation (3,9) are recovered, at least in part, with the transferrin receptor-bearing vesicles (Table 1). These studies suggested that plasma membrane functions are recognized and processed for externalization by a common mechanism. However, differences in the time frame for externalization of the different functions suggest independent targeting of individual proteins for externalization.

The transferrin receptor is lost from the cell during the interval when protein synthesis is still ongoing in peripheral sheep reticulocytes and in reticulocytes from other species (10). The question arose whether the transferrin receptor is still being replaced in these peripheral reticulocytes.

In agreement with Cox et al. (11) we showed that the transferrin receptor is synthesized de novo in peripheral sheep reticulocytes (Fig. 3). The data show that ^{35}S-methionine is incorporated into che transferrin receptor and that 20 μM hemin stimulates that incorporation. However, the rate of loss of the synthetic machinery is more rapid than the elimination of the pre-existing transferrin receptor. De novo synthesis of receptor virtually ceases after 3-4 hrs incubation of the cells in vitro (Fig. 3) whereas ∼90% loss of transferrin receptor from the cells requires ∼24-48 hrs (12).

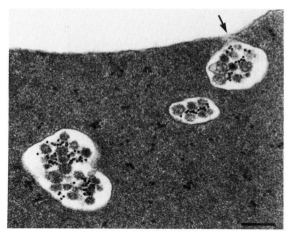

Fig. 1 Multivesicular bodies containing the transferrin receptor in sheep reticulocytes. Peripheral sheep reticulocytes were coated at 4°C with an antitransferrin receptor antibody followed by incubation at 4°C with a second antibody conjugated to colloidal gold as described (6). The cells were incubated at 37°C for 3 hr prior to fixation. Note that the colloidal gold is at the surface of the 50nm bodies inside the sacs (MVB). The arrow points to a fusion site between MVB and the plasma membrane. The bar represents 200 nm.

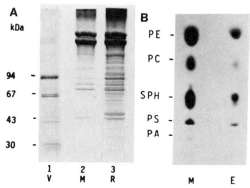

Fig. 2 Protein and lipid composition of exosomes. (A) Protein composition. Coomassie Blue stain of SDS-PAGE.
Lane 1- Total exosomes; 2- Erythrocyte membranes; 3- Reticulocyte membranes
(B) Lipid content. Lipids were extracted from membranes and 100,000 x g pellets and chromatographed by TLC. (From Johnstone et al. Vesicle Formation During Reticulocyte Maturation. J. Biol. Chem. 262, 9412-9420 1987).

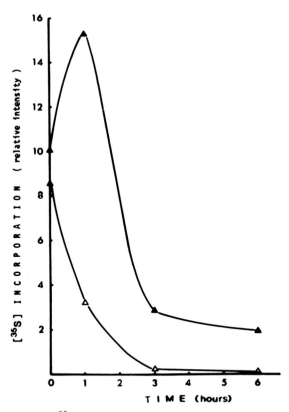

Fig. 3 Incorporation of ^{35}S-methionine into the transferrin receptor of sheep reticulocytes. The radioautograms of the immunoprecipitates were scanned with a laser densitometer. Closed symbols: 20μM hemin. Open symbols – No Hemin. (From Ahn and Johnstone. Maturation Associated Loss and Incomplete de novo Synthesis of the Transferrin Receptor in Peripheral Sheep Reticulocytes. J. Cell. Physiol. 140, 107–118. 1989).

Most of the newly formed receptor fails to reach the plasma membrane (13). Thus, labelled receptor cannot be removed by treatment of the cell surface with proteolytic enzymes, even after chase periods of 10 hrs or more. The data show (Fig. 4) that treatment of the cells with trypsin at 0°C or 37°C reduces the amount of immunoprecipitated, preformed receptor, but not the ^{35}S-labelled receptor. Moreover, the newly formed receptor is smaller (∼2 kD) than the pre-existing receptor (Fig. 5), presumably due to differences in carbohydrate processing. Prolonged incubation (4–5 hrs) does not alter the size of the receptor. Treatment of immunoprecipitates (native and ^{35}S-labelled) with TFMS (trifluoromethanesulfonate) which cleaves off all except the innermost carbohydrate residue, leaves ^{35}S-labelled and Coomassie Blue stained peptides of the same size, indicating that the peptide component of the receptor is complete (Fig. 5). Since treatment with endoglycosidase H also reduces the ^{35}S-labelled receptor (but not the pre-existing receptor) to the same size as TFMS treatment, the absence of complex oligosaccharides on the newly formed receptor is suggested. Cleveland mapping (14) detected an ^{35}S-labelled peptide whose migration did not correspond to that of an unlabelled peptide. After deglycosylation, the same peptide pattern on the gel was obtained with ^{35}S-labelled and unlabelled peptides. These data argue that only a single carbohydrate chain is incomplete on the new receptor.

A number of experiments suggest that the newly formed and preexisting receptors are in different cellular pools. (1) The two receptor populations respond differently to the presence of hemin and non-heme iron in the culture medium (13). Whereas hemin (20 μM) but not free iron increases the rate of loss of the pre-existing receptor (Fig. 6A), non-heme iron is more effective in increasing the loss of ^{35}S-labelled receptor (Fig. 6B). (2) In contrast with the pre-existing receptor, no detectable ^{35}S-labelled receptor is found in the released exosomes. It should be noted, however, that the newly synthesized receptor binds transferrin and can be shown to be a dimer on non-reducing gels (13). (3) To probe further the structural differences in the pre-existing and newly formed receptor, the cell lysates, containing ^{35}S-labelled receptor were centrifuged on Percoll gradients. The peaks of ^{35}S-labelled transferrin receptor migrated at a slightly higher density than the unlabelled receptor. The peak of radioactivity (13) was coincident with the peak of activity for galactosyl transferase, a Golgi enzyme.

The observations presented thus far suggest that the transferrin receptor is incompletely processed and lacking complex carbohydrate chains which are known to exist (at least on the human receptor). Evidence also exists for complex carbohydrate on the sheep receptor since the native receptor is more readily eluted with α-methyl mannoside from a Con A column than the ^{35}S-labelled receptor.

Table 1. Segregation of Plasma Membrane and Lysosomal Activities in Transferrin Receptor Containing Exosomes

Activity	Site of origin	% Removed by Anti Transferrin Receptor Containing Beads
[125]I-Transferrin Binding	Plasma Membrane & Endosomes	80–90%
Acetylcholinesterase	Plasma Membrane	40–50%
[3]H NBMPR-Binding	Plasma Membrane	40–50%
Phosphatase	Plasma Membrane	40–50%
N-acetylglucosaminidase	Lysosomes	~ 4%
ß-glucuronidase	Lysosomes	≤ 1%
Cathepsins	Lysosomes	Not Detectable

Exosomes from sheep reticulocyte cultures were bound to transferrin receptor-coated magnetic beads as described (8) and the activities associated with the beads are expressed as a percent of the total activity in the exosome fraction.

7

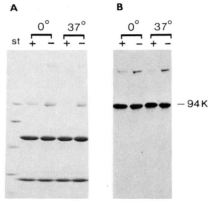

Fig. 4 The newly formed receptor is trypsin insensitive. After incorporation of ^{35}S–methionine by sheep reticulocytes, the cells were incubated with (+) 0.1% trypsin or without (−) for 30 min at 0° or 37°C. After isolation of membranes, the transferrin receptor was immunoprecipitated and the immunoprecipitates subjected to SDS PAGE and autoradiography. A. Coomassie Blue Stains B. Autoradiograms. The band at 94 kD is the transferrin receptor. (From Ahn and Johnstone. Maturation Associated Loss and Incomplete de novo Synthesis of the Transferrin Receptor in Peripheral Sheep Reticulocytes. J. Cell. Physiol. 140, 107–118 1989).

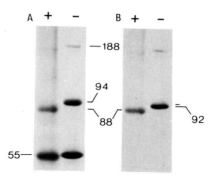

Fig. 5. The ^{35}S–labelled deglycosylated peptide is the same size as the native deglycosylated peptides. After ^{35}S–labelling of sheep reticulocytes and immuno-precipitation with anti-transferrin receptor antibody, a fraction of the immunoprecipitated material was treated with TFMS (+).
   A. Left panel: Change in mobility of the native (Coomassie Blue stained) receptor after TFMS treatment.
   B. Right panel: ^{35}S–labelled receptor from the same experiment. Note that after TFMS, a peptide of ∼88 kD is obtained in both cases. (From Ahn and Johnstone. Maturation Associated Loss and Incomplete de novo Synthesis of the Transferrin Receptor in Peripheral Sheep Reticulocytes. J. Cell. Physiol. 140, 107–118. 1989).

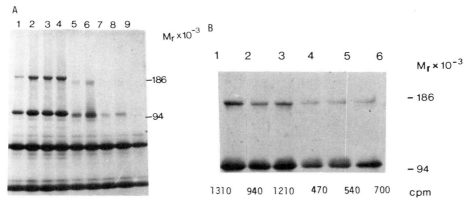

Fig. 6 Heme stimulates externalization of the native transferrin receptor. Iron stimulates the loss of ^{35}S-labelled receptor. After incubation of reticulocytes for 18 hr at 37°C with the additions below, vesicles and plasma membranes were harvested, immunoprecipitated and subjected to SDS-PAGE.

(6A). Left panel-Coomassie Blue stains. Lanes 1-4 Plasma membranes; 5-9 Vesicles. 1,6 + hemin 20 µM; 2,7 + SIH 100 µM; 3,8 FeSIH + 100 µM; 4,9 FeSIH + 100 µM + 1 mM DOH; 5 - No additions. Both the dimer (186kD) and the monomer (94kD) of the receptor are seen in the immunoprecipitates.

(6B). Right panel-audioradiographs immunoprecipitates of the ^{35}S-labelled transferrin receptor from the plasma membrane. Lane 1 - Initial labelling; 2-6 - 24 hr chase with unlabelled methionine; 2 - No additions; 3 + 0.5 mM PIH; 4 + 0.5 mM FePIH; 5 + 0.5 mM FePIH 1 mM DOH; 6 + 20 µM hemin;
The cpm in the sample (below each lane) represent the sum of the radioactivity in the monomer and dimer of the receptor. (From Ahn and Johnstone. Maturation Associated Loss and Incomplete de novo Synthesis of the Transferrin Receptor in Peripherial Sheep Reticulocytes. J. Cell. Physiol. 140, 107-118. 1989).

To further our understanding of the defective carbohydrate processing, reticulocytes were labelled with ^{3}H-mannose. The data show that ^{3}H-mannose can still be incorporated into total membrane proteins of peripheral sheep reticulocytes (Fig. 7A). Only a small fraction of the ^{3}H-label is found in the transferrin receptor, the majority residing in peptides which immunoblot with an antibody against Glycophorin A (Fig. 7B). Pronase digestion of the membrane bound proteins, followed by size separation of the ^{3}H-labelled glycopeptides, shows that the majority of the label ($\geq$90%) is in a residue containing eight or nine mannose residues. Inadequate incorporation of ^{3}H-mannose into the transferrin receptor itself precludes this type of analysis directly on the receptor. Assuming however, that ^{3}H-mannose incoporation into the receptor parallels the global incorporation of ^{3}H-mannose, these observations would suggest that the newly formed receptor is probably retained in the ER.

9

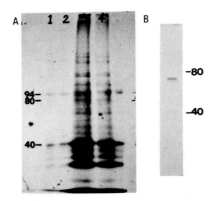

Fig. 7. Incorporation of ^{3}H–mannose into membrane proteins and the transferrin receptor of sheep reticulocytes.
Sheep reticulocytes were incubated with ^{3}H–mannose for 4 hr at 37°C. An aliquot of solubilized cell membrane was subjected to SDS–PAGE and fluorography. An aliquot of membranes was immuno-precipitated with antitransferrin receptor antibody and the immuno-precipitates subjected to SDS–PAGE and fluorography.
A – Left Panel: Lanes 1 & 2; duplicate immunoprecipitates (10 x more membrane than in 3 & 4); Lane 3; Tx–100 extract of total membranes; Lane 4; Tx–100 supernatant after immunoprecipitation of the transferrin receptor.
B – Right Panel: Immunoblot of plasma membranes with anti-Glycophorin A

Many studies of glycoprotein synthesis (15) have established that after removal of three glucose residues and a mannose residue, a glycopeptide with eight mannose residues leaves the ER and that further carbohydrate modification occurs in the Golgi. Thus our observations suggest that either the glycoproteins are retained in the ER or have not progressed beyond the cis Golgi.

To investigate these possibilities, attempts have been made to introduce a mouse liver microsomal fraction into lyzed reticulocytes (following the procedure of Magnani et al. (16) or to fuse reticulocytes with cultured cell lines. It was hoped that by introducing intracellular organelles, the completion of the receptor would be achieved and the reasons for lack of terminal receptor processing would be established. However, this has not proved possible.

Finally we have addressed the question whether incomplete processing and insertion of the transferrin receptor into the plasma membrane pertains to other newly synthesized membrane proteins. Glycophorin A is still synthesized by sheep reticulocytes. Like the transferrin receptor it does not appear to be inserted into the plasma membrane as judged by insensitivity of the ^{35}S–labelled peptide to chymotryptic digestion from the cell surface. These observations suggest that an early event in the maturation of the sheep reticulocyte is the loss of functions involved in translocating proteins from ER to Golgi or through the Golgi stacks so that membrane proteins synthesized late fail to be targeted to their natural sites.

DISCUSSION

Peripheral reticulocytes have long been used to study the synthesis of a soluble protein, namely hemoglobin. Recent studies have also shown that a transmembrane protein, lost during maturation, is made in peripheral sheep and rabbit reticulocytes. In the sheep, the synthesis in a 2 hr period is less than 0.1% of the pre-existing receptor. Since protein synthesis ceases by 4-6 hours of incubation of the peripheral reticulocytes, the newly formed receptor quantitatively makes little contribution to the total pool of receptor. Moreover, in sheep reticulocytes, the new receptor fails to be completed or targeted to the plasma membrane.

The receptor appears to be incompletely glycosylated. It is unlikely that the incomplete glycosylation is the reason for the lack of targeting to the plasma membrane. It has been shown in other systems (17,18,19) that inhibitors of late stage glycosylation, such as deoxymannojirimycin, do not prevent the transferrin receptor from appearing in the plasma membrane. Only treatment with tunicamycin has, in some instances (20), but not al (7), prevented targeting to the plasma membrane.

Our studies have suggested that the factors required to direct vesicular traffic between ER and Golgi or through the Golgi stacks are lost, resulting in retention of the partly glycosylated receptor in the internal membrane network. Glycophorin A, also synthesized by the peripheral sheep reticulocytes appears to suffer a similar fate.

The resealed reticulocyte might provide a useful assay system for studying factors required for translocation of plasma membrane proteins through the internal membrane network en route to the surface membrane.

Abbreviations:

PIH   pyridoxal isonicotinoyl hydrazone
SIH   salicylaldehyde isonicotinoyl hydrazone
DOH   4,6-dioxoheptonoate
TFMS   trifluoromethanesulfonic acid

Supported by grants from MRC (Canada) and National Institute of Health (USA).

REFERENCES

1. R. B. Zeidler and H. D. Kim, Pig Reticulocytes IV. In vitro maturation of naturally occurring reticulocytes with permeability loss of glucose, J. Cell Physiol. 112:360 (1982).

2. M. Inaba and Y. Maede, Na, K-ATP-ase in dog red cells. Immunological identification and maturation associated degradation by the proteolytic system, J. Biol. Chem. 261:16099 (1986).

3. S. M. Jarvis and J. D. Young, Nucleoside transport in sheep reticulocytes and fetal erythrocytes: A proposed model for the nucleoside transporter, J. Physiol.(Lond) 324:47 (1982).

4. R. M. Johnstone, M. Adam, J. R. Hammond, L. Orr and C. Turbide, Vesicle formation during reticulocyte maturation, J. Biol. Chem., 262:9412 (1987).

5. J. H. Jandl and J. H. Katz, The cell to plasma transferrin cycle, J. Clin. Investig. 42:314 (1963).

6. B. T. Pan, K. Teng, C. Wu, M. Adam and R. M. Johnstone, Electron microscopic evidence for externalization of the transferrin receptor in vesicular form in sheep reticulocytes, J. Cell. Biol. 10: 942 (1985).

7. C. Harding, J. Heuser and P. Stahl, Endocytosis and intracellular processing of transferrin and colloidal gold transferrin, Eur. J. Cell. Biol. 35:256 (1984).

8. R. M.Johnstone, A. Bianchini and K. Teng, Reticulocyte maturation and exosome release: transferrin receptor containing exosomes show multiple plasma membrane functions, Blood 74:1844 (1989).

9. C. M. Rennie, S. Thompson, A. C. Parker and A. Maddy, Human erythrocyte fraction in "Percoll" density gradients, Clin. Chim. Acta 98:119 (1979).

10. J. B. Lingrel H. Borsook, A comparison of amino acid incorporation into hemoglobin and ribosomes of marrow erythroid and circulating reticulocyte of severely anemic rabbits, Biochem. 2:304 (1963).

11. T. M. Cox, M. W. O' Donnell and I. M. London, Biosynthesis of the transferrin receptor in rabbit reticulocytes, J. Clin. Investig. 76:2144 (1985).

12. M. Adam, C. Wu, C. Turbide, J. Larrik and R. M. Johnstone, Evidence for a pool of non-recycling transferrin receptors in peripheral sheep reticulocytes, J. Cell. Physiol. 127:8 (1986).

13. J. Ahn and R. M. Johnstone, Maturation associated loss and incomplete de novo synthesis of the transferrin receptor in periferial sheep reticulocytes: Response to heme and iron. J. Cell. Physiol. 140:107 (1989).

14. D. W. Cleveland, S. G. Fischer, N. W. Kirschner and U. K. Laemmli, Peptide mapping by limited proteolysis in sodium dodecyl sulphate and analysis by gel electrophoresis. J. Biol. Chem. 252: 1102 (1972).

15. R. Kornfeld, S. Kornfeld, Assembly of asparagine linked oligosaccharides. Annual Rev. of Biochem. 54:631 (1985).

16. M. Magnani, L. Rossi, M. Bianchi, G. Serafini and V. Stocchi, Red blood cell loading with hexokinase and hexokinase inactivating antibodies. A new strategy for studying the role of enzymes in red cell metabolism and removal. Biomedica Biochimica Acta, 49:S149 (1990).

17. M. B. Omary and I. S.Trowbridge, Synthesis of human transferrin receptor in cultured cells, J. Biol. Chem. 256:12888 (1986).

18. M. D.Snider and O. C. Rogers, Membrane traffic in animal cells. Cellular glycoproteins return to the site of Golgi mannosidase, J. Cell. Biol. 103:265 (1986).

19. E. Ralton, H. J. Jackson, M. Zanoni and P. A. Gleeson, Effect of glycosylation inhibitors on the structure and function of the murine transferrin receptor, Eur. J. Biochem. 186:637 (1989).
20. C. Reckhow and C. Enns, Characterization of the transferrin receptor in tunicamycin-treated A431 Cells. J. Biol. Chem. 263:7297 (1988).

THE LOSS OF ENZYME ACTIVITY FROM ERYTHROID CELLS DURING MATURATION

David R. Thorburn[1,2] and Ernest Beutler[1]

[1]Department of Molecular & Experimental Medicine
Research Institute of Scripps Clinic
La Jolla, California, U.S.A.
[2]Now at the Murdoch Institute for
Research into Birth Defects
Royal Children's Hospital
Melbourne, Australia

It has been known for over thirty years that the activities of certain "age-dependent" enzymes decline during the circulatory life-span of erythroid cell (1). The kinetics of this decay have usually been assumed to be a simple exponential function, that is, occurring continually throughout the life of the cell. In this view, erythrocyte death was often thought to be determined by one or more critical enzymes decaying to a threshold level that was too low to sustain an adequate metabolic rate. Estimates of enzyme half-lives have been calculated from studies that fractionate erythrocytes into groups of different mean cell density (2,3). Although it is clear that red cell density increases with age, numerous recent studies have shown that there is not a simple correlation between age and density (4,5). This is because the major part of the change in density probably occurs very early in maturation, and there appears to be a substantial range of initial density. Density-fractionation and other in vitro methods are therefore unable to resolve cell preparation into fractions of uniform cell age, and cannot provide an accurate picture of enzyme decay.

IN VIVO STUDIES OF RETICULOCYTE MATURATION

Transient Erythroblastopenia

Given the problems of in vitro methods, we have investigated several in vivo systems that shed light on the patterns of loss of erythrocyte enzyme activity. Subjects with reticulocytosis have dramatically elevated activities of age-dependent enzymes, such as hexokinase, glucose 6-phosphate dehydrogenase (G6PD) and pyruvate kinase (6). Some surprising results were obtained, however, when subjects with depressed bone marrow function were

studied. Transient erythroblastopenia of childhood is a disorder in which
the marrow ceases to produce new erythroid cells for periods of up to
several months. Patients present very low reticulocyte counts and hemoglobin
levels, but the activities of most age-related enzymes are in the normal
range (Table 1) despite the fact that these patients have produced no new
cells for many weeks. Two exceptions to this behaviour are pyrimidine-5'-
nucleotidase (P5'N) and the metabolite creatine, both of which are
well below normal. Studies of bone-marrow transplant patients with
chemotherapy and radiation-induced aplasia gave data similar to those with
erythroblastopenia. Six days after induction of aplasia, P5'N and creatine
levels were low, while other enzyme activities were normal (5)

These results cannot be explained by assuming each enzyme decay as
a single exponential function, since this would predict significantly
low activities of all age-dependent enzymes in subjects with suppressed
erythropoiesis. Apart from P5'N, other age-dependent enzymes appear to have
nearly normal activity in old cells. There must therefore be more than one
mechanism of enzyme decay. The simplest model which could describe the
behaviour of enzymes like hexokinase is a biphasic one, in which there
is a rapid loss of activity early in the life-span (for example, during
reticulocyte maturation) but relatively little loss thereafter (7).

Reticulocyte Maturation in vivo

This biphasic decay model is supported by two recent experimental
studies of in vivo aged rabbit reticulocytes. In the first study (8),
rabbits were made anemic with phenylhydrazine. Packed cells from some of the
rabbits were then used to hypertransfuse the remaining animals, so that they
had a hematocrit of ~0.6, and a reticulocyte count of ~60%. The hematocrit
remained elevated for several weeks, suppressing the release of new
reticulocytes from the marrow. The reticulocyte count therefore fell
to subnormal levels within four days, allowing the study of changes in
age-dependent enzyme activities. Four enzymes were studied and each seemed
to show a slightly different pattern of behavior (Fig. 1). Hexokinase
activity declined by more than half over the first week, but remained fairly
constant thereafter. The activity of G6PD was stable for several days, but
then fell by about half over the next ten days, before stabilising. P5'N
decayed rapidly over the first week, but then continued to fall at a slower
rate. The age-independent enzyme lactate dehydrogenase remained constant
throughout the experimental period.

One possible source of concern with this experimental system was
that the reticulocytes were induced with phenylhydrazine, which may cause
abnormal proteolysis in reticulocytes (9). In order to address this issue,
and to study a larger number of enzymes, an alternative system was also
investigated. Rabbits were made anemic by repeated bleeding, and the
reticulocyte content was increased from 20-25 % to ~ 50% by density
fractionation. The reticulocyte-rich cells were then biotinylated and
transfused into untreated rabbits. At subsequent times, the cells
were recovered using an avidin support, and the activities of a panel

of 19 enzymes were determined (10). Hexokinase, G6PD, P5'N and lactate dehydrogenase showed essentially the same behaviour as in the hypertransfusion model. Three other enzymes (aldolase, pyruvate kinase and glutamate-oxaloacetate transaminase) showed a biphasic decay pattern similar to hexokinase, while the remaining enzymes were not strongly age-dependent.

As well as providing direct evidence for multiple patterns of enzyme decay, these two studies show that the initial phase of rapid decay of enzymes like hexokinase extends beyond the reticulocyte stage into young erythrocytes. Recent studies have shown that in both human (11,12) and rabbit (13) reticulocytes, hexokinase activity is distributed between two or three major isoenzymes. The activity of one isoenzyme is relatively constant in all cells, but the activity of the other(s) decreases with increasing age or density. Thus, a likely mechanism for determining how much of a particular enzyme is to be proteolysed, is to have multiple isoenzymes differing in their proteolytic susceptibility. Factors that may determine this susceptibility include intracellular location and intramolecular signals, such as the PEST, KFER and N-end hypotheses (reviewed in (14). Direct evidence that the N-end rule may function in reticulocyte proteolysis has come from studies of polymorphic variants of hypoxanthine phosphoribosyltransferase (15). Several systems that may be involved in the removal of execess reticulocyte enzymes have been identified. These include the lipoxygenase (16), and ubiquitin systems (17), proteases such as the multicatalytic protease (18), calpain (3), and cathepsin E (19), as well as physical expulsion of enzymes or organelles (20,21). The actual substrates of each system remain unclear.

HEXOKINASE DECAY IN CELL-FREE SYSTEMS

ATP-dependent Proteolysis

We have investigated the possible mechanisms involved in destruction of rabbit reticulocyte hexokinase. Magnani and co-workers have reported and ATP- and ubiquitin-dependent proteolysis of hexokinase in cell-free systems prepared from rabbit reticulocytes (22). We found that hexokinase activity was stable in stroma-free reticulocyte lysates in the presence or absence of ATP (23). ATP-dependent loss of activity could be observed, however, under certain conditions (Fig. 2). These were (i) storage of lysates at -20°C for several weeks prior to incubation, (ii) addition to fresh lysates of an $FeCl_3$/ascorbate free-radical generating system, or (iii) use of higher concentrations of fresh lysate in the incubation system (cytosol concentrations of $\geq 20\%$ of undiluted concentration, compared to 10% in the normal system). Hexokinase activity could be stabilised by addition of free radical scavengers either prior to storage (i), or during the incubation for (ii) and (iii) (not shown). It therefore appears that prior oxidation of hexokinase is a requirement for ATP-dependent proteolysis in stroma-free reticulocyte lysates. Moreover, ATP-dependent proteolysis of hexokinase was also found in stored lysates prepared from normal rabbit erythrocytes (23), raising some doubt as to whether this mechanism is involved in hexokinase proteolysis in vivo.

Table 1. Erythrocyte Components in Subjects with Transient Erythroblastopenia of Childhood

Enzyme	Enzyme Activity (IU/g of Hemoglobin; mean ± SD)		
	TEC subjects	Child controls	Adult controls
Hexokinase	1.8±0.5 (20)	2.0±0.6 (7)	1.8±0.4
G6PD	12.4±1.8 (19)	16.3±1.5 (7)	12.1±2.1
Pyruvate kinase	12.2±2.1 (20)	14.8±3.0 (7)	15.0±2.0
Aldolase	2.9±0.8 (20)	3.6±0.4 (7)	3.2±0.9
GOT	3.9±1.1 (20)	6.2±0.9 (7)	5.0±0.9
P5'N	40.1±18.0 (20)	149±37 (7)	138±18
LDH	216±37 (20)	214±31 (7)	200±27

Hematological	Concentration (mean ± SD)		
Parameters	TEC subjects	Child controls	Adult controls
Creatine	0.52±0.02 (9)	n.d.	1.31±0.31 (M) 1.50±0.25 (F)
Hemoglobin	5.9±1.9 (13)	11.3±2.5	15.5±1.3 (M) 14.0±1.3 (F)
Hematocrit	0.18±0.06 (8)	0.34±0.04	0.47±0.04 (M) 0.42±0.03 (F)

Cell preparation and assay were as described previously (36), and GOT was assayed in the presence of pyridoxal phosphate. The number of subjects is shown in parentheses. The age range of child controls was 1-5 years, except for hemoglobin and hematocrit, where it was 0-2 years. Abbreviations: TEC, transient erythroblastopenia; GOT, glutamate oxaloacetate transaminase; LDH, lactate dehydrogenase; n.d., not determined. The units of creatine, hemoglobin, and hematocrit are μmol/g of hemoglobin, g %, and 1/1, respectively. Control values were determined in our laboratory (6,36), except for hemoglobin and hematocrit (37).

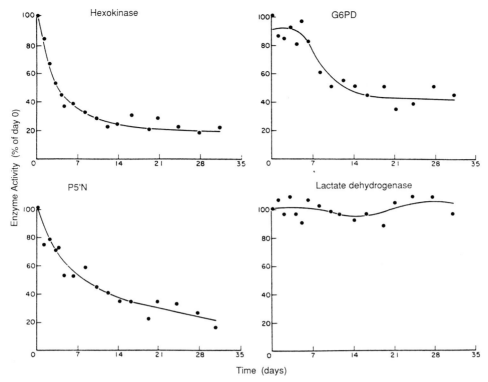

Fig. 1. Changes in red cell enzyme activities with time in hypertransfused
anemic rabbits. Enzyme activity is expressed as a percentage of the
specific activity on day zero. The initial samle was taken one hour
after completion of hypertransfusion, and each value represents the
mean of five independent experiments. The experimental details have
been reported previously (8), and the figure is modified from that
report. In this and all other experiments, white cells and platelets
were removed by cellulose filtration (38).

ATP-independent Proteolysis

Recently, we observed a substantial loss of hexokinase activity when
whole reticulocyte lysates (ie., containing organelles and membranes) were
incubated for 18 hours in the absence of ATP. This loss of activity was not
seen with whole erythrocyte lysates, stroma-free reticulocyte lysates, or
reticulocyte organelles and membranes (all greater than 80% of the initial
activity at 18h, not shown) and could be stabilised by addition of the
calcium chelator EGTA. Figure 3A shows that there is a biphasic dependence
of this loss of hexokinase activity on calcium concentration, with a maximal
loss at 0.2-0.5mM calcium. This biphasic dependence shows some similarity to
the calcium dependence of calpain, which is stimulated at low calcium levels
but inhibited at higher calcium concentrations due to increasing inhibition
by the endogenous inhibitor calpastatin (24). The actual (presumed) calcium
concentration at which decay of hexokinase activity is maximal is, however,
about 100 times higher than the expected value for maximal calpain activity
(25).

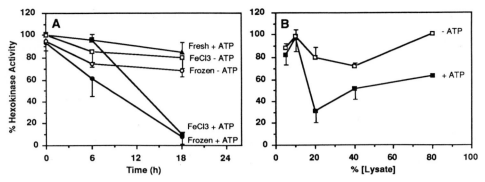

Fig. 2. Hexokinase activity in stroma-free lysates prepared from phenyl-hydrazine-induced rabbit reticulocytes. The incubations contained various amounts of reticulocyte lysate plus 50mM Tris-HCl, pH 7.5, 5mM MgCl , 1mM dithiothreitol, 25 U penicillin G and 25 μg streptomycin in a final volume of 0.5 ml ("-ATP") or the same components plus 2mM ATP, 10mM phosphocreatine and 12μg of creatine phosphokinase ("+ATP"). Incubations were at 37°C and hexokinase activity is expressed as a percentage of the specific activity measured in the fresh lysate. The data are the means ± SD of three or four independent experiments. (A) Samples contained a volume of hemolysate equivalent to 50μl of packed cells, corresponding to a final lysate concentration of 10% of that of undiluted cytosol. "Fresh" lysates were prepared on the day of the experiments, while "Frozen" lysates had been stored at -20°C for two to three weeks prior to the experiment. $FeCl_3$ lysates were fresh lysates containing a free radical generating system of 1μM $FeCl_3$ and 0.1mM sodium ascorbate. (B) Samples contained volumes of fresh lysate corresponding to between 5% and 80% of undiluted cytosol; all samples were incubated for 18h.

Western blotting of these lysates with a polyclonal antibody against rabbit reticulocyte hexokinase, showed that only about half of this loss of activity could be accounted for by loss of antigen (not shown). This phenomenon is therefore due in part to hexokinase inactivation. It can be inhibited partly, but not completely, by free radical scavengers such as glycerol (not shown). The lipoxygenase inhibitor salicylhydroxamic acid (SHAM) markedly stabilises hexokinase in this system (Fig. 3B), so hexokinase inactivation is probably due to oxidation by free radicals formed in the mitochondrial membrane. Reticulocyte lipoxygenase does not appear to be activated directly by calcium (16), although basophilic lipoxygenase shows a marked biphasic calcium dependence (26). Phospholipase A_2 , which is present in erythrocytes (27), is stimulated by calcium, and this enzyme releases free fatty acids from phospholipids. Since fatty acids are much better lipoxygenase substrates than phospholipids (16), this probably explains calcium stimulation of hexokinase inactivation. The reason that high calcium concentrations stabilise hexokinase activity is unclear, but may involve physical changes in membrane structure or fluidity. Varying the concentration of magnesium chloride in the range 0 - 4 mM has no effect on hexokinase activity in this system.

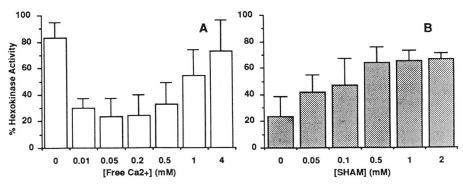

Fig. 3. Amount of hexokinase activity remaining after an 18h incubation at 37°C of whole lysate from sonicated rabbit reticulocytes. The system contained lysate corresponding to 30% of the concentration of undiluted cytosol, plus 50mM Hepes, pH 7.2, 1mM dithiothreitol, 1mM EGTA, and antibiotics as in Fig. 2, but no ATP or ATP-regenerating system. Various amounts of $CaCl_2$ were added to give estimated free Ca^{2+} concentrations of 0 to 4mM, calculated according to (39). Hexokinase activity is expressed as a percentage of the activity at 0h, and the data are means ± SD of four or five independent experiments. The incubations in (B) contained 0.2mM free Ca^{2+} and various concentrations of the lipoxygenase inhibitor salicylhydroxamic acid (SHAM).

Isoenzyme Profiles

Magnani and co-workers have separated rabbit reticulocyte hexokinase into three major isoenzymes (13) by anion exchange HPLC, and we obtained similar results using mono FPLC (Fig. 4). Two of the isoenzymes (I_a and I_b in their nomenclature) are cytosolic, and the I_a species is the major form present in mature cells. The predominant species present in reticulocytes is the I_a* isoenzyme, which is particulate, and presumably bound mostly to mitochondria. Analysis of the isoenzyme profiles during ATP-dependent and ATP-independent proteolysis might help to distinguish whether either or both systems are artefactual. The I_a isoenzyme would be expected to be preferentially spared by any proteolytic system that operates in vivo. Figure 4 shows the changes in hexokinase isoenzyme profiles during incubation of both types of hexokinase decay system. Stored stroma-free lysates have only a small amount of the I_a* isoenzyme, and when incubated with ATP, the activities of both cytosolic species decline, with the I_b species falling slightly faster. In whole lysates incubated without ATP the I_a* and I_b species both decay markedly, while the amount of the I_a species appears to be stable. The location of the main (I_a*) isoenzyme on the outer mitochondrial membrane would predispose it to free radical damage induced by lipoxygenase, since that is thought to be the predominant site of lipoxygenase activity. The isoenzyme profiles at the end of both incubations thus approach the profile expected for mature cells. FPLC analysis therefore suggests that either mechanism could be operating in vivo.

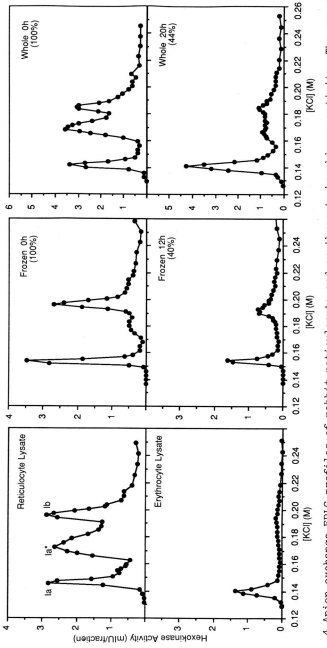

Fig. 4. Anion-exchange FPLC profiles of rabbit reticulocyte and erythrocyte hexokinase activity. The procedure was a modification of the method of Stocchi et al. (13), as described elsewhere (Thorburn and Beutler, submitted). Hexokinase activities were corrected so that each frame represents a volume of lysate equivalent to 10 mg of hemoglobin. "Frozen" samples were obtained from incubations (0 h and 12 h) of stored stroma-free reticulocyte lysate, equivalent to "Frozen + ATP" in Fig. 2A. "Whole" samples were obtained from incubations (0 h and 20 h) of whole reticulocyte lysate 0.2 mM free Ca^{2+}, as in Fig. 3. The percentage of zero-time hexokinase activity remaining in the incubated samples is shown in parentheses.

HEXOKINASE DECAY IN IN VITRO INCUBATIONS OF INTACT RETICULOCYTES

In order to determine the likely physiological roles of the two hexokinase decay systems, we have studied intact reticulocytes incubated for several days. The control system showed a slightly faster initial rate of hexokinase decay than observed in the in vivo systems ($\sim$40% of activity lost in 18 hours and $\sim$20% in 24 hours, respectively). This is probably due to the higher initial reticulocyte content ($\sim$95% compared with 50-60%) and correspondingly larger proportion of reticulocyte hexokinase in the in vitro system. The amount of hexokinase activity and antigen was determined after 18 hours incubation with a variety of compounds (Table 2). The activity and antigen values were in close agreement, suggesting that if hexokinase is inactivated in vivo, then rapid proteolysis ensues. Negligible amounts of hexokinase antigen were found in the 100,000 x g pellets of supernatants from the reticulocyte incubations (not shown). Expulsion from reticulocytes of vesicles containing mitochondrial-bound hexokinase is therefore unlikely to have been a significant contribution to the loss of hexokinase protein.

Cells incubated with 5% glycerol showed moderate stabilisation of hexokinase activity, as would be expected if hexokinase oxidation is a prerequisite for proteolysis. Incubation of cells with the lipoxygenase inhibitor SHAM stabilised hexokinase almost completely, suggesting that lipoxygenase activity is absolutely required for hexokinase decay. It is widely thought that the ubiquitin proteolytic system is largely responsible for proteolysis of mitochondrial proteins damaged by lipoxygenase (16). The marked stabilisation of hexokinase by ammonium chloride, chloroquine and 3-methyladenine suggests this is not the case for hexokinase. The former two compounds are inhibitors of lysosomal proteolysis, and 3-methyladenine is a specific inhibitor of autophagic sequestration (28). Lysosomal proteolysis therefore appears to be the major mechanism of hexokinase proteolysis during reticulocyte maturation. Lipoxygenase-induced oxidation (but not necessarily inactivation) of hexokinase and perhaps other mitochondrial proteins presumably allows recognition of the oxidised protein and/or mitochondria for autophagic uptake. This recognition could involve ubiquitination of the oxidised protein, since recent evidence suggests ubiquitination may be a signal for lysosomal uptake of proteins (29).

Lysosomes have been shown to be present in reticulocytes by both biochemical (30,31) and cytochemical (32) methods. The contribution of lysosomal proteolysis to reticulocyte maturation, however, has often been thought to be small (33, 34). Nevertheless, several studies have detected autophagic vacuoles in reticulocytes containing mitochondria and other organelles (21,32). Although the contents of some of these vacuoles may be expelled from the cells rather than digested, the results of the present study and the relatively late loss of lysosomal enzymes during reticylocyte maturation (32,35) suggest that lysosomes could contribute significantly to reticulocyte proteolysis.

SUMMARY

Erythrocyte enzyme activities in patients with reticulocytosis or transient erythroblastopenia show that loss of age-dependent enzyme activity

Table 2. Hexokinase Activity and Antigen in *in vitro* Incubated
         Reticulocytes

Sample	Activity	Antigen
0 h	28.6±9.0 (100%)	100%
18 h	59±8%	60±6%
+ 2 mM SHAM	101±5%	100±4%
+ 10 µM chloroquine	78±7%	81±8%
+ 5 mM NH$_4$Cl	86±8%	85±6%
+ 5 mM 3-methyladenine	90±12%	90±8%
+ 5% glycerol	74±3%	81±8%
42 h	51±9%	52±6%
66 h	44±6%	51±14%

Reticulocytes were prepared by bleeding and enrichment of the
reticulocyte content to 85%–95% as described previously (10).
Cells were indubated at 37°C in sealed containers at a hematocrit
of 0.002 in alpha medium, plus 25 mM Hepes, pH 7.4 at 37°C, 10 mM
NaHCO$_3$, an additional 20 mM glucose, 50 U/ml penicillin, 50 µM
streptomycin, and 5% foetal calf serum. Hexokinase activity at
0 h is expressed as IU/g of hemoglobin, and at subsequent times
as a percentage of the 0 h data. Hexokinase antigen was quantified
by Western blotting, as described elsewhere (Thorburn and Beutler,
submitted). The cells used for Western blotting had equal amounts
of lysate in each lane, corresponding to 100 ng of hexokinase
in the 0 h lysate; values are expressed as a percentage of the 0 h
data. All values are the means ± SD of three or four independent
experiments.

is not a simple exponential process occurring throughout the life-span of the cell. In vivo studies of reticulocyte maturation in rabbits indicate that there are multiple mechanisms of enzyme decay, and that proteolysis continues after the maturation of (morphologically recognisable) reticulocytes into young erythrocytes. Most reticulocyte hexokinase is degraded by lysosomal proteolysis, apparently triggered by an initial attack by lipoxygenase.

ACKNOWLEDGEMENTS

    This is publication No. MEM 6550 from Research Institute of Scripps Clinic. The work was founded by NIH grant HL25552, and DRT was supported by an Australian National Health & Medical Research Council C.J. Martin Fellowship and a Fulbright Fellowship.

REFERENCES

1. A. C. Allison and G. P. Burn, Enzyme activity as a function of age in the human erythrocyte, Brit. J. Hematol. 1:291 (1955)
2. C. Seaman, S. Wyss and S. Piomelli, The decline in energetic metabolism with aging of the erythrocyte and its relationship to cell death, Am. J. Hematol. 8:31 (1980)
3. E. Melloni, F. Salamino, B. Sparatore, M. Michetti, A. Morelli, U. Benatti, A. De Flora and S. Pontremoli, Decay of proteinase and peptidase activities of human and rabbit erythrocytes during cellular aging, Biochim. Biophys. Acta 675:110 (1981)
4. M. Morrison, C. W. Jackson, T. J. Mueller, T. Huang, M. E. Dockter, W. S. Walker, J. A. Singer and H. H. Edwards, Does cell density correlate with red cell age?, Biomed. Biochim. Acta 42:s107 (1983)
5. E. Beutler, How do red cell enzymes age? A new perspective, Brit. J. Haematol. 61:377 (1985)
6. E. Beutler and G. Hartman, Age-related red cell enzymes in children with transient erythroblastopenia of childhood and with hemolytic anemia, Pediatr. Res. 19:44 (1985)
7. E. Beutler, Biphasic loss of red cell enzyme activity during in vivo aging, Prog. in Clin. Biol. and Research 195:317 (1985)
8. A. Zimran, S. Torem and E. Beutler, The in vivo ageing of red cell enzymes: direct evidence of biphasic decay from polycythaemic rabbits with reticulocytosis, Brit. J. Haematol. 69:67 (1988)
9. S. Rapoport and W. Dubiel, The effect of phenylhydrazine on protein breakdown in rabbit reticulocytes, Biomed. Biochim. Acta 43:23 (1984)
10. A. Zimran, L. Forman, T. Suzuki, G.L. Dale and E. Beutler, In vivo aging of red cell enzymes: Study of biotinylated red blood cells in rabbits, Am. J. Hematol. 33:249 (1990)
11. M. Magnani, G. Serafini and V. Stocchi, Hexokinase type I multiplicity in human erythrocytes, Biochem. J. 254:617 (1988)
12. K. Murakami, F. Blei, W. Tilton, C. Seaman and S. Piomelli, An isozyme of hexokinase specific for the human red blood cell (HK), Blood 75:770 (1990)
13. V. Stocchi, M. Magnani, G. Piccoli and G. Fornaini, Hexokinase

microheterogeneity in rabbit red blood cells and its behaviour during reticulocytes maturation, <u>Mol. Cell. Biochem.</u> 79:133 (1988)

14. M. Rechsteiner, S. Rogers and K. Rote, Protein structure and intracellular stability, <u>Trends Biochem. Sci</u> 12:390 (1987)

15. G. G. Johnson, W. A. Kronert, S. I. Bernstein, V. M. Chapman and K. D. Smith, Altered turnover of allelic variants of hypoxanthine phosphoribosyltransferase is associated with N-terminal aminoacid sequence variation, <u>J. Biol. Chem.</u> 263:9079 (1988)

16. T. Schewe, S. M. Rapoport and H. Kuhn, Enzymology and physiology of reticulocyte lipoxygenase: Comparison with other lipoxygenases, <u>Adv. Enzymol.</u> 58:191 (1986)

17. A. Hershko, Ubiquitin-mediated protein degradation, <u>J. Biol. Chem.</u> 263:15237 (1988)

18. K. Tanaka and A. Ichihara, Involvement of proteasomes (multicatalytic proteinase) in ATP-dependent proteolysis in rat reticulocyte extracts, <u>FEBS Lett.</u> 236:159 (1988)

19. E. Ueno, H. Sakai, Y. Kato and K. Yamamoto, Activation mechanism of erythrocyte cathepsin E. Evidence for the occurence of the membrane-associated active enzyme, <u>J. Biochem.</u> 105:878 (1989)

20. R. M. Johnstone, A. Bianchini and K. Teng, Reticulocyte maturation and exosome release: Transferrin receptor containing exosomes shows multiple plasma membrane functions, <u>Blood</u> 74:1844 (1989)

21. N. A. Noble, Extrusion of partially degraded mitochondria during reticulocyte maturation, <u>in</u>: The Red Cell: Seventh Ann Arbor Conference", G.J. Brewer, ed., Alan R. Liss, Inc., New York, pp. 275-290 (1989)

22. M. Magnani, V. Stocchi, L. Chiarantini, G. Serafini, M. Dachà and G. Fornaini, Rabbit red blood cell hexokinase. Decay mechanism during reticulocyte maturation, <u>J. Biol. Chem.</u> 261:8327 (1986)

23. D. R. Thorburn and E. Beutler, Decay of hexokinase during reticulocyte maturation: Is oxidative damage a signal for destruction?, <u>Biochem. Biophys. Res. Commun.</u> 162:612 (1989)

24. E. Melloni, B. Sparatore, F. Salamino, M. Michetti and S. Pontremoli, Cytosolic calcium dependent proteinase of human erythrocytes: formation of an enzyme-natural inhibitor complex induced by Ca ions, <u>Biochem. Biophys. Res. Commun.</u> 106:731 (1982)

25. S. Pontremoli, E. Melloni, B. Sparatore, F. Salamino, M. Michetti, O. Sacco and B. L. Horecker, Binding to erythrocyte membrane is the physiological mechanism for activation of Ca -dependent neutral proteinase, <u>Biochem. Biophys. Res. Commun.</u> 128:331 (1985)

26. Y. Hamasaki and H. H. Tai, Calcium stimulation of a novel 12-lipoxygenase from rat basophilic leukemia (RBL-1) cells, <u>Biochim. Biophys. Acta</u> 793:393 (1984)

27. M. Seigneuret, A. Zachowski, A. Herman and P. F. Devaux, Asymmetric lipid fluidity in human erythrocyte membrane: new spin-label evidence, <u>Biochemistry</u> 23:4271 (1984)

28. P. B. Gordon and P. O. Seglen, Exogenous control of intracellular protein catabolism, <u>in</u>: Proteolytic enzymes: A practical approach, R.J. Beynon and J. S. Bond, eds., IRL Press, Oxford, pp. 201-210 (1989)

29. L. Laszlo, F. J. Doherty, N. U. Osborn and R. J. Mayer, Ubiquitinated protein conjugates are specifically enriched in the lysosomal system

of fibroblasts, FEBS Lett. 261:365 (1990)

30. S. Kornfeld and W. Gregory, The identification and partial characterization of lysosomes in human reticulocytes, Biochim. Biophys. Acta 177:615 (1969)

31. S. Yatziv, I. Kahane, P. Abeliuk, G. Cividalli and E. A. Rachilewitz, "Lysosomal" enzyme activities in red blood cells of normal individuals and patients with homozygous beta-thalassaemia, Clin. Chim. Acta 96:67 (1979)

32. G. Gronowicz, H. Swift and T. L. Steck, Maturation of the reticulocyte in vitro, J. Cell. Sci. 71:177 (1984)

33. F. S. Boches and A. L. Goldberg, Role for the adenosine triphosphate-dependent proteolytic pathway in reticulocyte maturation, Science 215:978 (1982)

34. S. M. Rapoport, in: The Reticulocyte, CRC Press, Florida, pp. 167-204 (1986)

35. L. Orr, M. Adam and R. M. Johnstone, Externalization of membrane-bound activities during sheep reticulocyte maturation is temperature and ATP dependent, Biochem. Cell. Biol. 65:1080 (1987)

36. E. Beutler, in: Red cell metabolism. A Manual of biochemical methods, 3rd edn., Grune & Stratton, Orlando (1984)

37. A. N. Mauer, in: Pediatric Hematology, McGraw Hill, New York, p. 5 (1969)

38. E. Beutler, C. West and K. G. Blume, The removal of leukocytes and platelets from whole blood, J. Lab. Clin. Med. 88:328 (1976)

39. U. Quast, A. M. Labhardt and V. M. Doyle, Stopped-flow kinetics of the interaction of the fluorescent calcium indicator QUIN 2 with calcium ions, Biochem. Biophys. Res. Commun. 123:604 (1984)

ROLE AND MECHANISM OF HEXOKINASE DECAY DURING

RETICULOCYTE MATURATION AND CELL AGING

Mauro Magnani, Luigia Rossi, Marzia Bianchi,
Giordano Serafini, Vilberto Stocchi

Istituto di Chimica Biologica "Giorgio Fornaini"
Università degli Studi di Urbino, Italy

Human red blood cells (RBC) as well as the erythrocytes of many mammals are removed from circulation at the end of their life by an impressively efficient biological mechanism that is not fully understood. It is commonly accepted that senescent erythrocytes are selectively recognized and removed from the circulation by spleen and liver macrophages (1), however, various mechanisms have been proposed to account for this selectivity (for a review see Blood Cells vol. 14, 1988). Among these, the immunologic mechanism of cellular removal is widely accepted although the precise identity of the senescent cell antigen(s) remains to be established and primary mechanism of red blood cell aging leading to immunoglobins and complement binding have yet to be elucidated.

Certainly, through its life-span, the erythrocyte remains dependent on preformed enzymes and the activity of a number of these declines during red cell aging (2), eventually leading to an age dependent metabolic impairment. This metabolic impairment could in turn alter the membrane surface properties to the extent that the erythrocyte is recognized as senescent (3)

This fascinating hypothesis accepted by a number of researchers has been supported only by non conclusive evidence that has been questioned by others. In this paper we will summarize the evidence obtained in our laboratory on the relationships between the erythrocyte metabolic properties and the binding of autologous immunoglobulins. Furthermore, since hexokinase has been identified as a regulatory step of red cell glycolysis and its decay found to be able to influence the red cell metabolic capacity, we will also report the data obtained on the mechanism of hexokinase decay during reticulocyte maturation.

Hexokinase and Glucose metabolism in cells of different ages

Glucose constitutes the main substrate of energy metabolism to both

Red Blood Cell Aging, Edited by M. Magnani and
A. De Flora, Plenum Press, New York, 1991

reticulocytes and erythrocytes (4-6). Quantitative measurements, however, have shown that glucose consumption in reticulocytes exceeds that of erythrocytes by 3-4 fold and undergoes a further decrease during cell aging (4-8). This decreased metabolic activity is not due to impairment of the glucose transport system (8) but, probably, to the decrease of many glycolytic enzymes (1,2,4). The hexokinase reaction (E.C. 2.7.1.1. HK) is the first step in red blood cell glycolysis (the only ATP-producing pathway of this cell) and is also considered a crucial rate limiting enzyme (9,10). In fact, compared with the other glycolytic enzymes, it shows the lowest activity, it is suppressed by its product, glucose 6-phosphate, and many other glycolytic intermediates, and cellular concentrations of MgATP are in the range of its Km (11). Furthermore, of all glycolytic enzymes, hexokinase shows the fastest decay during reticulocyte maturation and although at a much slower rate its activity continues to decrease also during cell aging (7,12,13). Direct evidence for the role of hexokinase decay in the age-dependent impairment of erythrocyte glycolysis has recently been obtained in our laboratory by loading mature human erythrocytes with homogeneous hexokinase. This procedure based on hypotonic hemolysis, isotonic resealing and reannealing of red blood cells allowed us to prepare erythrocytes with increased hexokinase activity and normal levels of all the other enzymes. These hexokinase-loaded erythrocytes were found to show a glycolytic activity similar to that found in young red blood cells (14) suggesting that glucose metabolism in human erythrocytes and its age-dependent decay is regulated by hexokinase.

Hexokinase decay and IgG binding

The major question, still open, is whether this age-dependent decay of hexokinase and glycolysis has any effect on red blood cell survival. In an attempt to answer this question we have compared the age-dependent decline of hexokinase activity and the amount of cells positive for surface immunoglobulins on red blood cell populations fractionated on discontinuous density gradients (15). As shown in fig 1, while most of the decline in hexokinase activity occurs in the first cell fractions (i.e. during reticulocyte maturation) the number of cells positive for surface immunoglobulins increases in the last fractions.

In other words the decline in hexokinase activity and, as shown above, of glycolytic rate occuring during reticulocyte maturation does not cause a significant binding of autologous IgG to the red blood cell membrane, instead this IgG binding becomes relevant during cell aging.

Subsequently we investigated the possibility that reducing the metabolic properties of erythrocytes to values similar to those found in the most dense cells could promote an autologous IgG binding. To this end we again manipulated the red cell metabolic properties by changing the hexokinase activity level.To do this we encapsulated monospecific hexokinase inactivating antibodies into human erythrocytes obtaining red blood cells with 20% residual hexokinase activity and 30% residual glycolytic rates. It is worth noting that, when incubated into autologous plasma, these cells become positive for surface immunoglobulins to a percentage of 66-70 % of total cells, allowing us to conclude that although the decay of hexokinase activity and glycolytic rates occurring during reticulocyte maturation does not influence the red cell immunological properties, a further reduction of the cell glycolytic rate can promote a relevant modification to the red cell membrane with exposition of new cell antigens recognized by autologous immunoglobulins.

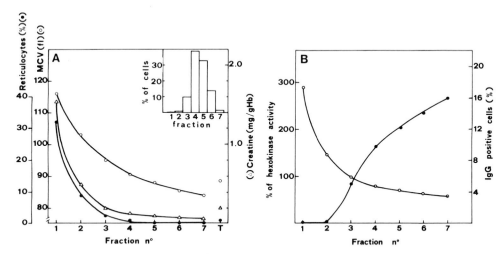

Fig. 1. A, Mean cell volume (MCV), creatine content and percentage of reticulocytes in red blood cell fractions obtained by density gradient centrifugation. Inset, percentage of cells in each fraction. B, Percentage of hexokinase specific activity and number of cells positive for surface immunoglobulins in the red blood cell fractions obtained as above. The 100% hexokinase specific activity was 0.98 U/g hemoglobin. Cells positive for surface immunoglobulins were determined after incubation with 1:250 solution of FITC-labelled affinity purified antibody to human IgG (K.P.L. Inc. Md) by flow cytometry. Readings have been performed by positioning the window at chanel 50.

Mechanism of hexokinase decay

The hexokinase specific activity, as shown in Fig. 1, undergoes a pronounced cell age dependent decay that is very fast during reticulocyte maturation and more slow afterwards. The specific activity in reticulocytes has been found to be 4-5 times that of corresponding mature red cells. We have previously shown by immunoprecipitation experiments that this maturation-dependent hexokinase decay is not due to accumulation of inactive enzyme molecules but to degradation of hexokinase (16). A cell-free system derived from rabbit reticulocytes was found to catalyze the decay of hexokinase activity and the degradation of the ^{125}I-labelled enzyme. This degradation was found to be ATP and ubiquitin dependent with an optimum at pH 7.5. This degradation also required a proteolytic fraction retained by DEAE-cellulose. We were also able to show that hexokinase should be covalently conjugated to ubiquitin in high molecular weight complexes before to influence the mitochondria stability in reticulocytes. In fact, salicylhydroxamate, an inhibitor of lipoxygenase, reduces the decay rate of hexokinase, while Ca^{2+} increases its decay rate. We have also shown that swelling of mitochondria by hypotonic solutions also enhances hexokinase degradation (22).

CONCLUSIONS

The primary mechanism(s) of red blood cell aging leading to removal of circulating senescent red blood cells by the immunological recognition have yet to be elucidated. For many years, based on a number of reported modifications occurring during red blood cell aging and involving the level of many enzymes as well as their kinetic and regulatory properties, researches have considered the age dependent metabolic decay of the red cell as one possible mechanism responsible for triggering red cell removal. In several cases this age dependent metabolic decline has been well documented (7,8) however other authors believe that the isolation method used to obtain aged red blood cells is not adequate to isolate senescent erythrocytes (23,24) and so many of the biochemical properties previously attributed to the senescent cells should be reconsidered (25). In this paper we present a different approach to this problem. First of all there is a general agreement on the fact that reticulocytes have much higher glycolytic rates than mature and old erythrocytes and that the fractionation procedures based on density separation of red cells in fractions of different ages are quite adequate to isolate reticulocytes from young and mature cells (26). In this paper we have shown that maturation of reticulocytes with a concomitant fast decay of many enzymatic activities and of a significant decrease of the red cell glycolytic rate is not associated with a significant increase of membrane bound immunoglobulins. However, we have also provided evidence for the role of hexokinase as a regulatory step of red cell glycolysis and shown that by a procedure of hypotonic hemolysis, isotonic resealing and reannealing it is possible to incapsulate into mature human erythrocytes significant amounts of homogeneous hexokinase or antibodies inactivating the endogenous enzyme. In this way red blood cells with glycolytic rates that mimic those found in young erythrocytes or those of old erythrocytes were obtained. A very interesting result was that erythrocytes with reduced metabolic abilities when incubated into autologous serum become positive for surface immunoglobulins in a very high percentage (65-70%) providing evidence for alterations of the membrane surface properties following a red cell metabolic impairment. The question that remains open is whether or not such a low glycolytic rate can be found in senescent red blood cells. Finally, we have shown that the maturation dependent decay of at least hexokinase is a process catalyzed by the ATP and ubiquitin dependent proteolytic system and that the intracellular distribution of this enzyme is a relevant feature for the regulation of this age dependent decay. It is worth noting that the ATP and ubiquitin proteolytic system itself becomes inactive during reticulocyte maturation providing the bases for a controlled degradation of proteins and organelles during cell maturation and aging. Furthermore studies are now needed to understand the molecular bases for enzyme decay during cell senescence when this proteolytic system is no longer active.

its degradation. The enzymes of the ubiquitin-protein ligase system are responsible for the formation of these aggregates. An important feature of this ATP and ubiquitin proteolytic system is that it is active only in reticulocytes (17). In fact we have shown that a cell-free system derived from mature erythrocytes, instead of the reticulocytes, is not able to catalyze either the conjugation of hexokinase to ubiquitin or its proteolysis.

Hexokinase intracellular distribution

In reticulocytes 50% of the total hexokinase activity is particulated while in mature erythrocytes all the hexokinase is soluble (18). The bound enzyme co-sediments with mitocondria and decreases very quickly during cell maturation and aging. Furthermore, rabbit reticulocyte hexokinase exists in distinct molecular forms designated Ia, Ia* and Ib (19, 20) which are separable by ion-exchange chromatography. Hexokinase Ib is the predominant enzyme in reticulocytes, it is present in soluble form and disappears during reticulocyte maturation. In contrast hexokinase Ia* is mitochondrial bound and its disappearance is parallel to the degradation of mitochondria during reticulocyte maturation.

In previous studies we have shown that when the bound enzyme is solubilized it is degraded by the ubiquitin ATP-dependent proteolytic system at the same rate of the soluble enzyme (21). In contrast, when hexokinase is bound it is not recognized at all by the same proteolytic system. To understand this behaviour that seems to have relevant implications for the age-dependent decay of red blood cell hexokinase and erythrocyte metabolism we have investigated in detail the role of hexokinase intracellular distribution on its degradation with respect to the selectivity of the reticulocyte proteolytic system.

Incubation of mitochondria with bound hexokinase and ^{125}I-ubiquitin in the presence of E_1, E_2 and E_3 (the enzyme of the ubiquitin conjugating system) shows a pronounced ATP-dependent conjugation of ubiquitin to several different mitochondrial proteins. However, by Western blotting experiments we have been able to show that the Mr of the mitochondrial bound hexokinase remains unchanged. In contrast the incorporation also of a single ubiquitin per hexokinase could have caused an appreciable difference in hexokinase electrophoretic mobility, in SDS-PAGE. Furthermore, purification of the mitocondrial bound hexokinase, subsequently to an incubation with ^{125}I-ubiquitin, ATP and the enzymes of the ubiquitin conjugating system did not reveal any significant incorporation of labelled ubiquitin, further supporting the above conclusion that the mithochondrial bound hexokinase is not degraded by the ATP and ubiquitin proteolytic system since it is not an ubiquitinable substrate. In contrast we previously showed that the soluble hexokinase can be ubiquitinated and subsequently degraded.

Factors affecting hexokinase intracellular distribution

Since during reticulocyte maturation mitochondria are degraded by the action of a specific lipoxygenase and by the ATP-dependent proteolytic system we expect an indirect effect on hexokinase decay of all factors able

ACKNOWLEDGEMENTS

Supported by C.N.R. target projects "Aging" and " Biotechnology and Bioinstrumentation".

REFERENCES

1. M. I. Berlin and P. D. Berk, The biological life of the red cell, in "The Red Blood Cell", D. Mac M. Surgenor, ed., Academic Press, New York (1975)
2. G. Fornaini, Biochemical modifications during the life-span of the erythrocyte, Ital. J. Biochem. 16:257 (1967)
3. H. H. Lutz, Red cell density and red cell age, Blood Cells 14:76 (1988)
4. G. J. Brewer, General red cell metabolism, in "The Red Blood Cell", D. Mac N. Surgenor, ed., Academic Press, New York (1975)
5. W. Siemens, M. Müller, R. Dumday, A. G. Holzhütter, J. Rathmann and S. M. Rapoport, Quantification of pathways of glucose utilization and balance of energy metabolism of rabbit reticulocytes, Eur. J. Biochem. 124:567 (1982)
6. W. Siemens, W. Dubiel, M. Müller and S. M. Rapoport, Accounting for the ATP-consuming processes in rabbit reticulocytes, Eur. J. Biochem. 139:101 (1984)
7. C. Seaman, S. Wyss and S. Piomelli, The decline in energetic metabolism with aging of the erythrocyte and its relationship to cell death, Am. J. Haematol. 8:31 (1980)
8. M. Magnani, E. Piatti, M. Serafini, F. Palma, M. Dachà and G. Fornaini, The age-dependent metabolic decline of the red blood cell, Mech. Ageing Dev. 22:295 (1983)
9. T. A. Rapoport, R. Heinrich, G. Jacobash and S. Rapoport, A linear steady-state treatment of enzymatic chains. A mathematical model of glycolysis of human erythrocytes, Eur. J. Biochem 42:107 (1974)
10. S. Minakami and H. Yoshikawa, Studies on erythrocyte glycolysis. II Free energy changes and rate limitings step in erythrocyte glycolysis, J. Biochem. 59:139 (1966)
11. M. Magnani, V. Stocchi, M. Dachà and G. Fornaini, Regulatory properties of rabbit red blood cell hexokinase at conditions close to physiological, Biochim. Biophys. Acta 804:145 (1984)
12. G. Fornaini, M. Magnani, A. Fazi, A. Accorsi, V. Stocchi and M. Dachà, Regulatory properties of human erythrocyte hexokinase during cell aging, Arch Biochem. Biophys. 239:352 (1985)
13. A. Zimram, S. Torem and E. Beutler, The in vivo ageing of red cell enzymes: direct evidence of biphasic decay from polycythemic rabbits with reticulocytosis, Br. J. Haematol. 69:67 (1988)
14. M. Magnani, L. Rossi, M. Bianchi, G. Fornaini, U. Benatti, L. Guida, E. Zocchi and A. De Flora, Improved metabolic properties of hexokinase-overloaded human erythrocytes, Biochim. Biophys. Acta 972:1 (1988)
15. M. Magnani, S. Papa, L. Rossi, M. Vitale, G. Fornaini and F. A. Manzoli, Membrane-bound immunoglobulins increase during red blood cell aging, Acta Haemat. 79:127 (1988)

16. M. Magnani, V. Stocchi, L. Chiarantini, G. Serafini, M. Dachà and G. Fornaini, Rabbit red blood cell hexokinase. Decay mechanism during reticulocyte maturation, J. Biol. Chem. 261:8327 (1986)
17. A. Hershko and A. Ciachanover, Mechanisms of intracellular protein breakdown, Annu. Rev. Biochem. 51:335 (1982)
18. M. Magnani, V. Stocchi, M. Dachà and G. Fornaini, Rabbit red blood cell hexokinase: intracellular distribution during reticulocytes maturation, Mol. Cell. Biochem. 63:59 (1984)
19. V. Stocchi, M. Magnani, F. Canestrari, M. Dachà and G. Fornaini, Rabbit red blood cell hexokinase. Evidence for two distinct forms and their purification and characterization from reticulocytes, J. Biol. Chem. 256:7856 (1981)
20. V. Stocchi, M. Magnani, G. Piccoli and G. Fornaini, Hexokinase microheterogeneity in rabbit red blood cells and its behaviour during reticulocytes maturation, Mol. Cell Biochem. 79:133 (1988)
21. M. Magnani, V. Stocchi, M. Dachà and G.Fornaini, Rabbit red bood cell hexokinase. Evidence for an ATP-dependent decay during cell maturation, Mol. Cell Biochem. 61:83 (1984)
22. M. Magnani, G. Serafini and V. Stocchi, Effects of Ca^{2+} and lipoxygenase inhibitors on hexokinase degradation in rabbit reticulocytes, Mol. Cell Biochem. 85:3 (1989)
23. T. Suzuki and G. Dale, Biotinylate erythrocytes: In vivo survival and in vitro recovery, Blood 70:719 (1987)
24. T. Suzuki and G. Dale, Senescent erythrocytes: Isolation of in vivo aged and their biochemical characteristics, Proc. Natl. Acad. Sci. U.S.A. 85:1647 (1985)
25. G. Dale and S. L. Moremberg, Time dependent loss of adenosine 5'-monophosphate deaminase activity may explain elevated Adenosine 5'-triphosphate levels in senescent erythrocytes, Blood 74:2157 (1989)
26. E. Beutler, Isolation of the aged, Blood Cells 14:1 (1988)

BEHAVIOUR OF GLUCOSE-6-PHOSPHATE DEHYDROGENASE DURING

ERYTHROID MATURATION

Maria D. Cappellini, Stefania Villa,
Alessandro Gaviraghi, Franco Martinez di Montemuros,
Dario Tavazzi, Daniela Panzeri and Gemino Fiorelli

Istituto di Medicina Interna e
Istituto di Scienze Biomediche Clinica Medica
Università di Milano, Monza Italy

INTRODUCTION

Glucose-6-phosphate dehydrogenase (G6PD; D-glucose-6-phosphate: $NADP^+$ oxidoreductase, EC 1.1.1 49) is the first enzyme of the hexose monophosphate pathway and provides NADPH essential for a variety of biosynthetic and detoxifying reactions (1).

Normal G6PD activity has been reported to decline during aging of erythrocytes (2), and this decay is even more pronounced for some variants of the enzyme (3, 4). The usual explanations for this event are 1) in vivo degradation of the enzyme which has been ascribed mainly to the erythrocyte proteolytic system, or 2) a lack of capacity for protein synthesis by erythrocytes, which are non-nucleated cells (5). Moreover, some variability of the enzyme in various tissues has been noted, and such differences have generally been attributed to post-translational modification of the enzyme (6).

Over 380 G6PD variants have been described in the past, suggesting a great number of alleles. However these predictions have so far been borne out by molecular analysis which has disclosed similar point mutations in different variants (7). All these observations raise several questions about G6PD behaviour not only during erythrocyte aging but also during erythroid differentiation and maturation. At present, most knowledge of the biochemistry of the enzyme stems from work on mature cells, and very little is known about the regulation of enzyme synthesis because of difficulties in obtaining human erythroid progenitors. This paper reports a study of G6PD activity and kinetic properties during erythroid maturation evaluated in normal and G6PD deficient erythroid precursors obtained from culture of peripheral blood in presence of erythropoietin.

MATERIALS AND METHODS

Peripheral blood was obtained from 6 normal adult males and from 6 adult males known to have a G6PD deficiency. The G6PD deficient variant of the 6 subjects was already characterized in our laboratory as Mediterranean variant on the basis of low activity and biochemical and kinetic properties.

Cell culture: mononuclear cells were separated by standard Lymphoprep method (Nyegaard and Co), washed at 4°C in Hanks' balanced salt solution (HBSS) (Flow Laboratories) and counted in isotonic saline containing 0.1% trypan blue dye. BFU-E cultures were set up according to Iscove's method (8) with minor modifications. Briefly, (3x15 /ml nucleated cell were plated in alpha medium containing 1% methylcellulose, 30% fetal calf serum, 1% deionized bovine serum albumin, 2×10^{-4} M beta-mercaptoethanol and 4 U/ml human recombinant erythropoietin (Cilag). The erythroid cultures were set up in 24-well plates; each well was covered completely with 0.5 ml light paraffin oil (9). The plates were incubated at 37°C in a fully humidified 5% CO_2 air atmosphere. BFU-E were counted, harvested and pooled on days 9 and 14 of culture. The collected cells were washed at least three times with alpha-medium to free them from serum and methylcellulose. The morphology was examined after May-Grunwald-Giemsa staining. The distribution of the different erythroid cells was calculated, based on cell size, cytoplasmic staining and nuclear condensation.

G6PD purification: the enzyme was purified from the erythroblasts obtained from BFU-E cultures and from erythrocytes by a two-step micromethod using affinity chromatography with 2'5' ADP-Sepharose 4B followed by an automated procedure of ion exchange chromatography with DEAE 5PW (not shown). The yield of enzymatic activity was estimated to be about 70% at the end of the procedure, with peaks of over 90% in the intermediate step in the case of normals, slightly less in the case of deficient variants.

Enzyme assays: G6PD activity was determined according to the WHO recommendations (10) at 25°C, using a Gilford model 2400-S spectrophotometer. The utilization of substrate analogues dNADP, 2dG6P and Gal6P (expressed as a percentage of the rate of G6P or NADP utilization) was tested according to Beutler (11). Michaelis constants (Km) for G6P and NADP were determined according to the WHO recommendations (12) in a pH 8.0 buffer containing 0.1 M Tris HCl. The G6P concentration ranged from 15 μM to 2 mM and the NADP concentration from 1 μM to 50 uM. All the kinetic assays were carried out at 25°C.

RESULTS

After 9 days of culture the colonies consisted mainly of early erythroblasts (90% basophilic cells) and after 14 days of culture they were mainly of late erythroblasts (95% orthochromatic cells).

G6PD activity, Km G6P and the percentage of substrate analogues utilization (dNADP, 2dG6P, Gal6P) in erythroid cells derived from cultured BFU-E and from erythrocytes of normal subjects and subjects with G6PD deficiency (Mediterranean variant) are shown in Table 1 and 2 respectively. The enzyme activity was significantly higher in normal and deficient erythroblasts versus the respective erythrocytes. In deficient erythroblasts the enzyme activity was about 25% of normal whereas in erythrocytes it was less than 10%. KmG6P ranged between 50 and 55 uM in both normal erythroid precursors and normal erythrocytes and between 25 and 28 μM in both deficient erythroid precursors and deficient erythrocytes. The percentage utilization of analogues did not change during erythroid maturation in normal cells but was significantly different in deficient BFU-E at day 9 versus day 14 of culture. At day 9, the behaviour was similar in normal and in deficient erythroblasts whereas at day 14 the deficient erythroblasts displayed the peculiar characteristics of G6PD deficient erythrocytes (Mediterranean variant).

Table 1. Biochemical Characteristics of G6PD (B) in Erythroid Precursors

| | Activity IU/10^9 cells | KmG6P μM | Utilization of | | |
			dNADP (%)	2DG6P (%)	Gal6P (%)
BFU-E (day 9)	3.9±0.2	50±4	65±6	9±2	10±1
BFU-E (day 14)	4.1±0.2	52±3	63±4	7±2	7±1
Erythrocytes	0.2±0.1	53±5	53±5	6±2	7±2

Table 2. Biochemical Characteristics of G6PD Mediterranean Variant in Erythroid Precursors

| | Activity IU/10^9 cells | KmG6P μM | Utilization of | | |
			dNADP (%)	2DG6P (%)	Gal6P (%)
BFU-E (day 9)	1.1±0.28	26±1	55±3	12±1	10±1
BFU-E (day 14)	1.0±0.23	27±1	267±25	48±11	37±6
Erythrocytes	0.01±0.01	27±1	237±3	42±7	31±3

Fig. 1-3 report the behaviour of NADP, 2dG6P and Gal6P during maturation of erythroid precursors and in erythrocytes from normal and deficient subjects.

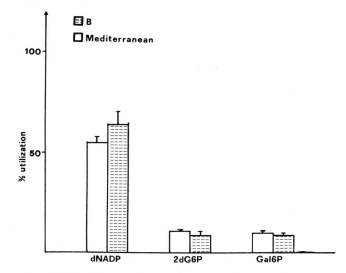

Fig. 1. Percent utilization of analogues in normal and G6PD deficient erythroid precursors (BFU-E, day 9 of culture).

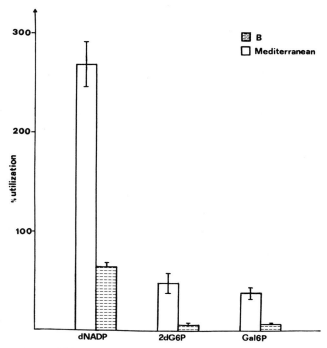

Fig. 2. Percent utilization of analogues in normal and G6PD deficient erythroid precursors (BFU-E, day 14 of culture).

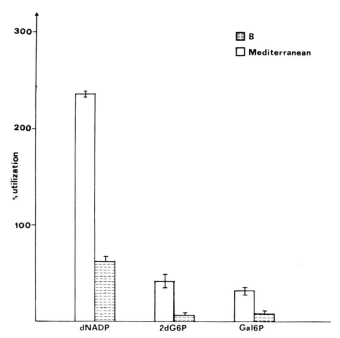

Fig. 3. Percent utilization of analogues in normal and G6PD deficient erythrocytes.

DISCUSSION

G6PD is considered a strongly age-related enzyme. Nevertheless apart from data on aging erythrocytes, virtually nothing is known about regulation of G6PD synthesis and turnover during erythroid maturation. This process cannot be easily investigated in mammalian cells because of difficulties in obtaining erythroid progenitors and precursors. To overcome this problem we produced erythroblasts in vitro by culturing peripheral BFU-E of normal subjects and of G6PD deficient subjects. This technique enables erythropoiesis to be studied at different stages of development. We observed that during erythroblast maturation the level of enzyme activity remained steady and was higher than in erythrocytes from both normal and deficient subjects. Similar results have been obtained in normal animals by Denton et al. (13) who studied the G6PD level in erythroid cells of bone marrow from phenylhydrazine-treated rabbits. They also found that the G6PD level remains steady during differentation from stem cells to dividing erythroblasts and then falls abruptly in reticulocytes. Nijhof et al. (14) have shown changes in activities and isoenzyme patterns of glycolytic enzymes in the mouse during erythroid differentiation in vitro using CFU-E (Colony Forming Unit-Erythroid) culture.

The utilization of substrate analogues in human erythroblasts has never been studied before. The different pattern that we observed in basophilic and orthochromatic deficient erythroblasts is rather surprising. The fact that early deficient erythroblasts had a substrate analogues utilization similar to that of normal erythroblasts and different from that of deficient late erythroblasts and erythrocytes, suggests that the enzyme undergoes some functional changes during cell maturation. The physiological significance of these postsynthetic changes is still unclear as are the mechanisms that regulate these changes. To some extent the different behaviour of G6PD behaviour of G6PD during erythroid differentation recalls the well known intertissue variability of G6PD. Such intertissue differences have generally been ascribed to postranslational modifications of the enzyme (15). Recently, the existence of multiple forms of human G6PD mRNA have been suggested which may produce multiple forms of enzyme protein. Moreover, the directly determined amino acid sequences of the amino end of the normal red cell G6PD does not exactly match the sequence deduced from cDNA (16). Hirono and Beutler (17) showed that these discrepancies could be due to alternative splicing of the same gene transcript. Such a mechanism could be hypothesized to explain the postsynthetic enzyme changes observed in different erythroid cells.

Sequencing of mRNA from red cells at different stages of development after the sensitive polymerase chain reaction technique is in progress.

ACKNOWLEDGMENTS

This work was supported by grant MPI 40% to G.F.

REFERENCES

1. L. Luzzatto and G. Battistuzzi, Glucose-6-phosphate dehydrogenase, Adv. Hum. Genet. 4:217 (1985).
2. A. C. Allison and G. P. Burn, Enzyme activity as a function of age in the human erythrocyte, Brit. J. Haematol. 1.1:291 (1955).
3. S. Piomelli, L. E. Corash, D. D. Davenport, J. Miraglia and E. L. Amorosi, In vivo lability of G6PD in GdA-and Gd Mediterranean deficiency, J. Clin. Invest. 47:940 (1968).
4. A. Morelli, U. Benatti, G. F. Gaetani and A. De Flora, Biochemical mechanims of glucose-6-phosphate dehydrogenase deficiency, Proc. Natl. Acad. Sci. USA 75,4:1979 (1978).
5. E. Beutler, Selectivity of proteases as a basis for tissue distribution of enzymes in hereditary deficiencies, Proc. Natl. Acad. Sci. USA 80:3767 (1983).
6. A. Kahn, P. Boivin, N. Vibert, D. Cottreau and J. C. Dreyfus, Post-translational modifications of human glucose-6-phosphate dehydrogenase, Biochimie 56:1295 (1974).
7. E. Beutler, Glucose-6-phosphate dehydrogenase: new percpectives, Blood 73:1397 (1989).
8. N. N. Iscove, F. Sieber and K. Winterhalter, Erythroid colony formation in cultures of mouse and human bone marrow: analysis of the requirement for erythropoietin by gel filtration and affinity chromatography on agarose-concanavalin, Am. J. Cell. Physiol. 83:309 (1974).

9. C. G Potter and M. D. Cappellini, Improved culture of BFU-E and CFC-GM by the use of an oil seal, Brit. J. Haematol. 54:153 (1983).
10. E. Beutler, K. G. Blume, P. C. Kaplan, G. W. Lohr, B. Ramot and W. M. Valentine, International Committee for standardization in hematology: recommended methods for red cell enzyme analysis, Br. J. Haematol. 35:331 (1977).
11. E. Beutler, Haemolytic anaemia in disorders of red cell metabolism, Plenum, New York (1978).
12. World Health Organization (WHO). Standardization of techniques for the study of G6PD. WHO, Tech. Rep. Ser. 366, Annex 7 (1967).
13. M. J.Denton, N. Spencer and H. R. V. Arnstein, Biochemical and enzymatic changes during erythrocyte differentation. The significance of the final cell division, Biochem. J. 146:205 (1975).
14. W. Nijhof, P. K. Wierenga, G. E. J. Staal and G. Jansen, Changes in activities and isoenzyme patterns of glycolytic enzymes during erythroid differentation in vitro, Blood 64:607 (1984).
15. L. Luzzatto and U. Testa, Human erythrocyte glucose-6-phosphate dehydrogenase: structure and function in normal and mutant subjects, Current topics in hematology, Alan R. Liss Inc. 1 (1978).
16. M. G. Persico, G. Viglietto, G. Martino, D. Toniolo, G. Paonessa, C. Moscatelli, R. Dono, T. Vulliamy, L. Luzzatto and M. D'Urso, Isolation of human glucose-6-phosphate dehydrogenase (G6PD) cDNA clones: primary structure of the protein and unusual 5' non coding region, Nucleic Acid Res. 14:2511 (1986).
17. A. Hirono and B. Beutler, Alternative splicing of human glucose-6-phosphate dehydrogenase messanger RNA in different tissue, J. Clin. Invest. 83:343 (1989).

MEMBRANES

KCl COTRANSPORT IN HbAA AND HbSS RED CELLS: ACTIVATION BY INTRACELLULAR

ACIDITY AND DISAPPEARANCE DURING MATURATION

J. Clive Ellory[1], Andrew C. Hall[1], Susan A. Ody[1],
Carlos E. Poli de Figueiredos[1], Susan Chalder[2]
and John Stuart[2]

[1]Department of Physiology
University of Oxford, Oxford
[2]Department of Haematology
University of Birmingham, Medical School, Birmingham

SUMMARY

Low intracellular pH was shown to be a potent activator of the KCl cotransport system in HbSS red cells, and in reticulocyte-rich fractions of HbAA red cells. Rheological experiments indicated that cell dehydration via the KCl cotransporter in response to low pH decreased the filterability of HbSS red cells. In vitro maturation experiments showed that the KCl cotransport system was rendered cryptic rapidly, in contrast to choline transport, and serine transport via system ASC, which disappeared much more slowly.

INTRODUCTION

Red cells from patients with sickle cell disease (HbSS) show large Cl-dependent K fluxes, consistent with the presence of a KCl cotransport system (1). In contrast, normal human red cells (HbAA) do not usually express significant Cl-dependent K^+ transport although certain manoeuvres (see Table 1) can reveal this pathway. If reticulocyte-enriched fractions are prepared by density separation from either normal donors, patients with haemochromatosis, or HbSS patients, increased activity of the KCl cotransport system is seen (2,3). This system is highly expressed in 'young' red cell fractions, but becomes cryptic with cell maturation (4). Its disappearance does not reflect loss of components from the membrane, unlike other membrane transport systems (5), since it is possible to reactivate KCl cotransport, even in mature and 'old' dense cells, by a variety of signals (Table 1). Of the treatments known to activate KCl cotransport listed in Table 1, reduced pH is the only one likely to play a role physiologically.

Originally, it was thought that activation of KCl cotransport by low pH_o was an indirect effect via an increase in cell volume, since lowering pH to 6.8 causes red cells to swell by about 10% (6). In fact we show here that KCl cotransport activation is a direct effect of the decrease in pH rather than a change in volume, and it is <u>internal</u>, not external pH which is important. We also look at the significance of incubation at low pH for HbSS red cells. It is known that low pH impairs the deformability of HbSS red cells. This could result either from a direct effect of acidity on the polymerization of haemoglobin, or from an increased cytoplasmic viscosity due to the loss of KCl and hence cell water. In this paper we show that activation of the KCl cotransport system contributes significantly to the loss of filterability seen in HbSS cells incubated at low pH. Finally, we looked at the rate of loss of KCl cotransport activity in reticulocyte enriched HbAA blood incubated <u>in vitro</u>, and compared the loss of activity of KCl cotransport with the rate of loss of amino acid transport activity via system ASC (17) and choline transport. The differences between the systems emphasizes the specific changes in particular transport systems during the complex process of red cell maturation.

METHODS

Blood

Venous blood was taken from healthy controls (HbAA) and with informed ethical consent from 10 out-patients with homozygous sickle cell disease (HbSS) who were in the asymptomatic steady-state. The blood was anticoagulated with lithium heparin (15 IU/ml blood). Fractions of blood enriched with reticulocytes were obtained either by Percoll-Isopaque continuous density gradient centrifugation (20), or by differential centrifugation in saline (8).

Table 1. Factors Activating KCl Cotransport in Human Erythrocytes[a]

Treatment	References
Albumin	Ellory & Hall (unpublished)
Hypotonicity	7,8
Reduced pH	9,10
Reduced Mg	10
High hydrostatic pressure	2
NEM treatment	11,12
Ghosting and resealing	13,14

[a]In order of increasing potency

Flux measurements

The cells from unfractionated whole blood were washed three times by centrifugation (3,000 x g, 10 min) using the following standard washing medium (mM); NaCl (145), KCl (15) glucose (5), MOPS (morpholinopropane sulphonic acid) (15) with the pH adjusted to 7.4 using NaOH. The influx of K^+ (using the congener ^{86}Rb) was measured at 37°C over 10 min at a haematocrit of about 5% and at either pH 7.4 or 6.8 in the appropriate medium (see below) with ouabain and bumetanide (both at 0.1 mM) present as previously described (2,8). In the experiments where the internal pH was varied at constant external pH, the following two solutions (A and B) were used. Solution A contained (in mM) NaNO$_3$ (160) and solution B contained Na glucuronate (160), with both also including KBr (15), glucose (5), MOPS (15) with both solutions being adjusted to pH 6.8 and 350 mOsm with 1N NaOH and sucrose addition respectively. Solution A contained mainly NaNO$_3$, the rapidly permeating anion allowing the pH$_i$ to follow pH$_o$ as predicted from the Donnan ratio for hydroxyl ions. Solution B contained mainly Na glucuronate, an impermeant anion which causes an internal alkalinization (15,16).

Both of these media must also contain a sufficiently high concentration of an anion capable of supporting KCl cotransport, but obviously not a concentration high enough to interfere significantly with the trans-membrane OH^- distribution. Bromide is a more effective activator of KCl cotransport than chloride (4), and therefore can be used at a lower concentration than Cl^-. To adjust the intracellular pH, two 5ml aliquots of whole HbSS blood from the donor were washed (20 volumes 3,000 x g, 5 min) in solutions A or B maintained ice-cold. The cell suspension were then incubated on ice with fresh medium for 30 min, washed again, resuspended for a further 30 min and centrifuged. The cells were then ready for use. Maintenance of a low temperature throughout was to limit the disappearance of KCl cotransport activity which has been shown to be temperature-dependent (4). K^+ influx was measured at 15 mM KBr with ^{86}Rb (1μCi/ml) and 37°C over 10 min in the two media (A and B) by washing, centrifugation (10,000 x g, 15 sec) and aspiration tecniques as previously described (17).

Measurement of internal and external pH

Samples from the cell suspensions used for the flux experiments were also taken for the measurement of pH$_i$ and pH$_o$. The pH was determined using a microcapillary electrode (Radiometer, Copenhagen, Denmark) with a water-jacket that maintained the sample temperature constant at 25°C. The extracellular pH was taken as that measured in the whole red cell suspensions (15). Intracellular pH and haemoglobin content were measured on the packed cells obtained by centrifuging cell suspensions (10,000 x g, 15 sec) and removing the supernatant. For the determination of haemoglobin content, the packed cells were resuspended to give about 50% haematocrit and aliquots added to Drabkin's reagent with the absorbance measured at 540 nm (18). The values for haemoglobin content can be used only for comparative estimations of haematocrit because of the problems associated with the unknown extracellular space in centrifuged HbSS red cells. The pH$_i$ was measured on red cell lysates obtained by repeatedly freezing samples of the cell pellet in liquid nitrogen (-196°C) and thawing in a water-bath at 37°C.

Rheological measurements

Whole blood was pre-filtered through Imugard IG 500 cotton wool, as previously described (19) to remove leucocytes, using a medium comprising (mM); NaCl (130), HEPES (40), pH 7.4 at 25°C. The cells were then fractionated on a Percoll-Isopaque continuous density gradient (20) and the red cells from the top half of the gradient removed and washed three times by centrifugation (3,000 x g, 5 min) in the above medium. Aliquots of the packed cells were suspended at a red cell count of $0.3 \times 10^{12}/l$ either in the above medium, or in a similar medium where KCl replaced NaCl, with glucose (5 mM) added and the pH adjusted to 7.4 or 6.8 and the osmolarity of all solutions was kept constant at 290 mOsm by the addition of sucrose. These aliquots (10 ml) were incubated for 15 h at 37°C in a Cyclical Gas Exchanger as previously described (20). Alternate cycles of oxygenation (2.5 min) and deoxygenation (8.5 min) were maintained throughout by gassing with O_2 and N_2. After 15 hr, the cells were fully oxygenated for 30 min before testing the rheological properties of the cells.

Sickle cell deformability was measured before and after the 15 hr incubation by filtration through 5 μm diameter polycarbonate membranes (Nuclepore Corporation, Pleasanton, California, U.S.A.) using a St. George's Filtrometer (Carri-Med Ltd., Dorking) as previously described (21). The results were expressed as an index of filterability adjusted for red cell count (22), the index representing the flow resistance of the cell suspension at a haematocrit of 3% to that of the suspending saline and was expressed in arbitrary units. An increase in the index reflects loss of filterability (i.e. loss of deformability). The mean cell volume (MCV) of oxygenated cells was calculated from the microhaematocrit obtained following centrifugation (13,000 x g, 5 min) and the red cell count measured by a Coulter Counter S Plus IV (Coulter Electronics Ltd., Luton). The haemoglobin content was determined as described above. Statistical significance was determined by the Wilcoxon signed rank test for paired data.

Maturation experiments

Red cells were incubated for up to 3 hr at 5% haematocrit in the standard saline medium. At appropriate time intervals, aliquots were withdraw, the cells washed twice in cold medium by centrifugation (3,000 x g, 5 min) and resuspendend for flux measurements. To measure KCl cotransport activity, the ^{86}Rb flux methodology was used, as described above; to measure Na-dependent serine transport, one aliquot of cells was washed and resuspended in a Na-free medium (same composition as the saline medium except KCl replaced NaCl). Serine influx was measured at an external serine concentration of 0.1 mM, with [14]C- labelled serine (0.1μCi/ml), over 5 min. The cell pellet was deproteinized and counted as previously described (4). Choline influx (using ^{14}C-choline) was measured over the range 2.5 to 100 μM using a similar procedure with 5 min fluxes in normal saline media as described above. Data were fitted to a simple Michaelis-Menten equation for the calculation of the apparent constants K_m and V_{max}.

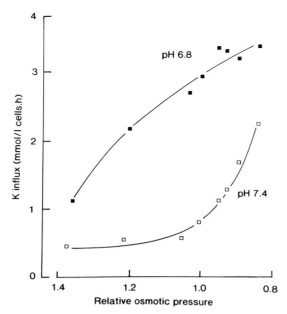

Fig. 1 Increased passive K transport in HbSS red cells induced by medium acidification and hypo-osmolarity. Results (means ±s.d.) shown are from one typical experiment of four donors with HbSS red cells.

RESULTS

Figure 1 compares the ouabain- and bumetanide- insensitive K influx in HbSS cells suspended in media of pH 6.8 and 7.4 as a function of osmolarity. It is apparent that lowering the pH increases K uptake at all osmolarities, including media that are markedly hypertonic. This suggests that low pH acts directly to activate the KCl cotransporter rather than acting indirectly via cell swelling. In these experiments both internal and external pH will change since the cells are suspended in conventional high chloride media. To distinguish whether internal or external pH was involved as the effective stimulus, cells were suspended in media where the bulk of the anion was impermeable glucuronate, conditions leading to intracellular alkalinization. In this case (Fig. 2), the KCl cotransport activity was defined as the fraction of K influx sensitive to H25 (after pre-treatment with bumetanide to eliminate NaKCl cotransport), which is a specific inhibitor of the KCl cotransporter under these conditions (23). This component was only significantly activated when the internal pH was lowered, no significant activity being measurable at an internal pH of 7.3, even though the external pH was 6.8. These media were made hypertonic to 350mOsm, by the addition of sucrose, to eliminate any influence of changing cell volume.

Figure 3 shows the filterability through 5 um pores of HbSS red cells suspended in isotonic NaCl or KCl media adjusted to either pH 7.4 or 6.8. In the NaCl medium at pH 6.8, loss of cell water as a result of activation of the KCl transporter can occur and there was a significant ($p < 0.01$) decrease in filter ability compared with the control (pH 7.4). The loss of cell water caused significant cell shrinkage (decrease in mean cell volume (MCV) from 99.5 to 90.5 fl; p <0.01) and an increase in mean cell haemoglobin concentration (MCHC) from 30.5 to 33.6 g/dl (p <0.01). In contrast, cells suspended in isotonic KCl cannot lose KCl when the volume-sensitive KCl cotransporter is activated because there is no gradient for KCl movement. In this case the smaller increase in filterability was not statistically significant and represents the direct effect of low pH on polymerization of sickle cell haemoglobin. It is therefore possible to distinguish the relative contributions of cell dehydration via KCl loss and the direct effect of pH on HbSS polymerization within cells.

Finally we have addressed the question of cell maturation and the factors which influence the rate at which the KCl cotransport system is rendered cryptic. Fig. 4 compares the effects of incubation in saline or 1% (w/v) bovine serum albumin in saline on the rate of loss of KCl cotransport in vitro. It is clear that the KCl cotransport system rapidly inactivates in saline. We have previously shown that autologous plasma can significantly reduce the rate of inactivation (24). The addition of albumin stimulates the initial activity of the transporter (cf Table 1) but the rate of inactivation is not decreased.

To compare the rate of loss of KCl cotransport with other membrane transport pathways, we have measured choline and amino acid uptake in reticulocyte-enriched and mature red cell populations. In 4 separate experiments on fractionated HbAA red cells from normal donors, the choline uptake rate under V_{max} conditions was 25.8 ± 3.7 and 38.8 ± 5.5 μmol/l cells.h in mature and reticulocyte-rich cell populations respectively. In a second series of experiments, on fractionated cells from 3 haemochromatosis patients respective V_{max} values were 22.1 ± 2.9 and 38.4 ± 4.2; in these experiments, K_m values were also obtained giving 8.5 ± 1.5 and 9.4 ± 1.8 μM respectively.

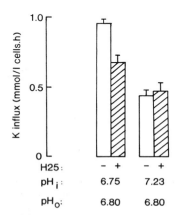

Fig. 2 Ouabain and bumetanide-insensitive K influx in a "reticulocyte-rich" fraction from an HbAA haemachromatosis patient, measured at constant external pH and two internal pHs, in the presence or absence of 0.1 mM H25, a KCl cotransport inhibitor. Results (means ±s.e.m. of triplicate samples) shown are from one typical experiment.

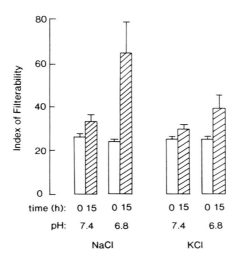

Fig. 3 Loss of 5μm pore filterability of HbSS red cells as a function of the
cation composition and pH of the suspending medium. An increase in
the index reflects a decrease of cell deformability. The composition
of the media and other experimental conditions are described in
Methods; results are means ± s.e.m. for 10 independent experiments.

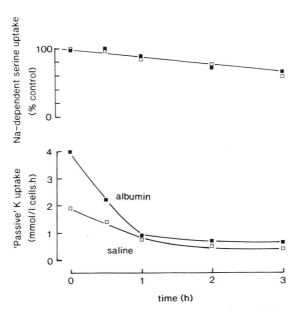

Fig. 4 The effect of _in vitro_ incubation on Na-dependent serine transport
(upper panel) or KCl cotransport in reticulocyte-rich HbAA red cells,
incubated in normal saline medium or the same medium supplemented
with 1% (w/v) bovine serum albumin.

The Na-dependent amino acid transport system ASC, which was measured in vitro in the same red cell samples as those used for the KCl transport experiments, indicated that this pathway was inactivated at a much slower rate than the KCl cotransporter (Fig. 4).

To investigate whether activation of the system by low pH alters the rate of loss of KCl cotransport, cells were incubated at either pH 6.8 or 7.4 for one hour, washed, and the transport activity measured under swollen and shrunken conditions (250 and 400mOsm respectively). No difference in transport rate was found, indicating that the rate of loss of transport activity was the same regardless of whether the system was operating or not (data not shown).

DISCUSSION

The data presented here show that: (1) internal acidity is a potent signal for activating the volume-sensitive KCl cotransporter in human red cells, (2) activation of this transporter causes HbSS red cells to shrink so that their filterability through 5 um pores is reduced, (3) that KCl cotransport is only present in "young" red cells, being cryptic in mature red cells, and (4) the rate of loss of KCl cotransport is different from the selective loss of other membrane transport systems, e.g. choline and serine transport.

The results in Figure 1 clearly demonstrate that lowering pH at all osmolarities studied activates the KCl transporter. The fact that low pH is effective even in shrunken cells (400mOsm) confirms the potency of this signal in activating the transporter. Brugnara et al. (25) have shown that KCl cotransport in the least dense fraction of HbSS cells separated on gradients shows a greater activity in acid solution than the older, denser cells, and Lew et al. (26) have proposed pH_i as the most likely physiological signal for dehydrating HbSS cells.

Due to their unusually high Cl-conductance red cells have an intracellular pH equivalent to the Donnan distribution of H^+ (or OH^-). However, by using an impermeant anion it is possible to vary internal and external pH independently over a certain range, allowing the question of which is effective in stimulating KCl cotransport to be addressed. One problem with such an approach is to maintain a sufficiently high halide ion concentration to activate the transporter, without increasing the conductance too much. We have managed to do this using Br^-, which has a much higher affinity for the KCl system in "young" human red cells (4). A second problem is the change in membrane potential which occurs in media with an impermeant anion. At high positive potentials, Halperin et al. (27), have shown activation of a voltage-sensitive cation conductance in human red cells. To avoid this problem we limited our experiment to conditions where the voltage only changes to +20mV, the threshold for opening this channel; further, we have used the potent KCl cotransport inhibitor H25 (23) to define the fraction of K transport occurring through this system. The result of such experiments clearly indicates that <u>internal</u> pH was the effector in this case medium supplemented with 1% (w/v) bovine serum albumin.

Having shown that internal pH, the principal trigger encountered physiologically, is effective in stimulating KCl cotransport, we next demonstrated that its activation has important consequences for sickle cells. Using the rheological measure of filtration through 5 um pores, we found a doubling of resistance to filtration following incubation at acid pH. In these experiments, a rheological incubation period of 15 hr was used to simulate the cumulative effects of multiple shorter periods of acidification in the life of a sickle cell. The relative importance of dehydratation rather than a direct effect of pH on HbSS polymerization was confirmed by incubating cells in isotonic KCl, conditions where the cells could not shrink by losing KCl, and where the effect of acid on filterability was much less. Transient acidification, such as occurs in muscle during exercise, is therefore potentially rheologically hazardous for HbSS cells.

Finally, we showed rapid inactivation of KCl cotransport activity when <u>young</u> red cells were incubated <u>in vitro</u> in saline, with less inactivation for cells in plasma (24). Albumin causes an immediate increase in activity of the transporter but cannot replace plasma in prolonging the maturation time for the KCl transport system. Presently it is not possible to identify the plasma constituent(s) responsible for this effect.

Incubation at pH 6.8 when KCl cotransport is fully operative, or at pH 7.4 when the KCl cotransporter is not active, did not affect the maturation rate. This indicates that it is irrelevant to the rate at which the transporter is rendered cryptic whether it is switched on, or not. It is also clear that the loss of KCl cotransport activity is different from the shedding of selective transporters identified by Johnstone and Teng (5), as shown in the present work by the slow loss of choline transport, and serine transport via system ASC. From the kinetic properties of the choline transport in reticulocyte-rich and mature cell fractions, it is evident that a relatively simple loss of transport units is involved rather than a kinetic alteration to the functioning of the transporter since there was no change in the K_m.

In conclusion we have demonstrated that KCl cotransport can play an important role in volume regulation in <u>young</u> red cells. The cumulative effect of recurring activation of this system by low plasma pH <u>in vivo</u> is likely to be a significant factor in regulating volume of this fraction of HbSS red cells, with the attendant danger of HbSS polymerisation and vascular obstruction.

ACKNOWLEDGEMENTS

This work was supported by the Welcome Trust and Action Research for the Crippled Child.

REFERENCES

1. M. Canessa, A. Spalvin and R. L. Nagel, Volume dependent and NEM-stimilated K, Cl transport is elevated in oxygenated SS, Sc and CC human red cells, FEBS letts, 200:197 (1986).
2. A. C. Hall and J. C. Ellory, The effects of high hydrostatic pressure on 'passive' monovalent cation transport in human red blood cells, J. Membrane Biol., 94:1 (1986a).
3. M. Canessa, M. E. Fabry, N. Blumenfeld, R.L. Nagel, Volume-stimulated Cl-dependent K efflux is highly expressed in young human red cells containing normal hemoglobin or HbS, J. Membrane Biol., 97:97 (1987).
4. J. C. Ellory and A. C. Hall, Human red cell volume regulation in hypotonic media, Comp. Biochem. Physiol. 90A: 533 (1988).
5. R. M. Johnstone and K. Teng, Membrane remodelling during reticulocyte maturation, News in Physiological Sciences 4:37 (1989).
6. V. L. Lew and R. M. Bookchin, Volume, pH and ion-content regulation in human red cells: Analysis of transient behaviour with an integrated model, J. Membrane Biol. 92:57 (1986).
7. D. Kaji, Volume-sensitive K transport in human erythrocytes, J. Gen. Physiol. 88:719 (1986).
8. A. C. Hall and J. C. Ellory, Evidence for the presence of volume-sensitive KCl transport in 'young' human red cells, Biochim. Biophys. Acta 858: 317 (1986).
9. J. C. Ellory, A. C. Hall and S. A. Ody, Is acid a more potent activator of KCl co-transport than hypotonicity in human red cells? J. Physiol., 420:149P (1989).
10. C. Brugnara and D. C. Tosteson, Cell volume, K transport and cell density in human erythrocytes, Amer. J. Physiol., 252:C269 (1987).
11. J. C. Ellory, P. B. Dunham, P. J. Logue and G. W. Stewart, Anion-dependent cation transport in erythrocytes, Phil. Trans. Roy. Soc. (London), B299:483 (1982).
12. P. K. Lauf, N. C. Adragna and R. P. Garay, Activation by N-ethylmaleimide of a latent KCl: flux in human red blood cells, Amer. J. Physiol. 246:C385 (1984).
13. P. B. Dunham and P. J. Logue, Potassium-chloride cotransport in resealed human red cell ghosts 250: C578 (1986).
14. J. R. Sachs, Volume-sensitive K influx in human red cell ghosts, J. Physiol. 92:685 (1988).
15. J. Funder and J. O. Weith, Chloride and hydrogen ion distribution between human red cells and plasma, Acta Physiol. Scand. 68:234 (1966).
16. K. Kirk, P. W. Kuchel and R. J. Labotka, Hypophosphite ion as a ^{31}P nuclear magnetic resonance probe of membrane potential in erythrocyte suspensions, Biophysical J. 54:241 (1988).
17. J. D. Young and J. C. Ellory, in "Red Cell Membranes - A Methodological Approach". Eds J. C. Ellory and J. D. Young, Academic Press, London.
18. J. V. Dacie and S. M. Lewis, "Practical Haematology", 5th ed. Churchill Livinstone, London.
19. J. Stuart, P. C. W. Stone, D. Bareford, N. M. Caldwell, J. E. Davies and S. Barr, Evaluation of leucocyte removal methods for studies of erythrocyte deformability, Clin. Hemorheol, 5:137 (1985).

20. M. N. Johnston, J. C. Ellory and J. Stuart, Bepridil protects sickle cells against the adverse rheological effects of cyclical deoxygenation, Br. J. Haematol, 73:522 (1989).
21. J. Dormandy, P. Flute, A. Matrai, L. Bogar, J. Mikita, G. D. O. Lowe, J. Anderson, S. Chien, E. Schmalzer and A. Herschenfeld, The new St. George's blood filtrometer, Clin. Hemorheol. 5:975 (1985).
22. Y. Y. Bilto, M. Player, M. J. West, J. C. Ellory and J. Stuart, Effects of oxpentifylline on erythrocyte cation content, hydration and deformability, Clin. Hemorheol. 7: 561 (1987).
23. J. C. Ellory, A. C. Hall, S. O. Ody, H. C. Englert, D. Mania and H. J. Lang, Selective inhibitors of KCl cotransport in human red cells, FEBS Letters, 262:215 (1990).
24. A. C. Hall, L. Bianchini and J. C. Ellory, Pathways for cell volume regulation via potassium and chloride loss, in: "Ion Transport", D. Keeling and C. Benham, ed.,Academic Press, London (1989), pp. 217-235.
25. C. Brugnora, T. Van Ha and D. C. Tosteson, Acid pH induces formation of dense cells in sickle erythrocytes, Blood, 74:487 (1989).
26. V. L. Lew, C. J. Freeman, O. E. Ortiz and R. M. Bookchin, A new hypothesis on the origin of irreversibly sickled cells (ISCs): Predictions from an integrated reticulocyte model, Clinical Research 36:567A (1988).
27. J. A. Halperin, C. Brugnara, M. T. Tosteson, T. Van Ha and D. C. Tosteson, Voltage-activated cation transport in human erythrocytes, Amer. J. Physiol. 257:C986 (1989).

CONFORMATIONAL CHANGES AND OXIDATION OF MEMBRANE

PROTEINS IN SENESCENT HUMAN ERYTHROCYTES

Augusta Brovelli, Maria A. Castellana,
Giampaolo Minetti, Giampiero Piccinini,
Claudio Seppi, Maria R. De Renzis
and Cesare Balduini

Department of Biochemistry
University of Pavia
Pavia, Italy

INTRODUCTION

Human red cells spend 120 days in the circulation and are then removed in an age-dependent manner (1). Since cell destruction is age-dependent, studies about red cell senescence focused on the mechanisms by which the aging of the cell leads to its destruction. The presence of autoantibodies on the surface of senescent cells produced the development of the autoimmune hypothesis for senescent cell removal from the circulation (2-4), and raised questions about the presence of senescence markers on the cell surface that permit such recognition and the mechanisms of their development during red cell life span. Studies on surface changes taking place during red cell senescence have been carried out mainly on density-separated red cells (5). A reduction in membrane surface area in the dense cell population is evident as a decrease in membrane cholesterol and phospholipid content (6,7) and in acetylcholinesterase activity and sialic acid content (8). Cell deformability decreases (9-12) and at the level of the membrane slight modifications of the covalent structure of some components have been described, produced by processes like oxidation (13-15), proteolysis (16, 17), glycation (18), methylation and transamidation (19), phosphorylation (20), and modifications of phospholipid asimmetry (21) and of topology and topography of proteins have been reported or hypothesized (22-25). Most of these modifications are effective in promoting autoantibody binding and/or phagocytosis in vitro, thus supporting a possible role of these mechanisms in determining recognition and removal of senescent cells. Investigations carried out with in vivo (26,27) and in vitro models (28,29) for red cell senescence and studies with mutant erythrocytes showed that oxidation plays a relevant role in determining surface properties of senescent cells and of many pathological cells with a decreased life span (30-32). Since the

Red Blood Cell Aging, Edited by M. Magnani and
A. De Flora, Plenum Press, New York, 1991

oxidative state of membrane proteins in human red cells of different age has not been investigated in detail in the past, we tried to quantitate the oxidative lesion the membrane proteins undergo during red cell life-span, in an attempt to understand what kind of membrane processes expressed in senescent red cells can be related to oxidation.

Diamide treatment of dog (26)and human (28) red cells has been used as a model of cell senescence. Diamide is an oxidant of thiol groups (33) and treatment of human red cells with micromolar diamide concentrations was shown to induce, as in senescent cells, the binding to the cell surface of anti-band 3 auto-antibodies and the complement-dependent phagocytosis of these cells (28). Since diamide treatment of red cells has been used to mimic cell aging at the membrane level, the oxidative processes concerning membrane proteins and occurring in diamide-treated red cells were compared to oxidative processes evidentiated in senescent red cell membrane.

METHODS

Separation of erythrocytes of different cell age and preparation of ghost membranes

Fresh human blood was collected in 3.8% (w/v) sodium citrate as anticoagulant. Erythrocytes were separated from plasma by centrifugation at room temperature at 1,700 g for 5 min. The cells suspended in 1 volume of 154 mM NaCl, 4.5 mM KCl, 5 mM phosphate, pH 7.4 (PBS) containing 0.2 mM phenylmethanesulphonyl fluoride (PMSF), passed through a mixture of α -and microcrystalline-cellulose as described by Beutler et al. (34) and then washed three times in PBS containing 0.2 mM PMSF. Erythrocytes of different cell density were prepared as previously described (35) by a modified version of Murphy's method (36). Filtered and washed erythrocytes were suspended in autologous plasma (hematocrit 60%) in the presence of 5 mM glucose and centrifuged for 1 h at 1,000 g at 20°C. The sedimented cells were divided into 8 fractions and the second, fourth and eighth fractions from the top were collected and submitted to the treatments described below. The creatine content of fractionated red cells (37,38) served as a cell age parameter. Washed erythrocytes were lysed in 10 volumes of 5 mM Tris/HCl buffer containing 0.2 mM PMSF and 1 mM ethylenediaminetetracetic acid, disodium salt (EDTA) as proteinase inhibitors to prepare ghost membranes in parallel: A) by the method of Marchesi and Palade (39) and B) by the same method, omitting the washing with 50 mM Tris-HCl buffer containing 0.5 M NaCl.

Treatment of middle-aged red cells with diamide (33)

Washed and packed middle-aged erythrocytes, corresponding to the fourth fraction, were incubated in PBS in the presence of 0.2, 0.8, 2.0 mM diamide at 37°C for 1 h by gentle agitation. Erythrocytes were collected by centrifugation at room temperature at 1,700 g for 5 min. and washed three times in PBS. An aliquot of cells was incubated without diamide and used as a control.

Evaluation of thiol distribution in membrane proteins using the thiol-reagent N-(7-dimethylamino-4-methyl-coumarinyl) maleimide (DACM) (40)

Washed and packed erythrocytes of different subpopulations were suspended in 1 volume of 6 mM DACM in PBS and processed as previously described (41). Erythrocytes were collected by centrifugation, washed three times in PBS and lysed to prepare ghost membranes (by the method B).

Treatment of different cell subpopulations with covalent inhibitors of the anion transporter band 3 protein

Washed and packed erythrocytes were divided into 3 aliquots: the first was suspended in 1 volume of 0.9 mM eosine maleimide (EM) (42) in PBS, the second in 1 volume of 0.9 mM eosine isothiocyanate (EITC) (43) in PBS and the third in 1 volume of 0.4 mM 4-4'- diisothiocyanostilbene-2-2'-disulfonic acid (DIDS) (44) in Krebs-Ringer phosphate buffer (KRp). The suspensions were incubated at 37°C for 1 h by gentle agitation in the dark. Erythrocytes were collected by centrifugation and the cells incubated in the presence of EM or EITC were washed three times in PBS. Red cells incubated in the presence of DIDS were washed five times in KRp containing 1% (w/v) bovine serum albumin, and one time ith PBS. Washed erythrocytes were lysed to prepare ghost membranes (by the method B).

Treatments of ghost membranes and SDS-polyacrylamide-gel electrophoresis (SDS-PAGE)

Ghost membranes prepared from red cells treated in situ with DACM_ were divided into 3 aliquots and denatured in reducing and unreducing conditions. An aliquot of ghost membranes denatured in unreducing conditions was treated again with 3 mM DACM, in order to label thiol groups which were not reactive in intact cells (45).
Ghost membranes prepared from red cells treated in situ with EM, EITC and DIDS were denatured in reducing conditions.
Electrophoretic separation, carried out according to Laemmli (46), was performed on polyacrylamide gel slabs (14 x 13.5 cm, 1.5 mm thickness) with running times of 4-5 h at 40 mA. The protein content of each sample was 60 μg. After electrophoresis the amount of the fluorochrome bound to each protein band was determined by quantitating the fluorescence intensity emitted at: $\lambda > 420$ nm (λ_{ex} =405 nm) for DACM, $\lambda > 570$ nm (λ_{ex} =546 nm) for EM and EITC and $\lambda > 400$ nm (λ_{ex} =366 nm) for DIDS, by scanning gels with a Camag TLC II densitometer. The same gels were then stained with Coomassie blue and scanned at 561 nm. The amount of fluorochrome bound to each protein band was evaluated by measuring the ratio between fluorescence intensity emitted by the fluorochrome and absorbance due to Coomassie blue staining.

Analytical methods

The assay of methionine sulfoxide (MetSO) content of membrane proteins was carried out on ghost membranes prepared with method A with buffers not containing PMSF. Ghosts were hydrolyzed in alkaline conditions as previously described (35) and MetSO was determined by ion exchange chromatography.

Thiol titration was performed on ghost membranes (prepared by method A) by spectrofluorometry after treatment of the sample with 0.1 mM DACM (40), as previously described (41).
Protein content was determined by the method of Lowry et al. (47) using serum albumin as a standard.
Hemoglobin (Hb) content of red cell suspensions was assayed with the Drabkin's reagent as described by Beutler (48).
Creatine content of red cells was assayed on density-separated red cell subpopulations as reported by Griffths (37).

RESULTS

Oxidation of membrane proteins in red cells of different age

Oxidative lesion of membrane proteins was studied in human erythrocytes of different age, density-separated by centrifugation in autologous plasma, and was evaluated on ghost membranes prepared by young, middle-aged and senescent cells, by assaying thiol and methionine sulfoxide groups, and in situ on intact cells, after treatment of red cell subpopulations with DACM, a coumarinyl derivative of maleimide, by measuring the amount of this thiol-reagent bound in situ by each membrane protein. Results obtained show that during aging of normal cells the oxidative state of membrane proteins increases (Tab. 1).

Table 1. MetSO and Thiol Group Content of Ghost Membranes Obtained from Erythrocytes of Different Density.

Cell	MetSO nmoles/mg protein n=7	t test[a]	Creatine µg/g Hb
young	25.41±5.97		190±40
middle-aged	36.91±6.79	$0.0005 < p \leq 0.005$	126±31
senescent	39.91±9.52	$0.0005 < p \leq 0.005$	62±19

Cell	% Thiol groups n=4	Creatine µg/g Hb
young	100	175±60
middle-aged	97±18	116±40
senescent	104±11	53±19

[a] Middle-aged and senescent cells were compared to the young ones; MetSO content of ghost membranes obtained from middle-aged and senescent cells did not differ significantly.

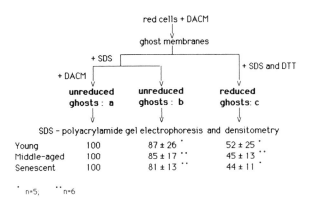

Fig. 1. Treatment of red cells with DACM and preparation of ghost
membranes for SDS–polyacrylamide gel electrophoresis: after
incubation of red cells with DACM ghost membranes were
prepared and divided into 3 aliquots. Sample a was used to
evaluate the total amount of thiols present (both accessible
in situ and after treatment of SDS–denatured ghosts with DACM),
samples b and c were used to evaluate the amount of thiols
accessible to the reagent after treatment of membranes in
unreducing and reducing conditions respectively. SDS: sodium
dodecylsulfate; DACM: N-(7-dimethyl amino-4-methyl-coumarinyl)
maleimide; DTT: dithiothreitol.

This was better shown by the assay of methionine sulfoxide residues
than by thiol titration, when studies were carried out on ghost membranes.
MetSO assay evidentiated a significant increase in the oxidation of
methionine from young to middle-aged and senescent erythrocytes,
whereas no significant difference appeared between middle-aged and senescent
cells (Tab.1). The amount of the sulfoxide in middle-aged erythrocytes
is 50% higher than the one found in membrane proteins of young cells, and
it does not increase further.

Thiol content of proteins determined in ghost membranes appeared
similar in all cell subpopulations studied (Tab. 1). In order to obtain
informations about the thiol state of individual membrane proteins, their
thiol state was assayed in intact cells by spectrofluorometry, using a
coumarinyl derivative of meleimide, the DACM, that after binding to free
thiols, becomes fluorescent. This reagent crosses the membrane and binds to
accessible thiols of integral and peripheral proteins. After ghost
preparation, membranes obtained from these cells were submitted to
SDS–PAGE: densitometry of fluorescent and Coomassie blue-treated gels
allowed us to quantitate the amount of DACM bound to each membrane
protein that was indicative of the amount of accessible thiol groups. The
amount of DACM bound was quantitated by measuring the ratio between
fluorescence intensity emitted by each protein and absorbance due to
Coomassie blue staining. Procedure adopted to evaluate the response of
membrane proteins to DACM treatment in intact cells is summarized in Fig. 1.
After incubation of cells with DACM and preparation of labelled ghosts, an

aliquot of membranes was treated with SDS and then again with DACM, to also label thiol groups which were not accessible in situ, and fluorescence emitted by each protein band after these treatments was taken as 100% value. After SDS-PAGE separation of ghost proteins, both in reducing and unreducing conditions, 85% thiol groups were found labelled in proteins after electrophoresis in unreducing conditions and 50% after electrophoresis in reducing conditions. No significant differences were observed in the three cell subpopulations studied. Since dithiothreitol used to prepare samples for electrophoresis in reducing conditions seems to displace part of the DACM bound by membrane proteins in intact cell, quantitation of DACM bound in intact cell by each protein was carried out on unreduced samples. Results obtained by calculating for each protein the ratio between fluorescence intensity emitted by DACM and absorbance due to Coomassie blue staining showed that membrane proteins bind the same amount of DACM in young and middle-aged erythrocytes.

On the contrary, in senescent red cells a significant decrease in the amount of DACM bound by band 3 protein, α-spectrin, 4.1 and 4.2 proteins was evident (Tab. 2). Similar percent recovery for each protein was found in middle-aged and senescent erythrocytes. Results obtained after densitometry of electrophoretograms obtained in unreducing conditions were very similar to those obtained in reducing conditions, previously reported (41), and are indicative of conformational changes of band 3 and of the main proteins of the membrane skeleton in senescent cells, or of oxidation of these proteins, or of both the processes.

<u>Functional analysis of band 3 protein in erythrocytes of different age</u>

Results obtained in studies about thiol distribution in individual membrane proteins in red cell subpopulations of different age support the

Table 2. Thiol Group Distribution in Membrane Proteins of DACM-treated Middle-Aged and Senescent Erythrocytes

Protein	F.I./O.D. %		t test	Modified thiols[c]
	middle-aged[a]	senescent[b]	n = 5	
Spectrin-	100	69±20	$0.010 < p \leq 0.025$	
				3-11
Spectrin-ß	100	76±31	n.s.	
Band 3	100	70±19	$0.010 < p \leq 0.025$	1-2
Band 4.1	100	69±25	$0.025 < p \leq 0.050$	2
Band 4.2	100	72±27	$0.025 < p \leq 0.050$	4
Actin	100	74±30	n.s.	1

[a] Creatine content: 109 ± 27 µg/g Hb.
[b] Creatine content: 53 ± 5 µg/g Hb.
[c] Estimated on the basis of the Cys content of each protein.

conclusion that conformational changes or oxidation of band 3 occur in senescent cells. To investigate if these processes affect the functional capacity of the protein in the anion translocation (49), functional analysis of band 3 protein was carried out on density-separated young, middle-aged and senescent cells, by measuring the binding to band 3 of three covalent fluorochrome inhibitors of the anion exchange: DIDS, EM and EITC, directed to different domains of the molecule (49). Intact cells of different age were treated with the fluorochrome, then ghost membranes were prepared from the labelled cells and submitted to SDS-PAGE. The amount of fluorochrome bound to band 3 was determined by densitometry of fluorescent gels and of the same gels after Coomassie blue staining, by the same procedure adopted to quantify the binding of DACM. Densitometry of fluorescent gels after electrophoresis of ghosts obtained from red cells treated with DIDS, EM, EITC is reported in Fig. 2 and shows that the three fluorochromes specifically bind to band 3. Quantitative data obtained after treatment of young, middle-aged and senescent cells with the three anion transport inhibitors are shown in Tab. 3: a decrease in the amount of DIDS and EITC bound by band 3 is evident in middle-aged red cells, while no difference between middle aged and senescent cells appears. Band 3 seems to bind the same amount of EM in young, middle-aged and senescent red cells.

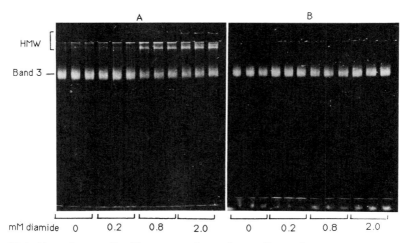

Fig. 2. Densitometry of fluorescent gels after electrophoresis of ghosts prepared from red cells treated with DIDS, EM and EITC. Beside band 3, a fluorescent band in the spectrin region, probably correponding to band 3 dimer, is evident in electrophoretograms of ghosts prepared from DIDS-treated cells.

Membrane processes in diamide-treated red cells

Oxidation of red cells with diamide has been used as a model of senescent cells (26,28), in particular to mimic cell aging at the membrane level. Since diamide treatment induces a decrease of reduced glutathione concentration, we supposed that after treatment of red cells with this reagent other phenomena besides thiol oxidation and membrane protein clustering could take place. Therefore after treatment of a middle-aged red cell subpopulation with 0.2, 0.8 and 2.0 mM diamide the following membrane processes were studied: membrane protein oligomerization, presence of band 3 in membrane protein aggregates, thiol oxidation in ghosts and in intact cells, oxidation of methionine to its sulfoxide in ghost membrane proteins. Oxidative processes concerning the membrane of senescent and diamide-treated cells were compared.

The presence of protein aggregates was analyzed by SDS-PAGE in reducing and unreducing conditions of ghosts prepared from diamide-treated cells, and was found on the membrane of red cells treated with 0.8 and 2.0 mM diamide. When red cells were treated with EM, a fluorochrome that specifically binds to band 3, the membrane protein aggregates appear to be fluorescent, indicating that band 3 participates in the formation of these protein clusters (Fig. 3). Quantitation of thiol groups in ghost membranes obtained from diamide-treated red cells showed a significant decrease of free thiols only in samples treated with 2.0 mM diamide (Tab. 4). Furthermore it is worth observing that a slight, but significant increase in the amount of thiols can be evidentiated in samples treated with 0.2 mM diamide.

Results of thiol distribution in membrane proteins in intact cells after diamide treatment cannot be described quantitatively due to the formation of membrane protein aggregates. Besides oxidation of thiols, oxidation of methionine residues is evident in membrane proteins of 0.2 mM and 2.0 mM diamide-treated cells (Tab. 4).

Table 3. Binding to Band 3 of Covalent Inhibitors of the Anion Exchange.

Cell	DIDS %	n	EM %	n	EITC %	n	Creatine μg/g Hb
young	100		100		100		159±52
middle-aged	84±16°	8	103±13	7	90±8°°	8	90±25
senescent	80±11°°	5	96±26	5	86±7°°	6	43±12

t test: °0.01 < p ≤ 0.05; °°p ≤ 0.01.
Middle-aged and senescent cells were compared to young cells: no differences were evident when senescent and middle-aged erythrocytes were compared.

DISCUSSION

In searching for parameters quantitating the oxidative state of membrane proteins in human erythrocytes during senescence, we observed that assay of MetSO content of ghost membranes is a good indicator of the oxidative process, since MetSO content of membrane proteins is 50% higher in middle aged and senescent red cells with respect to younger cells. Oxidation of methionine takes place early during red cell life-span in the circulation since it is already evident in middle-aged cells and reaches the same extent in middle-aged and senescent erythrocytes. A possible explanation of this behaviour could be the degradation of heavily oxidized proteins (50-52) and loss of membrane fragments containing the damaged material (41). In fact the cell size decreases during red cell life, and loss of membrane fragments takes place (6,53). Oxidation of methionine residues is expected to lead to conformationalchanges of proteins and partial loss of their function, and could finally trigger their degradation. In this case the assay of MetSO content of membrane proteins could also monitor, besides oxidation, the loss of membrane material through the red cell life in the circulation.

The evaluation of thiol group content of membrane proteins is not informative about oxidative processes, when performed on ghost membranes obtained from different red cell subpopulations. However when thiol state of membrane proteins was measured in intact cells, and individual membrane proteins were considered, after SDS-PAGE separation, a significant decrease of the amount of the thiol-reagent DACM bound to band 3 protein and to α -spectrin, 4.1 and 4.2 proteins was observed in senescent cells.

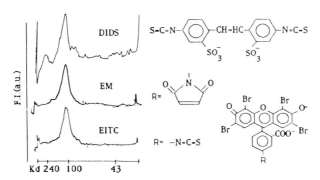

Fig. 3. SDS-polyacrylamide gel electrophoresis of ghost membranes prepared from red cells after treatment with 0.2, 0.8 and 2.0 mM diamide and incubation with 0.45 mM EM. Fluorescence emitted by EM bound to band 3 is evident on the gels. Electrophoresis was carried out in unreducing (A) and reducing (B) conditions. High molecular weight aggregates (HMW) disappeared after reducing treatment.

This can be indicative of oxidation or of conformational changes of these proteins in senescent erythrocytes, and from these processes a different topology or topography of membrane proteins can result. Two clinically important examples where oxidation has been demonstrated to interfere with supramolecular organization of membrane proteins and membrane stability are in sickle cell anemia and in cells stored under blood bank conditions. Protein 4.1 isolated from sickled erythrocytes contains less free cysteine, as well as less methionine and tyrosine. Oxidation of protein 4.1 apparently has functional consequences, since sickle protein 4.1 is impaired in its ability to bind to erythrocyte membranes (54). Spectrin isolated at different times during storage of blood exhibited a gradual decrease in its ability to bind to actin, that could be reversed to normal by reduction with dithiothreitol (55). A likely consequence of spectrin oxidation is loss of membrane surface area due to the formation of vesicles from areas of unsupported lipid bilayer (56).

Oxidation of methionine and cysteine could participate in a different way to the expression of membrane properties typical of senescent red cell, since the antioxidant equipment of the erythrocyte is differently effective against the two lesions. A methionine sulfoxide reductase, which reduces protein-bound methionine sulfoxide to methionine, occurs in numerous organisms and tissues (57,58). The presence of this enzyme has not been described in erythrocytes; therefore oxidation of methionine cannot be repaired by red cells, unless by degradation of heavily oxidized proteins (50-52). The accumulation of methionine sulfoxide in middle-aged cells could trigger degradation of proteins and lead to local instability of the membrane and loss of membrane fragments. Interestingly, comparing oxidative processes evident in physiologically aged cells and in diamide-treated cells, oxidation of methionine to its sulfoxide appears to be produced by milder oxidative conditions (0.2 mM diamide) than those determining a decrease of thiol group content in membrane proteins (2.0 mM diamide) in diamide-treated red cells, and to take place earlier than thiol modification during red cell circulating life.

Table 4. MetSO and Thiol Group Content of Ghost Membranes Prepared from Diamide Treated Erythrocytes

Diamide mM	MetSO %	n	t test	Thiol groups %	t test n=8
0.0	100			100	
0.2	177±52	5	$0.010 < p \leq 0.025$	107±6	$0.005 < p \leq 0.01$
0.8	136±40	4	n.s.	102±7	n.s.
2.0	132±24	4	$0.025 < p \leq 0.050$	76±9	$p \leq 0.0005$

An increase of diamide concentration leads to a relative decrease of sulfoxide, and since it is reasonable to suppose the possibility of protein degradation and of membrane fragmentation in diamide-treated cells, this hypothesis is now under investigation. Red cells are well equipped to maintain the reduced state of thiol groups, through the glutathione system. In senescent cells an impairment of the redox systems (5) could lead to oxidation of some thiols in membrane proteins. In fact in senescent cells band 3 protein, α-spectrin, 4.1 and 4.2 proteins appear to have a decreased thiol content or to have undergone conformational changes resulting in a decreased accessibility of their thiol groups to the reagent DACM. Band 3 protein is involved in the expression on the senescent cell surface of the neoantigen promoting IgG binding and clearance by macrophages (2-4,59,60). To relate structural properties of this molecule to its function of anion transporter in young, middle-aged and senescent cells, a functional analysis was carried out. Although the protein aggregation induced by the anion transport inhibition and eventual differences in the labelling time-course deserve further investigation, results so far obtained showed a decrease in the binding of DIDS and EITC in middle-aged and senescent cells, and no differences in the EM binding. Methionine oxidation and functional impairment of anion binding sites seem to be concomitant. Although we have at present no data about a possible relationship between oxidation of methionine and functional impairment of band 3, oxidation of this molecule could be the earliest event leading to the senescent antigen production (23,24,27).

Results obtained in these studies about the mechanisms of development of senescence markers during the red cell life-span suggest several considerations. If one studies the presence of senescence markers on the cell surface that permit recognition of aged red cell for removal, comparison between membrane properties of middle-aged and senescent cells has to be done, and modification of thiols we observed in senescent cells seems to be a senescence marker. If one investigates the mechanisms producing the appearance of senescence markers, investigation about the membrane processes taking place during all the cell life span in the circulation could give more information. Studies carried out so far show that membrane fragmentation is a membrane process taking place throughout the cell life-span, and our studies have shown that oxidation of methionine and a functional impairment of band 3 are already expressed in middle-aged red cells. Oxidation and proteolysis of oxidized proteins are among the processes leading to this membrane fragmentation. Loss of membrane fragments is an aging process and immunological mechanisms proposed for the clearance of senescent cells could operate in the removal of membrane fragments as well (61). A senescent cell binds IgGs irreversibly, and is a cell no longer able to undergo loss of plasma membrane. Why is this limit acheived? What happens on the cell surface when a cell is no longer able to repair its membrane by losing damaged regions? The answer to these questions could help to understand how senescent markers develop.

ACKNOWLEDGMENTS

This work was supported by grants to A.B. from MURST and from CNR (Target Project Biotechnology), and by CNR (Target Project Ageing) to C.B.

REFERENCES

1. G. S. Eadie and I. W. Brown, Red blood cell survival studies, Blood 8:1110 (1953).
2. M. M. B. Kay, Mechanism of removal of senescent cells by human macrophages "in situ", Proc. Natl. Acad. Sci. USA 72:3521 (1975).
3. M. M. B.Kay, Role of physiologic autoantibody in the removal of senescent human red cells, J. Supramol. Struct. 9:555 (1978).
4. H. U. Lutz and G. Stringaro-Wipf, Senescent red cell-bound IgG is attached to band 3 protein, Biomed. Biochim. Acta 42:S117 (1983).
5. M. R. Clark, Senescence of red blood cells: progress and problems, Physiol. Rev. 68:503 (1988).
6. M. P. Westerman, L. E. Pierce and W. N. Jensen, Erythrocyte lipids: a comparison of normal young and normal old populations, J. Lab. Clin. Med. 62:394 (1963).

7. C. C. Winterbourn and R. D. Batt, Lipid composition of human red cells of different ages, Biochim. Biophys. Acta 202:1 (1970).

8. N. S. Cohen, J. E. Ekholm, M. G. Luthra and D. J. Hanahan, Biochemical characterization of density-separated human erythrocytes, Biochim. Biophys. Acta 419:229 (1976).

9. M. R. Clark, N. Mohandas and S. B. Shohet, Osmotic gradient ektacytometry: comprehensive characterization of red cell volume and surface maintenance, Blood 61:899 (1983).

10. G. B. Nash and H. J. Meiselman, Red Cell and ghost viscoelasticity. Effects of hemoglobin concentration and in vivo aging, Biophys. J. 43:63 (1983).

11. R I. Weed, The importance of erythrocyte deformability, Am. J. Med. 49: 147 (1970).

12. A. R. Williams and D.R. Morris, The internal viscosity of the human erythrocyte may determine its life-span in vivo, Scand. J. Haematol. 24:57 (1980).

13. L. M. Snyder, L. Leb, J. Piotrowski, N. Sauberman, S. C. Liu and N. L. Fortier, Irreversible spectrin-haemoglobin crosslinking in vivo: a marker for red cell senescence, Brit. J. Haematol. 53:379 (1983).

14. H. Q. Campwala and J. F. Desforges, Membrane-bound hemichrome in density-separated cohorts of normal (AA) and sickled (SS) cells, J. Lab. Clin. Med. 99: 25 (1982).

15. S. K. Jain, Evidence for membrane lipid peroxidation during the in vivo aging of human erythrocytes, Biochim. Biophys. Acta 937:205 (1988).

16. M. Morrison, K. S. Au and L. Hsu, Are the red cell proteases a clock mechanism which turns on a signal of senescence?, Biomed. Biochim. Acta 46:S79 (1987).

17. M. M. B. Kay and J. R. Goodman, IgG antibodies do not bind to band 3 in intact erythrocytes; enzymatic treatment of cells is required for IgG binding, Biomed. Biochim. Acta 43: 841 (1984).

18. S. P. Sutera, R. A. Gardner, C. W. Boylan, G. L. Carrol, K. C. Chang, J. S. Marvel, C. Kilo, B. Gonen and J. R. Williamson, Age-related changes in deformability of human erythrocytes, Blood 65:275 (1985).

19. J. R. Barber and S. Clarke, Membrane protein carboxyl methylation increases with human erythrocyte age, J. Biol. Chem. 258:1189 (1983).

20. G. Fairbanks, J. Palek, J. E. Dino and P. A. Liu, Protein kinase and membrane protein phosphorylation in normal and abnormal human erythrocytes: variation related to mean cell age, Blood 61: 850 (1983).

21. R. A. Schlegel, L. McEvoy, M. Weiser and P. Williamson, Phospholipid organization as a determinant of red cell recognition by the reticuloendothelial system, in: "Red blood cells as carriers for drugs – Potential therapeutic applications", C. Ropars, M. Chassaigne and C. Nicolau eds., Vol. 67, Pergamon Press, Oxford (1987).

22. E. Schweizer, W. Angst and H. U. Lutz, Glycoprotein topology on intact human red blood cells reevaluated by cross-linking following amino group supplementation, Biochemistry 21:6807 (1982).

23. H. U. Lutz, R. Flepp and G. Stringaro-Wipf, Naturally occurring autoantibodies to exoplasmic and cryptic regions of band 3 protein, the major integral membrane protein of human red blood cells, J. Immunol. 133:2610 (1984).

24. P. S. Low, S. M. Waugh, K. Zinke and D. Drenckhahn, The role of hemoglobin denaturation and band 3 clustering in red blood cell aging, Science 227: 531 (1985).

25. M. M. B. Kay, G. J. C. G. M. Bosman and C. Lawrence, Functional topography of band 3: specific structural alteration linked to functional aberrations in human erythrocytes, Proc. Natl. Acad. Sci. U.S.A. 85: 492 (1988).

26. G. J. Johnson, D. W. Allen, T. P. Flynn, B. Finkel and J. G. White, Decreased survival 'in vivo' of diamide-incubated dog erythrocytes, J. Clin. Invest. 66:955 (1980).

27. M. M.B. Kay, G. J. C. G. M. Bosman, S. S. Shapiro, A. Bendich and P. S. Bassel, Oxidation as a possible mechanism of cellular aging: vitamin E deficiency causes premature aging and IgG binding to erythrocytes. Proc. Natl. Acad. Sci.USA 83: 2463 (1986).

28. P. Arese, F. Bussolino, R. Flep, P. Stammler, S. Fasler and H. U. Lutz, Diamide enhances phagocytosis of human red cell in a complement and anti band 3 antibody-dependent process, Biomed. Biochim. Acta 46: S84 (1987).

29. M. Beppu, A. Mizukami, M. Nagoya and K. Kikugawa, Binding of anti-band 3 autoantibody to oxidatively damaged erythrocytes, J. Biol. Chem. 265:3226 (1990).

30. B. H. Rank, J. Carlsson and R.P. Hebbel, Abnormal redox status of membrane-protein thiols in sickle erythrocytes, J. Clin. Invest. 75: 1531 (1985).

31. P. S. Becker, J. S. Morrow and S. E. Lux, Abnormal oxidant sensitivity and -chain structure of spectrin in hereditary spherocytosis associated with defective spectrin-4.1 binding, J. Clin. Invest. 80:557 (1987).

32. P. Arese and A. De Flora, Pathophysiology of hemolysis in glucose-6-phosphate dehydrogenase deficiency, Semin. Hematol. 27:1 (1990).

33. N. S. Kosower, E. M. Kosower and B. Wertheim, Diamide, a new reagent for the intracellular oxidation of glutathione to the disulfide, Biochem. Biophys. Res. Commun. 37.593 (1969).

34. E. Beutler, C. West and K. G. Blume, The removal of leukocytes and plateles from whole blood, J. Lab. Clin. Med. 88:328 (1976).

35. A. Brovelli, C. Seppi, G. Pallavicini and C. Balduini, Membrane processes during 'in vivo' aging of human erythrocytes, Biomed. Biochim. Acta 42:S122 (1983).

36. J. R. Murphy, Influence of temperature and method of centrifugation on the separation of erythrocytes, J. Lab. Clin. Med. 82: 334 (1973).

37. W. J. Griffiths, The determination of creatine in body fluid and muscle and of phosphocreatine in muscle, using the autoanalyzer, Clin. Chim. Acta 9:210 (1964).

38. J. Fehr and M. Knob, Comparison of red cell creatine level and reticulocyte count in appraising the severity of hemolytic processes, Blood 53:966 (1979).

39. V.T. Marchesi and J. E. Palade, The localization of Mg-Na-K-activated adenosine triphosphatase on red cell ghost membranes, J. Cell Biol. 35 :385 (1967).

40. K. Yamamoto, T. Sekine and Y. Kanaoka, Fluorescent thiol reagents – Fluorescent tracer method for protein SH groups using N-(7-dimethylamino-4-methyl-coumarinyl) maleimide. An application to the proteins separated by SDS-polyacrylamide gel electrophoresis, Anal. Biochem. 79:83 (1977).

41. A. Brovelli, C. Seppi, A. M. Castellana, M. R. De Renzis, A. Blasina and C. Balduini, Oxidative lesion to membrane proteins in senescent erythrocytes, Biomed. Biochim. Acta 49:S218 (1990).

42. E. Nigg, M. Kessler and R. J. Cherry, Labeling of human erythrocyte membranes with eosin probes used for protein diffusion measurements -Inhibition of anion transport and photooxidative inactivation of acetylcholinesterase, Biochim. Biophys. Acta 550:328 (1979).

43. T. Chiba, Y. Sato and Y. Suzuki, Characterization of eosin 5-isothiocyanate binding site in band 3 protein of the human erythrocyte, Biochim. Biophys. Acta 897: 14 (1987).

44. M. K. Ho and G. Guidotti, A membrane protein from human erythrocytes involved in anion exchange, J. Biol. Chem. 250: 675 (1975).

45. C. Seppi, M. A. Castellana, G. Minetti, G. Piccinini, C. Balduini and A. Brovelli, Evidence for membrane protein oxidation during 'in vivo' aging of human erythrocytes, Mech. Age. Dev. in press.

46. U. K. Laemmli, Cleavage of structural proteins during the assembly of the head of bacteriophage T4, Nature 227:680 (1970).

47. O. H. Lowry, N. J. Rosebrough, A. L. Farr and R. J. Randall, Protein measurement with the Folin phenol reagent, J. Biol. Chem. 193:265 (1951).

48. E. Beutler, The preparation of red cells for assay, in "Red cell metabolism – A manual of Biochemical methods" 3rd edition, Grune and Stratton, New York (1984).

49. D. Jay and L. Cantley, Structural aspects of the red cell anion exchange protein, Ann. Rev. Biochem. 55:511 (1986).

50. R. L. Levine, C. N. Oliver, R. M. Fulks and E. R. Stadtman, Turnover of bacterial glutamine synthetase: oxidative inactivation precedes proteolysis, Proc. Natl. Acad. Sci. USA 78:2120 (1981).

51. A. J. Rivett, Preferential degradation of the oxidatively modified form of glutamine synthetase by intracellular mammalian proteases, J. Biol. Chem. 260:300 (1985).

52. K. J. A. Davies and A. L. Goldberg, Proteins damaged by oxygen radicals are rapidly degraded in extracts of red blood cells, J. Biol. Chem. 262:8227 (1987).

53. M. Morrison, A. W. Michaels, D. R. Phillips and S. Choi, Life span of erythrocyte membrane protein, Nature 248:763 (1974).

54. R. S. Schwartz, A. C. Rybicki, R. Health and B. H. Lubin, Protein 4.1 in sickle erythrocytes – Evidence for oxidative damage, J. Biol. Chem. 262:15666 (1987).

55. L. C. Wolfe, A. M. Byrne and S. E. Lux, Molecular defect in the membrane skeleton of blood bank-stored red cells, J. Clin. Invest. 78:1681 (1986).

56. G. M. Wagner, D. T-Y. Chiu, J-H. Qju, R. H. Heath and B. H. Lubin, Spectrin oxidation correlates with membrane vesciculation in stored RBCs, Blood 69:1777 (1987).

57. N. Brot, L. Weissbach, J. Werth and H. Weissbach, Enzymatic reduction of protein-bound methionine sulfoxide, Proc. Natl. Acad. Sci. USA 78:2155 (1981).

58. A. Spector, R. Scotto, H. Weissbach and N. Brot, Lens methionine sulfoxide reductase, <u>Biochem. Biophys. Res. Commun.</u> 108:429 (1982).
59. M. M. B. Kay, S. R. Goodman, K. Sorensen, C. F. Whitfield, P. Wong, L. Zaki and V. Rudloff, Senescent cell antigen is immunologically related to band 3, <u>Proc. Natl. Acad. Sci. USA</u> 80:1631 (1983).
60. M. M. B. Kay, Localization of senescent cell antigen on band 3, <u>Proc. Natl. Acad. Sci. USA</u> 81:5753 (1984).
61. H. Mueller and H. U. Lutz, Binding of autologous IgG to human red blood cells before and after ATP-depletion - Selective exposure of binding sites (autoantigens) on spectrin-free vesicles, <u>Biochim. Biophys. Acta</u> 729:249 (1983).

SELECTIVE BINDING OF MET–HEMOGLOBIN TO ERYTHROCYTIC MEMBRANE:

A POSSIBLE INVOLVEMENT IN RED BLOOD CELL AGING

Bruno Giardina[1], Roberto Scatena[1], Maria E. Clementi[1],
Maria T. Ramacci[2], Franco Maccari[2], Loredana Cerroni[1]
and Saverio G. Condò [1]

[1] Dipartimento di Medicina Sperimentale e Scienze
Biochimiche II Università di Roma – Roma, Italy
[2] Istituto di Ricerca sulla Senescenza
Sigma Tau, Pomezia, Italy

INTRODUCTION

It is well known that in vivo and under normal physiological conditions intraerythrocytic hemoglobin may exist in three different forms represented by oxygenated, deoxygenated and partially oxidized hemoglobin (1–4). Apart from the first two derivatives whose relative proportions are continuously changing during the oxygenation deoxygenation cycle, met–hemoglobin is normally present at a steady–state level of about 1%.

Because of the very high concentration of the intraerythrocytic hemoglobin it was obvious to investigate its interaction with the plasmic membrane of the cell. Some of these studies, performed in the last decade, have clearly demonstrated that while spectrin binds preferentially the oxygenated derivative of hemoglobin (5), the cytoplasmic fragment of band 3 (the anion transport protein of the erythrocyte), actin and tubulin bind preferentially to deoxyhemoglobin very likely at the level of the 2,3–DPG binding site (6–8). However these reversible interactions still remain of uncertain physiological significance.

On the other hand, in spite of the reversibility of the hemoglobin membrane interactions outlined above, it is well known that it remains difficult to prepare hemoglobin–free red cell membranes. The apparent paradox of this observation could be explained taking into account the presence of a given percentage of met–Hb; thus we have to consider that about 3% of the total hemoglobin is cycled to met–Hb each day (9) and that this derivative may play an important role in the initiation and development of lipid peroxidation after coming into contact with the membrane lipids.

Red Blood Cell Aging, Edited by M. Magnani and
A. De Flora, Plenum Press, New York, 1991

In this perspective we have investigated the binding of met-Hb to the red cell membrane trying to put in evidence a preferential binding of this derivative besides that already known in the case of hemichromes (10-12).

MATERIALS AND METHODS

Heparinized or citrated blood was obtained from volunteer normal individuals, from patients heterozygous for G6PD deficiency and from blood transfusion units.

Erythrocytes were isolated by centrifugation at 500g for 10 min at 5°C. The supernatant and buffy coat were removed and the packed cells were washed three times by suspension in isotonic phosphate buffer (5mM sodium phosphate plus 0.15M NaCl at pH 8.0), followed by centrifugation, as above. Washed erythrocytes were hemolyzed with 60 volumes of cold hypotonic phosphate buffer (5mM sodium phosphate, 0.1mM PMSF, 1mM Na EDTA, pH 8.0).

The packed red ghosts were washed with 5mM sodium phosphate buffer (pH 8.0) until supernatant was completely hemoglobin free, twice with 5mM sodium phosphate plus 0.15M NaCl at pH 8.0 in order to release possibly remaining Hb still bound to the membrane by ionic interactions, and finally with 5mM sodium phosphate (pH 8.0) to remove remaining salts.

Hemoglobin solutions were prepared as usual (1). The α and β chains were obtained using the procedure developed by Bucci and Fronticelli (13). The α and β p-mercuribenzoate chains were demercurated by β-mercaptoethanol treatment (1) on a Sephadex G-25 column. Samples of horse heart myoglobin were obtained from Sigma Chemical Co. as the Type III lyophilized reagent in the ferric state. Spectrophotometric measurements were carried out using a Kontron 860 Uvikon spectrophotometer.

RESULTS

Red cell membranes, prepared as described in the Materials and Methods section, result to be constituted by two different fractions, the major of which is represented by the so called white ghosts i.e. hemoglobin-free membranes. The remaining fraction is characterized by a given amount of hemoglobin still bound to the membrane-matrix as may be evidenciated spectrophotometrically after treatment with toluene (14). In accordance with a previous study (15) we have named this latter fraction red button. It may be important to outline that, as we will show later on, the ratio white ghosts/red button is variable being correlated to the age of the sample and/or to the occurence of oxidation processes.

Fig. 1 shows typical absorption spectra, in the visible region (from 350 to 700 nm),of a suspension of red cell membranes from two different preparations referring respectively to fresh (panel A) and old (30-40 days; panel B) human erythrocytes.

These absorption spectra, relative to the hemoglobin which is still bound to the membrane after a complete washing cycle, clearly indicate that erythrocytic membranes from old cells are characterized by a higher content of hemoglobin.

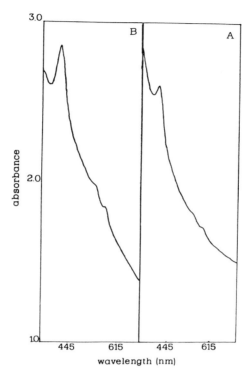

Fig. 1. Absorption spectra, from 350 to 700 nm, of a red cell membrane suspension in 5 mM sodium phosphate buffer at pH 8.0. Spectra refer respectively to old (30-40 days; panel A) and fresh (panel B) human erythrocytes.

We may greatly decrease (by about 50%) the amount of hemoglobin linked to the membranes by addition to the washing buffer of 1 mM sodium azide or potassium cyanide and by increasing the pH towards values higher than 8.0 (Fig. 2).

Since N_3^-, CN^- and OH^- are all ligands of the ferric form of hemoglobin (1) the release of Hb from the membranes brought about by addition of these ions strongly suggests that almost 50% of the red button is constituted by met–Hb or better by the various oxidation intermediates deriving from the progressive oxidation of the molecule. For the remaining 50% of the red button hemoglobin appears to be irreversibly bound to the membrane matrix very likely in the form of hemo or hemichromes (10-12).

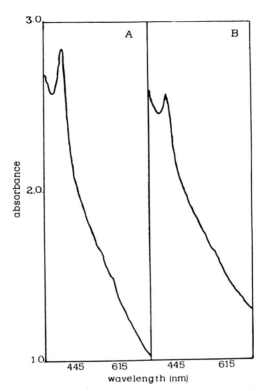

Fig. 2. Absorption spectra, from 350 to 700 nm, of the red cell membrane suspension in 5 mM sodium phosphate buffer at pH 8.0. Spectra refer to old (30-40 days) human erythrocytes before (panel A) and after (panel B) addition to the washing buffer of the 1mM sodium azide.

That partially oxidized hemoglobin molecules may reversibly be bound by the erythrocytic membranes is further supported by a number of experiments in which artificially oxidized (30%) hemoglobin solutions have been incubated with hemoglobin-free membranes (white ghosts). The optical spectra of the partially oxidized hemoglobin solution before and after the addition of the white ghosts are reported in Fig. 3. As evident from the relative absorbance values at 630, 576 and 540 nm the % of met-Hb present in solution decreases significantly (by about 20%) upon addition of the white ghosts indicating that partially oxidized Hb may be preferentially bound by the erythrocytic membrane even in the presence of oxygenated hemoglobin. Of course this is paralleled by the partial conversion of the white ghosts into the red button. Moreover the latter may be re-converted into white ghosts upon addition of the ferric ligands described above.

The same type of experiment performed by using isolated α and ß chains of human hemoglobin A as well as horse myoglobin demonstrated (see Fig. 4) that while oxidized chains are preferentially bound by the membranes, their partner subunits (i.e. ß chains) as well as horse myoglobin do not display any significant preferential interaction.

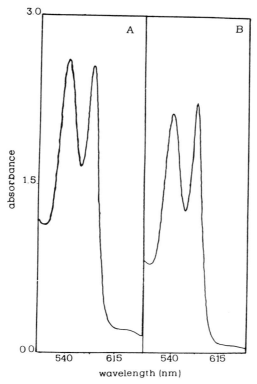

Fig. 3. Absorption spectra, from 500 to 650 nm, of partially oxidized human hemoglobin A solution before (panel A) and after (panel B) incubation with white ghosts.

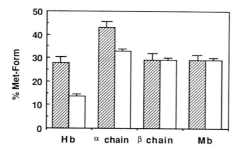

Fig. 4. Percentage of met-form present in different hemoprotein solutions before (shaded bars) and after (white bars) incubation at 8°C with white ghosts.

The possible role of the free sulphydryl groups of the protein in
this interaction has been investigated by using reagents such as
ß-mercaptoethanol and parachloromercuribenzoate (PMB) which are known to
specifically react at the level of the -SH of cysteine residues. Hence,
addition of mercaptoethanol to the red button does not induce any
significant release of the hemoglobin bound to the membrane.

Moreover the above described behaviour of the isolated chains (either α
or ß) is not altered when their -SH groups (α104, ß93 and ß112) are blocked
by PMB.

As far as the age of the erythrocytes is concerned, a set of
preliminary experiments performed on blood sample kept at 4°C for several
weeks revealed an increase of the met-Hb bound to the red cell membrane as a
function of the storage time. The results of these experiments are reported
in Fig. 5 in terms of the optical density change observed in the Soret band
on membrane samples obtained from erythrocytes after different storage time.

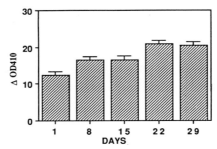

Fig. 5. Optical density change at 410 nm of membrane suspensions in 5 mM
sodium phosphate buffer at pH 8.0 as a function of red cells in
vitro age. The storage time of total blood samples at 8°C is
reported in abscissa.

This finding is paralleled by results obtained by inducing a
progressive oxidation of the intraerythrocytic hemoglobin with increasing
concentration of hydrogen peroxide (16). An increase of the % of met-Hb
results in a increase of the % of red button and hence of the hemoglobin
bound to the membrane.

The behaviour obtained in the case of normal red blood cells is
supported by analysis of erythrocytes from patients affected by diseases in
which an oxidative stress may play a significant physiopathological role
such as G6PD deficiency or iatrogenic anemias secondary to drugs or
hemodialysis where a marked increase of the bound met-Hb has been observed
(Fig. 6).

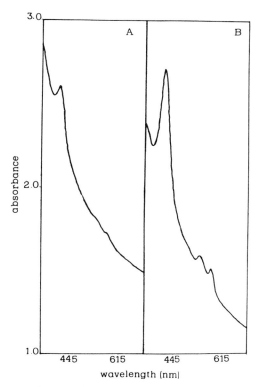

Fig. 6. Absorption spectra, from 500 to 700 nm, of the red cell membrane
        suspension in 5 mM sodium phosphate buffer at pH 8.0. Spectra refer
        respectively to normal (panel A) and G6PDH deficient (panel B)
        erythrocytes.

DISCUSSION

    The interaction of hemoglobin with the erythrocyte membrane is
interesting, both as a model system to study how cytoplasmic components
may interact with a plasma membrane and as a part of the current effort to
develop a more integrated physical description of the human erythrocyte. In
this perspective our study clearly indicates that besides oxy-and
deoxyhemoglobin also the intermediates of its progressive oxidation may be
selectively bound by the erythrocytic membrane. This binding has been
shown to be abolished or at least greatly decreased by addition of ligands
of the ferric form such as azide, cyanide and OH^- . The case of the latter
ligand is obviously linked to the pH dependence of the observed
interactions. Thus the binding tends to vanish for pH values higher than the
pK of the acid-alkaline transition of met-Hb (pK=8.1).

    As far as the molecular mechanism of the interaction is concerned the
effect of ligands of the ferric form led us to think of a major role of the
positive charge carried by the oxidized heme. However this did not appear to
be the case as indicated by the experiments performed by using the oxidized

81

isolated α and ß chains of human HbA as well horse myoglobin. In fact, of all these molecular species only the α chains displayed a significant interaction with the plasma membrane. Hence the positive charge of the heme iron is certainly important but it has to be coupled with some still unknown structural characteristics at the level of the protein moiety which seem to play a crucial role in determining the selectivity of the binding.

On the whole the results presented in this note suggest that hemoglobin might interact at the cytoplasmic surface of the erythrocyte membrane through its α chains. This observation is again in full agreement with the well known fact that the oxidation rate of α chains in hemoglobin tetramers is much faster than that of the ß chains (17) and that for this reason the intermediate oxidized hemoglobin in vivo is present mainly as $\alpha^{+3}\beta^{+2}$ rather than $\alpha^{+2}\beta^{+3}$ valency hybrid (4).

Finally it is very suggestive that we observed an increase of the met-Hb bound to the membrane as a function of the age of the erythrocytes as well as in cases of G6PD deficiencies. Therefore we may suggest that the interaction of met-Hb with the plasma membrane may be of great physiological significance in relation to the enzymatic reducing mechanisms that maintain the hemoglobin molecule in its active state. In other words the plasma membrane could be considered as playing an active role in the reduction of the molecule at least since it may contribute to keeping in place the hemoglobin molecule may be even in the right stereochemical situation.

However in old or abnormal erythrocytes where the reduction systems may be altered (18,19) met-Hb instead of being readily reduced can slowly be transformed into hemichromes which in turn may react with band 3 or spectrin leading to a decrease in deformability of the cells. Hence the irreversible crosslinking processes occurring in vivo and thought to be at the basis of those mechanisms which lead to the elimination of old erythrocytes (20-22) are part of the chain of reactions which start from the reversible and preferential interaction of met-Hb with the plasma membrane of the cell. It may be relevant to recall that in the case of the crosslinks between spectrin and hemoglobin it has been demonstrated (23,24) that the α chains are ten times more reactive than the respective ß chains.

ACKNOWLEDGEMENTS

This was supported by C.N.R. and Sigma Tau (P.F. "Invecchiamento").

REFERENCES

1. E. Antonini and M. Brunori, "Hemoglobin and myoglobin in their reactions with ligands", A. Neuberger and E. L. Tatum, North-Holland Publ. Co., Amsterdam (1971).
2. M. F. Perutz, Species adaption in a protein molecule, Mol. Biol. Evol. 1:1 (1983).
3. A. F. Riggs, The Bohr effect, Ann. Rev. Physiol. 50:181 (1988).
4. A. Tomoda, M. Takeshita and Y. Yoneyama, Analysis of met-form

hemoglobin in glucose-depleted human red cells, FEBS lett. 88:247 (1978).

5. M. Devogel, J. Leonis and J. Vincetelli, The alteration of the functional properties of human hemoglobin by spectrin, Experientia 33:1429 (1977).

6. G. Chetrite and R. Cassoly, Affinity of hemoglobin for the cytoplasmic fragment of human erythrocyte membrane band 3, J. Mol. Biol. 185:639 (1985).

7. B. R. Premachandra, Interaction of hemoglobin and its component alfa and beta chains with band 3 protein, Biochemistry 25:3455 (1986).

8. I. Lebbar, F. Stetzkowski-Marden, O. Mauffret and R. Cassoly, Interactions of actin and tubulin with human deoxyhemoglobin, Eur. J. Biochem. 170:273 (1987).

9. C. E. Moya, S. Shah and T. M. Sodeman, The erythrocyte, in: "Sodeman's Pathologic Physiology", 7th edition W. B. Saunders, Baltimore (1985).

10. P. S. Low, S. M. Waugh, K. Zinke and D. Drenckhahn, The Role of hemoglobin denaturation and band 3 clustering in red blood cell aging, Science 227:531 (1985).

11. P. S. Low, Structure and function of the cytoplasmic domain of band 3: center of erythrocytes membrane-peripheral protein interactions, Biochim. Biophys. Acta 864:145 (1986).

12. S. M. Waugh, J. A. Walder and P. S. Low, Partial characterization of the copolymerization reaction of erythrocyte membrane band 3 with hemichromes, Biochemistry 26:1777 (1987).

13. E. Bucci and C. Fronticelli, A new method for the preparation of α and β subunits of human hemoglobin, J. Biol. Chem. 240:PC551 (1965).

14. A. Tsuneshige, K. Imai and I. Tyuma, The binding of hemoglobin to red cell membrane lowers its oxygen affinity, J. Biochem. 101:695 (1987).

15. R. Scatena, S. G. Condò, M. E. Clementi, M. Corda, M. T. Sanna, M. G. Pellegrini and B. Giardina, Methemoglobin and enzymatic reduction systems: physiological and pathological implications, Ital. J. Biochem. (1990) in press.

16. M. Sharabani, B. Plotkin and I. Aviram, Lipid peroxidation in red blood cell membranes, Cell. Molec. Biol 30:329 (1984).

17. A. Mansouri and K. H. Winterhalter, Non equivalence of chains in hemoglobin oxidation and oxygen binding, Biochemistry 13:3311 (1974).

18. M. Takeshita, M. Tamura, T. Yubisui and Y. Yoneyama, Exponential decay of cytochrome b_5 and cytochrome b_5 reductase during senescence of erythrocytes: relation to the increased methemoglobin content, J. Biochem. 93:931 (1983).

19. E. Hegesh, J. H. Hegesh and A. Kaftory, Congenital methemoglobinemia with a deficiency of cytochrome b_5, N. Engl. J. Med. 314:757 (1986).

20. C. Rice-Evan and E. Baysal, Iron-mediated oxidative stress in erythrocytes, Biochem. J. 244:191 (1987).

21. M. R. Clark, Senescence of red blood cells: progress and problems, Physiol. Rev. 68:503 (1988).

22. T. Suzuki and G. L. Dale, Membrane proteins in senescent erythrocytes, Biochem. J. 257:37 (1989).

23. L. M. Snyder, L. Leb, J. Piotrowski, N. Sauberman, S. C. Liu and N. L.

Fortier, Irreversible spectrin–hemoglobin crosslinking in vivo: a marker for red cell senescence, Bri. J. Haematol. 53:379 (1983).

24. N. Shaklai, B. Frayman, N. Fortier and M. Snyder, Crosslinking of isolated cytoskeletal proteins with hemoglobin: a possible damage inflicted to the red cell membrane, Biochim. Biophys. Acta 915:406 (1987).

DIABETES MELLITUS AND RED BLOOD CELL AGING: A STRUCTURAL

AND FUNCTIONAL STUDY

Laura Mazzanti[1], Rosa A. Rabini[1],
Roberto Staffolani[1], Emanuela Faloia[2],
Roberto De Pirro[2], Armanda Pugnaloni[3],
Gian Paolo Littarru[1], Graziella Biagini[3]

[1]Institute of Biochemistry
[2]Institute of Endocrinology
[3]Institute of Anatomy
 University of Ancona
 Ancona, Italy

INTRODUCTION

Physiological insulin concentrations modulate the allosteric properties of membrane-bound enzymes, such as rat erythrocyte acetylcholinesterase (AchE) and Na^+/K^+ ATPase (1). Modifications in both AchE and Na^+/K^+-ATPase enzymatic activities have been observed in human erythrocyte membranes obtained from patients affected by Insulin-dependent-Diabetes Mellitus (IDDM) (2,3).

The process of red blood cell (RBC) aging produces a number of alterations in the composition of the plasma membrane as well as and in the function and activity of membrane-bound enzymes, which are partly similar to those reported in IDDM (4). It is not clear if modifications present in erythrocyte membranes from diabetic patients are a direct consequence of alterations induced by diabetes mellitus at the plasma membrane level or if they are the consequence of an accelerated aging process induced by the disease. The aim of the present study is to investigate the influence of aging on the RBC acetylcholinesterase and Na^+/K^+-ATPase activity both in normal subjects and in diabetic patients. The protein structural pattern was also investigated using the freeze-fracturing technique.

PATIENTS AND METHODS

10 insulin-dependent diabetic patients (24-43 years) and 20 healthy subjects (26-47 years) were studied. All diabetic patients showed a good metabolic control.

Red Blood Cell Aging, Edited by M. Magnani and
A. De Flora, Plenum Press, New York, 1991

Erythrocytes were isolated from total blood by means of elution on microcristalline cellulose: alpha-cellulose (1:1) columns as described (5) and washed three times in Hepes Buffered Stock Solution (HBSS) diluted 1:20. RBCs were separated into five subpopulations on Percoll/BSA density gradients according to the previously described method (6) with minor modifications. Erythrocyte membranes from each fraction were obtained as described (7). AchE activity was measured in triplicate on erythrocyte plasma membranes according to previously described method (8). The enzymatic activity is expressed as µmol acethylthiocholine hydrolysed/min/mg membrane protein. Protein concentration was determined by the Lowry method (9) using albumin as a standard. The Na^+, K^+ activated Mg^{2+} dependent ATPase activity was determined on erythrocyte membranes by the Kitao method (10). Inorganic phosphate (Pi) hydrolyzed from reaction was measured according to Fiske and Subbarow (11). Enzyme activity was expressed as difference in organic phosphate released in the presence and absence of 10 mmol/l ouabain. The ATPase activity assayed in the presence of ouabain was subtracted from the total Mg^{2+} dependent ATPase activity to calculate the activity of the ouabain-sensitive Na^+, K^+ -ATPase. The results are expressed as umol Pi/(mg membrane proteins x 60 min). Erythrocyte membrane suspensions as obtained after biochemical treatment were slightly fixed in glutharaldehyde 1.2% for 15 minutes in phosphate buffer pH 7.4 to prevent the possible clustering of IMP particles (12) and treated with a solution of 27% glycerol in phosphate buffer as cryoprotectant to avoid eutectic formations. One drop of suspension was placed on gold plates and quickly frozen in liquid freon 22, then transferred into liquid nitrogen. Specimens were fractured by a Balzers 360M device at -113°C without etching; shadowed with 2.2 nm Carbon-platinum replica at 45 degrees, followed by a carbon deposition at 60 mA for 50 sec at 2×10^{-6} Torr. The replicas were cleaned with chlorox at 20°C for 1 hour, rinsed in bidistilled water, and mounted on nickel-copper grids. Observations were carried out with a Philips 301 E.M. at 80 KV. Fractured planes with microvillous membranes were printed at a final magnification of 85 Kx . Photographs were analyzed by Image Analysis System Quantimet 920 (Cambridge instr.) for quantitative investigations. The intramembrane particles (IMP), appearing as black dots with white shadows, were measured along the perpendicular axis in the direction of the shadowing by scanning with a high resolution Chalnicon TV-camera. Detection of intramembrane particles was carried out using proper gray levels into unit circular areas of 15,000 nm area. The parameters measured were: number of IMP per unit area (i.e. 0.015 nm), mean diameter of IMP and clustering index (V factor calculated as Variance/mean of number of IMP per unit area) (13) which means that the IMP were clustered for numerical values of factor V greater than 1, while they are distributed in regular Poisson pattern for factor V less than 1. Statistical analysis was performed to calculate the significance level of measurements as previously described (14).

RESULTS

Plasma membranes from normal subjects decreased their AchE activity during aging, while membranes from diabetic patients showed increased activity (Table 1). The comparison of values obtained in normal and diabetic subjects by means of the Student's test shows that mature, but not young, erythrocytes had different AchE activity.

Table 1. AchE and Na$^+$/K$^+$-ATPase Enzymatic Activities in Early Young (Fraction 1) and Old (Fraction 5) Red Blood Cell Plasma Membrane Obtained from Controls (C) and Diabetic Subjects (D).

	Fraction 1		Fraction 5	
	AchE	Na$^+$/K$^+$-ATPase	AchE	Na$^+$/K$^+$-ATPase
C (n=10)	1.48 ±0.20	1.62 ±0.06	0.60 ±0.15	0.59 ±0.04
D (n=10)	1.36 ±0.19	1.05 ±0.06	3.63 ±0.16 °	0.27 ±0.04 °

°p<0.001

The enzymatic activities are expressed for AchE: µmol/mg prot/min; for Na$^+$:K$^+$-ATPase:µmol Pi/mg prot/min.

Table 2. Number of IMP, Mean Diameter and V Factor of Membranes Obtained from Early Young (Fraction 1) and Oldest (Fraction 5) Erythrocyte in Both Controls and Diabetic Patients.

	Fraction 1			Fraction 5		
	NoIMP /µm^2	Mean diameter (A)	V	No IMP /µm^2	Mean diameter (A)	V
C (n=10)	3472 ±1131	93.25 ±0.47	1.54 ±0.11	3139 ±9	91.71 ±0.5	1.80 ±0.12
D (n=10)	3759 ±1356	102.24 ±2.2 °	1.99 ±0.14 °	4163 ±1224 °	96.54 ±0.40 °	1.43 ±0.30

°p < 0.001   °°p<0.05

Table 1 also shows that aging produces a progressive reduction in membrane Na^+, K^+ ATPase activity both in normal and diabetic subjects. Particularly, RBC plasma membranes obtained from normal subjects showed higher enzymatic activity in comparison to diabetic subjects in all the fractions tested (data not shown).

Table 2 shows that intramembrane particles (IMPs) have a significantly higher mean diameter in diabetic patients in respect to the controls in all the fractions tested. Factor V increased in the early erythrocytes in diabetics in comparison to controls, while it was decreased in the oldest RBC fraction in diabetics.

DISCUSSION

RBC aging provoked a progressive reduction of membrane Na^+, K^+ ATPase activity in both normal and diabetic subjects. This observation is in agreement with previous studies performed on fractionated RBCs from normal subjects and on unfractionated RBCs from diabetic patients (3,4). It is clear from the present data that all RBC fractions obtained from diabetic subjects showed lower Na^+/K^+ ATPase activity in comparison with controls, thus pointing out that this membrane defect appears at any RBC age; it is tempting to suggest that this phenomenon is indipendent of the effect of diabetes on RBC aging.

AchE activity decreased during aging in normal subjects but increased in mature RBCs from diabetic patients. Diabetes mellitus does not affect this activity in early young RBCs; this phenomenon may not be ascribed to the effect of the disease state on the plasma membrane but probably it comes from an effect appearing during aging.

The morphometric analysis showed that IMPs of RBC membranes decrease their diameter during aging both in controls and diabetic subjects; particularly IMPs have higher diameter in diabetics than in controls in every fraction studied. The increased diameter of IMPs in diabetics in respect to the controls may be at the basis of the alterations observed in the activities of membrane-bound enzymes.

REFERENCES

1. R. N. Farias, Membrane cooperative enzymes as a tool for the investigation of membrane structure and related phenomena, Adv. Lipid Res. 17:251 (1980).
2. I. Testa, R. A. Rabini, P. Fumelli, E. Bertoli and L. Mazzanti, Abnormal membrane fluidity and acetylchelinesterase activityin erythrocytes from insulin-dependent diabetic patients, J. Clin. Endocrinol Metab. 67 (6):1129 (1988).
3. L. Mazzanti, R. A. Rabini, I. Testa and E. Bertoli, Modifications induced by diabetes on the physico-chemical and functional properties of erythrocyte membrane, Eur. J. Clin. Invest. 19:84 (1989).
4. M. R. Clark, Senescence of red blood cells: Progress and Problems, Physiol. Rev. 68:503 (1988).

5. E. Beutler, C. West and K. G. Blume, The removal of leukocytes and platelets from whole blood, <u>J. Lab. Clin. Med.</u> 88:328 (1976).

6. A. Camagna, L. Rosseti, R. De Pirro, M. Di Franco, R. Lauro, P. Samoggia, P. Caprari and G. Salvo, Characterization of differences in insulin receptors from young and old red blood cells, <u>J. Endocrinol. Invest.</u> 10:371 (1987).

7. G. W. Burton and K. U. Ingold, K. E. Thompson, An improved procedure for the isolation of ghost membranes from human red blood cells, <u>Lipids</u> 16:946 (1981).

8. G. L. Ellman, D.K. Courtney, V. Jr. Andres and R. M. Featherstone, A. new and rapid colorimetric determination of acetylcholinesterase activity, <u>Biochem. Pharmacol.</u> 7:88 (1961).

9. O. H. Lowry, M. Y. Rosenburg, A. L. Farr and R. T. Randal, Protein measurement with the Folin phenol reagent, <u>J. Biol. Chem.</u> 193:265 (1951).

10. T. Kitao and K. Hattori, Inhibition of erythrocyte ATPase activity by aclacynomycin and reverse effects of ascorbate on ATPase activity, <u>Experentia</u> 39:1362 (1983).

11. C. Fiske and Y. Subbarow, The colorimetric determination of phosphorus, <u>J. Biol. Chem.</u> 193:375 (1925).

12. J. H. Willison and A. J. Rowe, Replica, shadowing and freeze-etching techniques, in Pratical Methods in Electron Microscopy, Audrey M. Glauert, ed., North-Holland, Amsterdam (1980).

13. S. W. Last, L. J. Tertoolen and J. G. Bluemink, Quantitative analysis of the numerical and lateral distribution of intermembrane particles in freeze-fractures of biological membranes, <u>Eur. J. Cell. Biol.</u> 23:273 (1981).

14. J. Roos, M. Robinson and R. L. Davidson, Cell fusion and intramembrane particle distribution in polyethylene glycol-resistant cells, <u>J. Cell Biol.</u> 96:909 (1983).

METABOLISM

CHARACTERIZATION OF SENESCENT RED CELLS FROM THE RABBIT

George L. Dale, Robert B. Daniels, Joshua Beckman
and Shannon L. Norenberg

Department of Molecular and Experimental Medicine
Research Institute of Scripps Clinic
La Jolla, CA 92037

INTRODUCTION

The mammalian erythrocyte survives a multitude of insults during its circulating lifespan including oxidant attack, calcium influxes, repeated deformation and glycation among others (1). Nevertheless, the majority of red cells survive and apparently function well for the entire pre-programmed time period which represents their lifespan. Neither the mechanism which determines the time frame of the lifespan nor the signal that triggers the removal of the senescent cell by macrophages is known (1). The lack of knowledge concerning this fundamental biological process can clearly be attributed to a single underlying problem, the difficulty of reliably isolating aged red cells (2). The majority of investigators in this field have utilized a variety of physical techniques for isolating aged cells (1) based upon assumptions as to changes which may occur with red cell aging, for example, an increase in cellular density. As a result, there have been literally thousands of reports documenting the changes in red cell properties as a function of cellular density with the assumption that these findings reflect changes with age. It has now, however, become increasingly clear that density fractionation is not capable of producing a sufficiently pure population of aged erythrocytes to allow any biochemical characterization of age-dependent changes (3-6).

Three techniques, however, are currently available which do reliably isolate aged erythrocytes (7-9). The first of these was introduced by Ganzoni et al. (7) in 1971 and utilizes an ingenious procedure of hypertransfusing mice to suppress erythropoiesis. The method is quite successful and has served as the gold standard for the field since the technique's introduction even though the method is limited to rodents. The second technique relies on a disease in humans, transient erythroblastopenia of childhood. With this disorder, erythropoiesis ceases totally with a concomitant aging of the preexisting red cells. At times after interruption of red cell synthesis, the patient can be bled for samples of aged red

Red Blood Cell Aging, Edited by M. Magnani and
A. De Flora, Plenum Press, New York, 1991

cells (8). The problems associated with this model are its rarity and the uncertainties associated with the timing of the interruption in erythropoiesis. In addition, once a diagnosis is made, the patient is routinely transfused, thereby eliminating the possibility of any further samples. The final method is a recently introduced one which involves biotinylation of rabbit red cells (9). In this model, the biotinylated cells have been shown to survive normally in vivo and to be selectively recoverable by an avidin support. This last technique has allowed a more careful dissection of temporal changes which occur during the red cell aging process (10,11) and has, thereby, helped to delineate subtle changes which had been previously undetected.

The purpose of this report is to document some recent findings with the biotinylation model. It should be emphasized that these studies were carried out in rabbits so that conclusions concerning similarities with other species should be drawn carefully; however, work with other valid aging models has shown a remarkable conservation of red cell aging characteristics between species (8,10-15).

MATERIAL AND METHODS

Recovery of Aged Red Cells

New Zealand White rabbits were treated with phenylhydrazine and biotinylated in vivo as previously detailed (5,10). Briefly, 3 Kg. rabbits were injected subcutaneously with 7.5 mg/Kg phenylhydrazine on three consecutive days and then ten days later were biotinylated by intravenously infusing 15.4 mg N-hydroxysuccinimidobiotin (NHS-biotin) per Kg body weight (5). After in vivo aging, the biotinylated cells were recovered by panning on plastic Petri dishes coated with an avidin-biotinylated-gelatin matrix (11). Hemoglobin levels were quantitated with Drabkins reagent (16).

Quantitation of IgG

Total levels of IgG were determined with an enzyme-linked immunospecific assay (ELISA). Specifically, polystyrene microtiter wells were coated with 0.1 ml of 5 μg/ml goat anti-rabbit IgG (GARG) in PBS for four hours at room temperature. After blocking the wells with bovine serum albumin and washing with 0.05% Tween-20 in PBS, the hemolysate (300 μg of hemoglobin) in 0.5% Tween-20/PBS was applied to the wells and incubated overnight at 4°. After this binding step and a subsequent wash as above, the wells were incubated with 0.1 ml of 400 ng/ml biotinylated-GARG in PBS containing 1 mg/ml BSA. After four hours at room temperature, the plate was washed and the amount of bound, biotinylated-GARG was quantitated with an ABC kit from Vector Laboratories, Burlingame, CA.

Enzyme Assay

AMP deaminase was assayed as previously reported (17) with the following exception: for the assay of effectors, the hemolysate in question was dialyzed against 10 mM HEPES; pH 7.0 for at least three hours at 4° to allow endeogeneous modulators of the enzyme to be removed. The actual assay with effectors was performed in 10 mM HEPES, pH 7.0 rather than the phosphate buffer previously described (17). Effectors were added at the final concentrations listed below.

Acetylcholinesterase was assayed as previously described (16). The only change was that washed stroma were used for the assay rather than whole hemolysate. This change was necessary since the activity of this enzyme in rabbits is considerably lower than that in human red cells; and with the lower activity, the blank rate for the complete hemolysate becomes a significant problem. Pyrimidine 5'-nucleotidase was measured by the method of Torrance et al. (18).

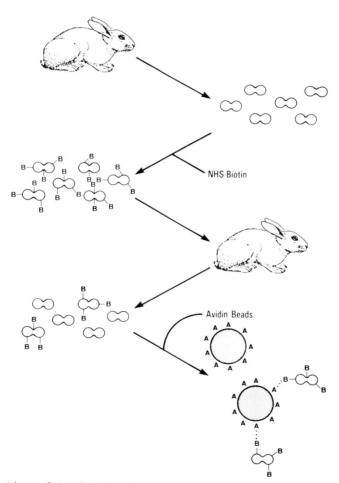

Fig. 1. Isolation of Aged Red Cells. Rabbit erythrocytes were biotinylated with NHS-biotin and reinfused into the animal. After prolonged *in vivo* circulation, the biotinylated cells were selectively recovered by their affinity for an avidin support. Reprinted with permission.

RESULTS AND DISCUSSION

Red Cell Aging Model

The general scheme used to study red cell aging in the rabbit is shown
in Figure 1. With this technique, rabbit red cells are biotinylated either
in vitro or in vivo (5, 9). The degree of substitution is quite modest in
that only 10,000 to 20,000 molecules of biotin are measurable on the surface
of each cell. Previous studies have shown that the biotinylated red cells
have a normal in vivo lifespan (9), and this observation is the best
available evidence that derivatization of the red cells does not adversely
affect the cells. After in vivo aging, the biotinylated cells are recovered
by panning on an avidin-coated Petri dish (11). As mentioned above, this
system has now been used to examine a variety of cellular parameters in aged
red cells (10, 11, 14), and detailed below are several additional properties
of these senescent erythrocytes.

Adenine Nucleotide Metabolism

Previously we have shown that the adenine nucleotide pool size of aged
red cells is increased (10). This rather surprising finding was one of the
first made with the biotin system and has subsequently been confirmed in the
human model of red cell aging, transient erythroblastopenia of childhood
(13). The increase in total ATP levels as the red cell ages appears to be
due to a loss of the enzyme AMP deaminase as a function of cell age (11)
(Figure 2).

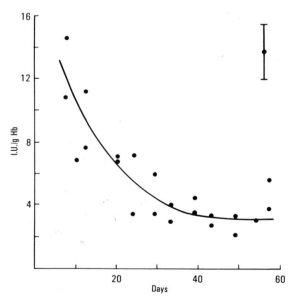

Fig. 2. AMP Deaminase Activity in Aged Red Cells. Biotinylated red cells
were recovered at various times after derivatization and assayed for
AMP deaminase activity (International Units/g hemoglobin). A total
of 23 samples from nine different animals were assayed; control
enzyme activity for AMP deaminase was 9.67 ± 1.76 I.U./g Hb (mean ±
1 SD; n=23). Reprinted with permission.

Adenine nucleotide metabolism in the red cell is unusual in that the level of adenine nucleotides is controlled only at the catabolic side of the pathway and not at the synthetic side (19, 20); therefore the loss of catabolic capacity in the face of a continued and unregulated synthetic rate results in the accumulation of ATP as the cell ages. A similar situation is seen in a genetically determined deficiency of erythrocyte AMP deaminase which results in an accumulation of ATP (21).

We have now examined the kinetics of the residual AMP deaminase enzyme found in aged red cells and compared this to control enzyme (Table 1). The rather surprising finding is that the aged red cell enzyme is kinetically normal except for one parameter, the ability to be stimulated by potassium

Table 1. Effectors of AMP Deaminase as a Function of Red Cell Age

Effector	Red Cell AMP Deaminase Activity (μmoles/min/g Hb)		
	Control	Aged	Young
None	1.9 ± 1.1	2.2 ± 1.2	2.0 ± 0.6
ATP	14.2 ± 2.1	15.2 ± 5.7	19.9 ± 2.8
KCl	36.5 ± 11.2	13.4 ± 3.9*	35.7 ± 0.6
2,3–DPG	1.3 ± 1.2	1.6 ± 1.4	1.6 ± 0.5
	(n=5)	(n=4)	(n=2)

Aged red cells, 45, 50, 56 and 58 days old, were isolated and assayed for AMP deaminase in the presence and absence of several effectors. The concentrations of effectors were 1 mM ATP, 150 mM KCl and 1 mM 2,3–DPG. The young cells were biotinylated and recovered 7 to 10 days later.
* $p < 0.05$

chloride. The other effects give similar results for both aged and control enzymes. However, the ability of the aged enzyme to be stimulated by KCL is diminished to approximately 35% of normal which is approximately the same magnitude of residual enzyme activity which was originally reported for AMP deaminase in aged red cells (11). A reasonable question arising from data is what are the synergistic effects of multiple stimulators or inhibitors? Additional experiments demonstrated that KCl is the dominant effector for AMP deaminase in that the same loss of activity in the aged samples is found when a cocktail of activators is used as when just KCl stimulation is tested (data not shown).

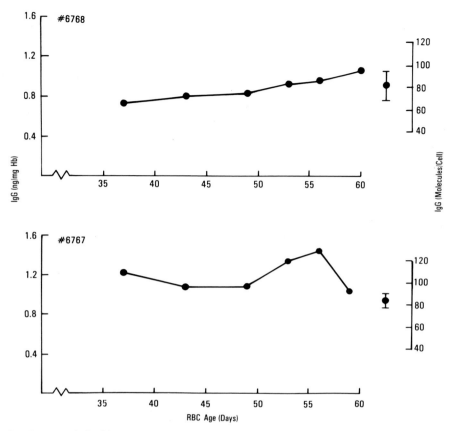

Fig. 3. Immunoglobulin Content of Aged Red Cells. Aged, biotinylated red cells were isolated as detailed. The absolute IgG content was determined with an ELISA assay. The control value for each individual rabbit is shown at the right side of the graphs. Calculation of IgG molecules per cell was based on an MCH of 23 pg.

Immunoglobulin Content of Aged Red Cells

The possible role of surface bound immunoglobulin in determining red cell senescence has been a popular proposal for several years (22-25). The large variety of theories in this field can be distilled into the following brief summary: during red cell aging, a cryptic or nascent antigen is exposed either as a function of time or damage to the cell. In addition, there is a housekeeping antibody present in the plasma which recognizes this senescent antigen, binds to the aging cell and eventually serves as the signal for recognition of the aged cell by the reticuloendothelial system. This theory has been tested in one valid model of red cell aging by Singer et al. (23) using the Ganzoni hypertransfusion model. These investigators did find elevated levels of IgG on the aged, mouse erythrocytes as measured by a phagocytosis assay.

We have now monitored the absolute levels of IgG on aging rabbit red cells using a quantitative ELISA assay. These results shown in Figure 3 indicate that the rabbit erythrocytes did not accumulate significant amounts of IgG during the aging process. In fact, the total level of IgG never exceeded 200 IgG molecules per cell for the two animals presented nor for any of the six total animals which have now been studied. The significance of these modest levels of surface-bound IgG is difficult to quantitate, but a variety of studies have suggested that from 200 to 2000 molecules of IgG are required to promote phagocytosis of red cells (26-28). However, the most important observation from the current experiments is that none of the six animals studied displayed an increasing level of IgG as the erythrocytes aged which casts doubt on the role of immunoglobulin accumulation in red cell senescence in the rabbit.

Acetylcholinesterase Activity of Aged Red Cells

One recurring question in the field of red cell aging concerns the state of the cell's membrane. Previously we have monitored the state of several skeletal proteins (14) and individual lipid classes (29) but have not found any dramatic changes. For the present experiments, we have measured the level of acetylcholinesterase (AChE) in control and aged red cells. The interest in AchE is two fold: first, the absence of AchE in paroxysmal nocturnal hemoglobinuria has heightened interest in this enzyme as a marker for a class of ecto enzymes which are anchored into the membrane via phosphatidylinositol tails (30). Secondly, AChE has been used repeatedly as an age-dependent marker of red cells (1), and the validity of that assumption has not been tested with a valid model of red cell aging.

The data presented in Table 2 indicate that aged red cells from the rabbit have levels of AChE which average approximately 70% of control activity. This apparent drop of 30% in activity may not actually represent an age-dependent change as much as a maturation change since reticulocytes have a higher content of membrane per cell and are known to lose AChE during maturation. A more detailed analysis of time-dependent changes in AChE should clarify this situation; in either case it is clear that AChE is not markedly age-dependent in contrast to earlier reports (31).

Enzyme Activity Decay During Red Cell Aging

Earlier we reported that many enzyme activities which had previously been used as markers of red cell age were, in fact, rather stable through out the cell's lifespan (10). Similar findings have been reported for human red cells using the transient erythroblastopenia of childhood model (8). However, one interesting observation made by Beutler and Hartman was that pyrimidine 5'-nucleotidase (P5N) was actually rather age-dependent in human red cells (8). To further examine this question, we have quantitated the level of P5N in aging red cells from the rabbit, and the data shown in Figure 4 demonstrate that this particular enzyme is also age-dependent in rabbits. In addition, the decay pattern of the P5N activity is quite similar to that seen for AMP deaminase (11) (Figure 2) in that the loss of activity is not linear with time but rather there is a rapid loss during the first 30 days of the cell's lifespan with a more gradual loss from then until the death of the cell at 60 days.

Table 2. Acetylcholinesterase Activity of Aged Red Cells from the Rabbit

Rabbit	Red Cell Age (days)	AChe Activity (nmol/min/mg)
6412	45	11.7
6415	46	11.8
6414	53	15.0
6412	54	13.8
Control	–	18.7 ± 2.6 (n=4)

Acetylcholinesterase activity was measured on purified stroma and expressed as nmole of product per minute per mg stromal protein. Control activity is expressed as mean ± 1 SD; the average for the aged samples was 13.1 ± 1.6 nmol/min/mg which is 70% of the control value.

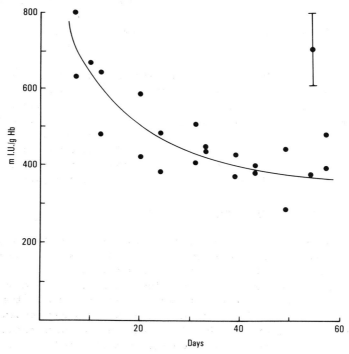

Figure 4. Pyrimidine 5'-Nucleotidase Activity of Aged Erythrocytes. Aged red cells were isolated as detailed in Figure 2 and assayed for pyrimidine 5'-nucleotidase activity. A total of 22 aged red cell samples from 7 to 57 days old were assayed; the control activity was 527 ± 99 mIU/g Hb (maen ± 1 SD; n=19).

SUMMARY

The above data continue to demonstrate the metabolic well being of the aged red cell as it is isolated from rabbits. The abundance of ATP, the absence of surface-bound IgG and a variety of other observations at this time lead to the tentative conclusion that the senescent red cell is amazingly healthy. Many investigators have predicted that the red cell is removed from the circulation as a metabolically exhausted effete cell. There is currently no evidence to support this other than a decrease in deformability of the cells with time (32), but it is not clear that this decline in deformability is sufficient to keep the cell from circulating. In either case, many of the previously proposed causes of cellular removal are clearly incorrect for the rabbit, and it is now time to focus on new directions for observing either cellular impairment or perhaps the presence of a cellular clock which is independent of the cell's metabolic state.

Another point which should be addressed is the reliability of the biotinylation model in rabbits as it relates to red cells in other species. So far several observations in aged red cells isolated with valid models have been reproduced across species boundaries including the rise in ATP (10,13), the fall in AMP deaminase activity (11,13), the shift in the 4.1a to 4.1b protein ratio (14,15,33), the stability of a number of glycolytic enzymes (8,10,12), and the instability of pyrimidine 5'-nucleotidase activity (8) (Figure 4). To this point, the rabbit has been a reliable model of red cell aging and one with implications for other species.

REFERENCES

1. M. R. Clark, Senescence of Red Blood Cells, Progress and Problems. Physiol. Rev. 68:503 (1988).
2. E. Beutler, Isolation of the aged, Blood Cells 14:1 (1988).
3. M. Morrison, C. W. Jackson, T. J. Mueller, T. Huang, M. E. Docktor, W. W. S. Walker, J. A. Singer, and H. H. Edwards, Does red cell density correlate with red cell age, Biomed. Biochim. Acta 42:107 (1983).
4. M. R. Clark, L. Corash, and R. H. Jensen, Density distribution of aging, transfused human red cells, Blood 74:217A (1989).
5. G. L. Dale and S. L. Norenberg, Density fractionation of rabbit erythrocytes results in only a slight enrichment for aged cells, Biochim. Biophys Acta, in press (1990).
6. M. G. Luthra, J. M. Friedman, and D. A. Sears, Studies of density fractions of normal human erythrocytes labeled with iron-59 in vivo, J. Lab. Clin. Med. 94:879 (1979).
7. A. M. Ganzoni, R. Oakes, and R. S. Hillman, Red cell ageing in vivo, J. Clin. Invest. 50:1373 (1971).
8. E. Beutler and G. Hartman, Age-related red cell enzymes in children with transient erythroblastopenia of childhood and with hemolytic anemia, Pediatr. Res. 19:44 (1985).
9. T. Suzuki and G. L. Dale, Biotinylated erythrocytes: In vivo survival and in vitro recovery, Blood 70:791 (1987).

10. T. Suzuki and G. L. Dale, Senescent erythrocytes: The isolation of in vivo aged cells and their biochemical characteristics, Proc. Natl. Acad. Sci. USA 85:1647 (1988).

11. G. L. Dale and S. L. Noremberg, Time-dependent loss of adenosine 5'-monophosphate deaminase activity may explain elevated adenosine 5'-triphosphate levels in senescent erythrocytes, Blood 74:2157 (1989).

12. A. M. Ganzoni, J. P. Barras, and H. R. Marti, Red cell ageing and death Vox Sang. 30:161 (1976).

13. D. E. Paglia, W. N. Valentine, M. Nakatani, and R. A. Brockway, AMP deaminase as a cell-age marker in transient erythroblastopenia of childhood and its role in the adenylate economy of erythrocytes, Blood 74:2161 (1989).

14. T. Suzuki and G. L. Dale, Membrane proteins in senescent erythrocytes, Biochem. J. 257:37 (1989).

15. T. J. Mueller, C. W. Jackson, M. E. Docktor, and M. Morrison, Membrane skeletal alterations during in vivo mouse red cell aging. Increase in the Band 4.1a:4.1b ratio, J. Clin. Invest. 79:492 (1987).

16. E. Beutler, in Red Cell Metabolism, A Manual of Biochemical Methods, New York, Grune and Stratton, 1983.

17. G. L. Dale, Radioisotopic assay for erythrocyte adenosine 5'-monophosphate deaminase, Clin. Chim. Acta 182:1 (1989).

18. J. D. Torrance, D. Whittaker, and E. Beutler, Purification and properties of human erythrocyte pyrimidine 5'-nucleotidase, Proc. Natl. Acad. Sci. USA 74:3701 (1977).

19. F. Bontemps, G. Van den Berghe, and H. G. Hers, Pathways of adenine nucleotide catabolism in erythrocytes, J. Clin. Invest. 77:824 (1986).

20. F. L. Meyskens and H. E. Williams, Adenosine metabolism in human erythrocytes, Biochim. Biophys. Acta 240:170 (1971).

21. N. Ogasawara H. Goto, Y. Yamada, I. Nishigaki, T. Itoh, and I. Hasegawa, Complete deficiency of AMP deaminase in human erythrocytes, Biochem. Biophys. Res. Commun. 122:1344 (1984).

22. M. M. B. Kay, Mechanism of removal of senescent cells by human macrophages in situ, Proc. Natl. Acad. Sci. USA 72:3521 (1975).

23. J. A. Singer, L. K. Jennings, C. W. Jackson, M. E. Dockter, M. Morrison, and W. S. Walker, Erythrocyte homeostasis: Antibody-mediated recognition of the senescent state by macrophages, Proc. Natl. Acad. Sci. USA 83:5498 (1986).

24. H. U. Lutz, S. Fasler, P. Stammler, F. Bussolino, and P. Arese, Naturally occurring anti-band 3 antibodies and complement in phagocytosis of oxidatively-stressed and in the clearance of senescent red cells, Blood Cells 14:175 (1988).

25. M. Magnani, S. Papa, L. Rossi, M. Vitale, G. Fornaini, and F. A. Manzoli, Membrane-bound immunoglobulins increase during red blood cell aging, Acta Haematol. 79:127 (1988).

26. W. F. Rosse, Quantitative immunology of immune hemolytic anemia. II. The relationship of cell-bound antibody to hemolysis and the effect of treatment, J. Clin. Invest. 50:734 (1971).

27. B. Zuppanska, E. Thompson, E. Brojer, and A. H. Merry, Phagocytosis of erythrocytes sensitized with known amounts of IgG1 and IgG3 anti-Rh antibodies, Vox Sang. 53:96 (1987).

28. M. O. Jeje, M. A. Blajchman, K. Steeves, P. Horsewood, and J. G. Kelton, Quantitation of red cell associated IgG using an immunoradiometric assay, Transfusion 24:473 (1984).
29. G. L. Dale, S. L. Norenberg, and R. B. Daniels, Phospholipid and cholesterol content of senescent erythrocytes. submitted.
30. W. F. Rosse, Phosphatidylinositol-linked proteins and paroxysmal nocturnal hemoglobinuria, Blood 75:1595 (1990).
31. E. Kamber, A. Poyiagi, and G. Deliconstantinos, Modifications in the activities of membrane-bound enzymes during in vivo ageing of human and rabbit erythrocytes, Comp. Biochem. Physiol. 77B: 95 (1984).
32. G. L. Dale and N. Mohandas, manuscript in preparation.
33. Y. Ravindranath, F. Brohn, and R. M. Johnson, Erythrocyte age-dependent change of membrane protein 4.1: Studies in transient erythroblastopenia, Pediatr. Res. 21:275 (1987).
34. This is publication number 6506-MEM from the Research Institute of Scripps Clinic. Partial support was provided by grant AG 08545 from the National Institutes of Health.

103

THE MECHANISM OF ENZYME DECLINE IN THE RED BLOOD CELL

DURING THE "IN VIVO" AGING PROCESS

Sergio Piomelli and Carol Seaman

Division of Pediatric Hematology Oncology
Columbia University College
of Physicians & Surgeons
New York, U.S.A.

While it is recognized that the human red blood cell (RBC) has a finite life-span of approximately 120 days, the mechanism that controls its ultimate removal from the circulation remains largely unknown. A better understanding of the process of RBC senescence can be obtained by the study of cohorts of cells of progressively increasing age. Our laboratory has utilized for this purpose separation of RBC into groups of progressively increasing specific gravity utilizing buoyant density gradients (1,2). This technique has been validated by experiments with ^{14}C-glycine cohort-labelled RBC that have demonstrated a progressive movement of the label from the top to the bottom with cell age (1) as well as by in vivo experiments of RBC survival that have demonstrated that the lighest RBC have the longest survival and the heaviest ones, the shortest (3). On the basis of these data a mathematical model has been designed to interpret, in terms of in vivo survival, the changes observed in parameters of RBC of progressively increasing gravity (4). Studies from our laboratory suggested that one of the mechanisms for the removal from the circulation of senescent RBC is a progressive decline of metabolic activities (5).

Recently Beutler hypothesized that the only decline of activities occurs at the reticulocyte-RBC transition (6,7). His hypothesis implies that all age-dependent activities of the RBC decline in a biphasic manner and the progression observed with density is artifactual. He argues that, in order to understand the mechanism of RBC aging, it is necessary to isolate the very oldest RBC. To quote some of his views: "...the role of declining red cell metabolism on aging and destruction of normal red cells is best approached experimentally by examining red cells just before they are removed from the cirulation..." This concept appears to us at least simplistic. Even if it were possible to achieve this goal, those few RBC would not be those in whom the process of aging is most advanced, but, instead would most likely represent a selected group of survivors, much as the very oldest human beings alive are indeed those in whom the process of aging has been slower, and thus appear younger.

In fact, in recent experiments by Dale, in Beutler's own institution, the very oldest RBC isolated by biotinilation had a higher then normal concentration of ATP (8). These results can only be interpreted to indicate that only RBC capable of maintaining a high level of ATP can survive to 120 days. The oldest RBC thus are those in whom aging did not occur. If a high ATP level is a requisite for long survival, toward the end of the life span the only surviving cell would be those that maintained their ATP best. To quote Beutler again: "...the cells being examined are still circulating.... They are, as it were, down but not yet out."

It is not necessary to isolate specific groups of cells, to understand the mechanism of cellular aging. Appropriate conclusions can be reached by simply shifting the age distribution of the whole RBC population in a predictable and reproducible manner.

Without ever visiting the faraway stars, astronomers have a clear understanding of their nature and life span, by indirect evidence gathered through spectral analysis. Without ever isolating the youngest and oldest red cells, biochemists can understand the nature of aging and life span of the RBC, through analysis of density gradients.

For many years, studies of RBC aging have been based on the analysis of differences between cells of different density techniques, on the assumption that RBC density is correlated with age. Our laboratory introduced the technique of centrifugation on discontinuous density gradients, that, on the one hand allows the various RBC to reach their specific buoyant density and on the other permits isolation of the different layers without remixing, by slicing the nitrocellulose ultracentrifuge tube (1,2).

The RBC population is normally distributed with respect to density. Within the total distribution, each cohort of RBC of the same age is similarly normally distributed. As the cohort ages, its mean density increases, but its variance remains constant. We have validated this hypothesis in vivo by following in time the distribution in the gradient of narrow cohorts of rabbit RBC labelled with ^{14}C-glycine (a non-reutilized isotopic label that, being incorporated in the hemoglobin, is not subject to elution). Moreover, we have demonstrated that in vivo light RBC have a much longer survival than heavy RBC (3).

The relative distribution of cohorts of RBC of minimal age (young), maximal age (old), and intermediate ages in discontinuous density gradients is predictable:
 - When layers of different densities are isolated, each contains RBC of every age, but in different proportions.
 - The upper layers (light cells) have a preponderance of younger RBC and the lower (dense cells) of the older ones.
 - The majority of RBC are in the layers near the mean density, and layers at the extreme levels contain very few RBC.

- At approximately 4.1 standard deviations from the mean, two minuscule fractions contain exclusively very young and very old RBC, respectively. Those fractions are however so minute they cannot, in practice, be isolated.
- By density (or any other technique) it is practically impossible to isolate exclusively the youngest or oldest RBC. However, as the age distribution of the RBC in the gradient is predictable, it is easy to compute the mean age of each fraction, by their position in the gradient.

On these premises we developed a mathematical model to correlate position in the gradient with RBC age (4). This is based on the recognition that, as shown by our in vivo experiments, the increase in density with RBC age is exerted on the entire population. Hence, the absolute RBC density is not appropriate to describe cell age, that is instead best represented by the cumulative distribution function (CDF - expressed as percentage distance from the top). As the change in density effects the entire normally distributed RBC population, the probits of the CDF are the appropriate units to use, since these are based on the normal distribution (each probit unit, in fact, corresponds to one standard deviation from the mean). Hence, on the abscissa, the probit of the mean distance from the top of each fraction (expressed as CDF) bears a precise relation to its mean RBC age. On the ordinate, since age-dependent parameters decay in a random fashion, their logarithms are the appropriate representation. Therefore, when the probits of the CDF are plotted on the abscissa and the logarithms of the metabolic parameter on the ordinate, the slopes of those age-dependent RBC parameters that decline in a continuous and random manner become linear and it is possible to measure their life-spans.

To define with precision the relation between position of the fractions in the gradient and their mean cell age, it is necessary to establish a definite correlation between the probit scale and the RBC age scale. This can be achieved by an understanding of the normal distribution and the knowledge of RBC population characteristics. Since the distribution of RBC with regard to density as well as with regard to age is normal, it is also symmetrical. Thus, fractions located at the center must have a mean age equal to half the life-span of the RBC (= 60 days, in the case of human RBC). At two symmetrical points at the tails of the distribution, the mean RBC age is respectively minimal (<1 day), at the extreme left and maximal ($\sim$120 days), at the extreme right. To identify those levels, it is necessary to follow the rate of decline in the gradient of cells of minimal age (or the corresponding rate of increase of cells of maximal age). As the distribution is symmetrical, it is enough to define only one of these intercepts (4).

Although it is not currently possible to identify RBC of maximal age, those of minimal age can be easily identified, since they are the reticulocytes. As the reticulocyte percentage in normal blood is 0.8%, the average age of the reticulocyte is approximately one day. Hence, the intercept, on the ordinate, of the slope of decline in the gradient of the reticulocytes at a value of 100% ($\log_{10} = 2$) corresponds on the abscissa with the position in the gradient (probit) where the mean age of the RBC =<1 day. This point has been identified experimentally by intercept on the ordinate of the steep rate of decline of reticulocytes at the adscissa value of 0.9 Probit (= 4.1 σ below μ). A corresponding point on the other extreme of the abscissa represents the position in the gradient where the mean age of the RBC = ∼120 days. Therefore:

- at 4.1 standard deviations below the mean the average RBC age is =<1 day,
- at the center of the gradient the average RBC age is = 60 days,
- at 4.1 standard deviations above the mean the average RBC age is =∼ 120 days.

Since this model is based on precise criteria, its validity can be experimentally demonstrated. In fact, each slope of decline of an age-dependent parameter should yield at the level of probit 5 (= μ) a value equal to that of the unfractionated RBC population; a population of RBC consisting exclusively of reticulocytes should yield for each RBC parameter a value of the same order of magnitude as that predicted by the intercept on the ordinate of its slope in the gradient when the abscissa value is 0.9 probit.

This model has been validated by the demonstration that in all cases studied, the level of activity at the center of the gradient was equal to that of the unfractionated population. In addition, when RBC fractions consisting <u>exclusively</u> of 100% reticulocytes were isolated from the blood of individuals with increased reticulocytes, the observed specific activity of glucose-6-phosphate dehydrogenase (G6PD) reached exactly the value predicted by the model. It is clear that the model described yields precise results, at least for those RBC parameters that decline continuously.

Beutler hypothesized that metabolic changes in the RBC take place only at the reticulocyte-mature RBC transition, and therefore, for age-dependent parameters of the RBC the activity, in the reticulocytes should be several times (>100) greater than for mature RBC. He argued, to support this unsubstantiated hypothesis, that the progressive rate of metabolic activities observed in density gradient separated RBC is artifactual, and only reflects the decline in percentage of reticulocytes. This view implies that metabolic RBC activities decline with age in a biphasic fashion, the largest changes taking place at the reticulocyte-mature RBC transition, with only minor decline afterwards.

This hypothesis is directly contradicted by the experimental data on G6PD from our laboratory mentioned above. This dispute can also be easily resolved by the simple, appropriate experiments detailed in this report. We took advantage of the fact that different isozymes may have different rates of decline. Our recent work has shown that the hexokinase (HK) activity of the RBC results from two independent isozymes, HK_I and HK_R (8). The HK_R isozyme is RBC-specific, is predominant in the reticulocytes, and decays at an extremely rapid rate. On the other hand, the second isozyme, HK_I, which is common to several other tissues, declines at a much slower rate. Therefore, a comparison of the rate of decline in the gradient of HK with those of other age-dependent enzymes of the RBC offers the opportunity to verify if the buoyant density gradient technique can distinguish a biphasic rate of decline from a progressive one. The activity of HK was measured simultaneously with that of three other age-dependent enzymes: pyruvate kinase (PK), pyrimidine-5'-nucleotidase (P5N) and glutamic-oxalacetic transaminase (GOT), in RBC separated on discontinuous gradients of arabino-galactane. These were designed to yield top fractions containing 5-10% or less of the total RBC population, in order to maximize the opportunity to observe a biphasic rate of decay.

MATERIALS AND METHODS

RBC suspensions

These were prepared by filtering at 4°C defibrinated blood diluted in buffered saline with glucose (BSG) through two double layers of Whatman #2 filter paper, followed by 2 repeated washings in BSG. This technique yields RBC suspensions totally free of platelets and essentially devoid of leukocyte contamination while retaining all reticulocytes (1). The purification of RBC with this method is at least as efficient (and probably more so) than the use of columns of α-cellulose and microcrystalline cellulose (1,9).

Density Gradients

These were prepared as described previously (2). For these experiments, two ml of packed RBC suspension were layered over 4 layers of arabino-galactane ranging in density from 1.097 to 1.12 respectively in a cellulose nitrate tube (1.6 x 10.2 cm). The total volume of the gradient was 12 ml, thus providing a gradient/RBC suspension ratio of 6/1. The tube was centrifuged at 60,000 x g for 45 minutes in the SW27 swinging bucket in a Spinco ultracentrifuge. The resulting five RBC layers were isolated by slicing the tube, and washed with BSG (or with plain saline for assays of P5N). The cumulative distribution function was determined by measuring the Hgb content of each layer and expressing it as percentage of the total Hgb (4). The RBC from each layer were immediately frozen in small aliquots to be used for measurements of enzyme activities.

Enzyme measurements

HK was measured by NADPH production, through a link with G6PD, and PK by oxidation of NADH through a link with lactic dehydrogenase as previously described (5). P5N was measured by inorganic phosphate generation using a linked indicator system that produces uric acid (10). GOT was measured by production of oxaloacetate followed by oxidation of NADH through a link with malic dehydrogenase (11). Reticulocyte counts were performed after supravital stain with new methylene blue. At least 1000 cells were counted.

RESULTS

The results were plotted as previously described, with the age dependent parameter on the ordinate and the probit of the cumulative distribution function on the abscissa (4). For each enzyme, in order to assess whether the in vivo decline was linear or biphasic, the data were analyzed by polynominal regression. The β coefficients were evaluated for each degree of the model (12). If the data were curvilinear, then the coefficients should also be significant by t-test for the second (or higher) degree polynominal, besides those for the first degree. If there is no deviation from linearity, the β coefficients for x^2 would be non-significant.

By these criteria, both PK and GOT showed a simple log-linear rate of decline. However, in the case of both HK and P5N, a second degree polynomial regression model indicated substantial deviation from linearity (Table 1). Therefore, in these cases, two separate slopes were computed. A biphasic decline most likely would be asymmetric, with an initial steeper slope reflecting changes at the reticulocyte to mature RBC transition. Therefore, the first slope was computed from the data in the layers with probit = <4 (these layers contain 15% of the total cell population) and the second slope was computed from the data from the layers with probit = >4 (the remaining 85% of the total cell population).

Thus, for HK and P5N it appears clear that there are two distinct rates of decline (one for the fractions containing less than 15% of the cells and one for the others). For HK the apparent t½ of the initial slope was computed at 15 days and for the second one at 51 days; for P5N the apparent t½ of the initial slope was computed at 11 days and for the second one at 34 days. The other enzymes measured clearly declined linearly, with a single slope; the t½ for PK was computed at 31 days and for GOT at 30 days.

Table 1. Rates of Decay of Selected RBC Enzymes

Enzyme	First t½ (days)	Second t½ (days)	ß-coeff. for x^2	P
HK	15	51	0.031	0.002
P5N	11	34	0.033	0.029
PK	31	--	0.009	0.134
GOT	30	--	0.014	0.144

DISCUSSION

Experiments with cohort labelling of newly formed RBC using non-eluting isotopes, such as [14]C- or ^{15}N-labelled glycine that becomes an integral part of the hemoglobin molecule, have shwon that initially the circulating radioactivity declines with a negligible slope. These findigs have been interpreted as demonstrating that in the initial part of the RBC lifespan there is little random cell loss (in the case of the human RBC, the loss is essentially negligible). This initial phase, after a period of time variable among species, is then followed by another phase of extremely rapid exponential rate of disappearance of radioactivity. These findings have been interpreted to indicate that, at a fixed point in the cell life-span, a very marked random loss takes place when all residual cells are removed from the circulation. The midpoint of this fast rate is taken as the average lifespan. All these experiments therefore indicate that, although in some species there is also a significant component of random loss, the RBC has a defined and finite lifespan (1,14).

In has appeared reasonable to most investigators that the mechanism of RBC removal results from the progressive alterations of intrinsic characteristics, until a point is reached when the RBC becomes extremely vulnerable and is removed from circulation (15). Thus the rather precise lifespan reflects a progressive process of aging. This view has been recently challenged by Beutler who suggested that removal of the RBC is totally unrelated to senescence, but instead results from a _signal_ that takes place, while the cell is still viable and fully functional (16).

Beutler's suggestion contradicts the very large body of experimental observations that demonstrate a progressive decline of important constituents and metabolic activities of the RBC. Since most of these data are based on studies of RBC separated on the basis of density, that appears to be a function of age (17), Beutler found it necessary to challenge as well the validity of the relationship between cell density and cell age, in order to support his hypothesis (6). He theorized that, since metabolic change only take place at the reticulocyte-mature RBC transition, the linear rates of change observed in density separated RBC are artifactual. The present study demonstrates instead that, when a biphasic rate of decline occurs _in vivo_, this is clearly demonstrated by the buoyant density gradient analysis.

In our laboratory, in addition to the current observation on HK (9), a very distinct biphasic rate of decline has been reported for other situations, such as for instance, the extremely rapid rate of disappearance of the protoporphyrin base from the red blood cells of patients with erythropoietic protoporphyria, as contrasted to the slow rate of decline of Zn-protoporphyrin (18). Biphasic rates of decline have been observed in density gradients also by others (19).

The observed biphasic rate cannot be an artefact due to leukocyte or platelet contamination of the lightest layers, since our "pure red cells suspension" are essentially free of these elements. In fact, in the case of enzymes, such as PK, that are present in the leukocytes at an extremely high ratio of activity relative to the red blood cells, the slope is absolutely linear.

Our results clearly demonstrate that, when RBC are carefully separated on the basis of their buoyant density, any biphasic rate of decline is apparent in the gradient derived plots. Thus, the progressive single linear rate of decline observed in the gradient for most enzymes reflects a real single exponential rate of disappearance _in vivo_. These observations are in keeping with previous reports from our laboratory that demonstrated that, when suspensions of RBC consisting of nearly 100% reticulocytes could be isolated from reticulocyte-rich blood, their enzymatic activities were found in the range predicted by a single exponential slope of decay (4) and not in the much higher range that would be required to satisfy Beutler's hypothesis (6,8,16).

The current studies therefore confirm the utility and validity of density fractionation in the study of RBC senescence. The view that this process results from a continued and progressive alteration of RBC components and activities remains, at present, in agreement with the current experimental evidence (20-23).

CONCLUSIONS

When changes take place at the reticulocyte-RBC transition, these are clearly reflected in the density gradient by a biphasic slope. The progressive metabolic decline of the RBC with in vivo aging demonstrated by buoyant density gradient separation remains the hypothesis most valid and consistent with the current available experimental data.

ACKNOWLEDGEMENTS

These studies were supported by Grant # DK26793-11 from the Institute for Diabetes, Digestive and Kidney Diseases, and Grant # HL28381-08 from the Institute for Heart, Lung and Blood Diseases, National Institutes of Health, Bethesda, MD.

REFERENCES

1. S. Piomelli, G. Lurinsky, L. R. Wasserman, The mechanism of red cell aging. I. Relationship between cell age and specific gravity evaluated by ultracentrifugation on a discontinuous density gradient, J. Lab. Clin. Med. 69:659 (1967).
2. L. M. Corash, S. Piomelli, H. C. Chen, C. Seaman, E. Gross, Separation of erythrocytes according to age on a simplified density gradient, J. Lab. Clin. Med. 84: 147 (1974).
3. S. Piomelli, C. Seaman, J. Reibman, A. Tytun, J. Graziano, N. Tabachnik, L. Corash, Separation of younger red cells with improved in vivo survival: A new approach to chronic transfusion therapy, Proc. Natl. Acad. Sci. (USA) 75:3474 (1978).
4. S. Piomelli, L. M. Corash, D. D. Davenport, J. Miraglia, E. L. Amorosi, In vivo lability of G6PD in Gd^{A-} and $Gd^{Mediterranean}$ deficiency, J. Clin. Invest. 47:940 (1968).
5. C. Seaman, S. Wyss, S. Piomelli, The decline in energetic metabolism with aging of the erythrocyte and the mechanism of cell death, Am. J. Hemat. 8:31 (1980).
6. E. Beutler, G. Hartman, Age-related red cell enzymes in children with transient erythroblastopenia of childhood and hemolytic anemia, Pediatr. Res. 19:44 (1985).
7. G. Dale, Characterization of senescent red cells from the rabbit, This symposium.

112

8. E. Beutler, Annation: How do red cell enzymes age? A new perspective, Br. J. Haematol. 61:377 (1985).

9. K. Murakami, F. Blei, W. Tilton, C. Seaman and S. Piomelli, An isoenzyme of hexokinase specific for the human red blood cell (HKR), Blood 75:770 (1990).

10. E. Beutler, C. West, K. G., The removal of leukocytes and platelets from whole blood, J. Lab. Clin. Med. 88:329 (1976).

11. C. R. Zerez, K. R. Tanaka, A continuous spectrophotometric assay for pyrimidine-5'-nucleotidase, Analyt. Biochem., 151:282 (1985).

12. E. Beutler, Red Cell Metabolism. A manual of biochemical methods 3rd Edition. Orlando, F.L., Grune and Stratton, Inc., pp. 87-89 (1984).

13. R. J. Freund, R. C. Littell, SAS System for Regression. Cary, NC, SAS Institute, Inc., pp. 103-108 (1986).

14. D. Shemin, D. Rittenberg, The life span of the human red blood cell, J. Biol. Chem. 166:627 (1946).

15. S. B. Bull, Red cell senescence, Blood Cells, 1:1 (1988).

16. E. Beutler, The relationship of red cell enzymes to red cell life-span, Blood Cells, 14:69 (1988).

17. M. R. Clark, S. B. Shoet, Red cell senescence, Clin. Haematol., 14:223 (1985).

18. S. Piomelli, A. A. Lamola, M. Poh-Fitzpatrick, C. Seaman, L. Harber, Erythropoietic protoporphyria and Pb intoxication: The molecular basis for difference in cutaneous photosensitivity. I. Different rates of diffusion of protoporphyrin from the erythrocytes, both in vivo and in vitro, J. Clin. Invest. 56:1519 (1975).

19. B. M., R. A. Fisher and H. Harris, Post-translational alterations of human erythrocyte enzymes, in Markert C.L. (ed): Isozymes, Vol. I, New York, Academic Press, pp. 781-795 (1975).

20. S. Piomelli, C. Seaman, L. Corash, How do red cell enzymes age? Hypothesis and facts, Brit. J. Haematol., 64:407 (1986).

21. S. Piomelli, Commentary to: E. Beutler's The relationship of red cell enzymes to red cell life-span, Blood Cells, 14:81 (1988).

22. S. Piomelli, Commentary to: D. Danon and Y. Marikowsky: The aging of the red blood cell. A Multifactor process, Blood Cells, 14:16 (1988).

23. C. Seaman: Commentary to: S. Landaw: Factors that accelerate or retard red blood cell senescence, Blood Cells, 14:63 (1988).

THE RELATIONSHIP BETWEEN THE BLOOD OXYGEN TRANSPORT AND THE

HUMAN RED CELL AGING PROCESS

Michele Samaja[1], Ermanna Rovida[2], Roberto Motterlini[3]
and Massimo Tarantola[3]

[1]Dipartimento di Scienze e Tecnologie Biomediche
Università di Milano
[2]Istituto di Tecnologie Biomediche Avanzate del C.N.R.
[3]Istituto Scientifico San Raffaele, Milano, Italy

SUMMARY

We have studied the relationship between the in vivo aging process of the human red cell (RBC) and its main function, the transport of O_2 from the lungs to the tissues. This study included several approaches. First, we observed that the affinity for O_2 in young RBCs was lower than in old RBCs (p<0.0005) due to different intracellular concentration of 2,3-diphospho-glycerate, main effector of hemoglobin. Second, we explored whether there are some subgroups of the healthy human population with altered RBC age distribution: females in the age range 25-35 exhibited significantly younger RBCs (p < 0.0005) and lower RBC-O_2 affinity (p < 0.01) than other groups. Correspondingly, the RBC-O_2 affinity in female blood was significantly lower (p < 0.002) than in male blood. Third, we correlated by two independent methods the lowered RBC-O_2 affinity to a more efficient O_2 delivery to the tissues by two independent methods: 1) calculating the size of the cardiac output increase required to sustain the tissue oxygenation after an increase of the RBC affinity for O_2 ; and 2) monitoring the enhanced cardiac function in isolated rat hearts perfused with RBCs at low O_2 affinity. Finally, comparing some hematologic findings relevant for the O_2 transport in two healthy populations with different RBC age distributions, such as age-matched females and males, it appeared that the low RBC-O_2 affinity in females is an adaptive response to their lower [Hb] . All these approaches agreed in indicating that the lowered RBC-O_2 affinity, and by reflect the younger RBC average age, appears associated to more favourable conditions of tissue oxygenation. We conclude that the human RBC aging process is not only a suitable model to study the cellular and tissutal aging mechanisms, but may also have a considerable influence on some physiological and pathological patterns in humans.

Red Blood Cell Aging, Edited by M. Magnani and
A. De Flora, Plenum Press, New York, 1991

INTRODUCTION

The human red blood cell (RBC) is designed for the specific purpose to transport oxygen from the lungs to the tissues. Nearly all the metabolic, physiological and physical properties of the RBC are aimed to the optimization of this function. For example, the RBC can accomodate large amounts of hemoglobin (Hb), a protein with O_2-carrying properties. In addition, the subcellular structures necessary for the aerobic metabolism, such as the mitochondria and the nuclei, lack in the mature RBC for the dual purpose of accomodating more Hb and allowing greater flexibility in the passage of the RBC through narrow capillaries. Finally, the presence of 2,3-diphosphoglycerate (DPG), intermediate of the RBC glycolytic pathway and major allosteric effector of Hb (Benesch and Benesch, 1967), allows the RBC to regulate its own affinity for O_2 by intrinsic mechanisms. Indeed, erythrocytic DPG regulates the RBC-O_2 affinity in two ways: by its allosteric control on Hb and by lowering the intracellular pH through the Gibbs-Donnan effect (Samaja and Winslow, 1981). Due to the synergism of the two effects and to the high intracellular concentration of DPG, it becomes perhaps the most important regulator of the RBC-O_2 affinity.

The lack of genetic material and of energy-producing compartments gives the human RBC the unique characteristics of a well-defined 120-day life span (Eadie and Brown, 1953), equilibrium set point between the production of new cells by the bone marrow and their removal from the circulation by the spleen. Indeed, human spleen removes selectively from the circulation the oldest RBCs only (Jandl and Cooper, 1983; Mentzer and Clark, 1983), in contrast to other species where the RBC removal is randomized (Clark, 1988). It is still matter of discussion if the primary cause of the RBC aging process is to be searched in the events leading to the metabolic depression, in the stressing factors present in the circulation, or both. Whatever the cause, we believe that it is to be clarified whether the RBC in vivo aging is linked to the main function of the RBC, i.e., the O_2 transport, and whether or how it participates to the organism adaptive responses to environmental and endogenous stimuli. We will here review our approach to this problem.

The O_2 affinity in young and old RBCs

Previous studies have shown that the $[DPG]/[Hb]$ ratio and the P_{50} (PO_2 at which half of Hb is bound to O_2 and useful index of the RBC-O_2 affinity) are higher in the young than in the old RBC fraction (Haidas et al., 1971); Schmidt et al., 1987). Recently, we have confirmed these data (Samaja et al., 1990a) with an approach by which it was inferred that DPG alone does not fully explain the lower RBC-O_2 affinity in the young RBCs, but another unknown intracellular factor must be involved. Nonetheless, DPG always remains the most important allosteric effector of Hb in the RBC. The decreased concentration of DPG in the senescent RBC is related mainly to the age-induced change of the activity of key regulatory enzymes that operate along the glycolytic pathway, but the possibility that the level of DPG may also be set by direct effect of some other factors, for example acute exposure to an hypoxic environment, cannot be ruled out (Mairbaurl et al., 1990).

The RBC age distribution in humans

In the search of a relationship between the RBC aging process and the transport of O_2 to the tissues, it is essential to find a model where the two variables can be easily correlated. For this purpose, one needs a suitable method to characterize the average RBC age distribution in a sample of blood. Several features have been proposed as suitable markers for the RBC senescence, including the metabolic depletion (Seaman et al., 1980), the changes in the RBC deformability and shape (Linderkamp and Meilsen, 1982), and the altered activity of some enzymes (Kadlubovski and Agutter, 1977). The striking correspondence between the age of the RBC and its density is of particular interest because it allows both preparative and analytical studies. The same peculiarity was used to measure the RBC age distribution in some hematological disorders (Nakashima et al., 1973). It was found that iron-deficiency anemia, pyruvate kinase deficiency and polycythemia vera were associated with an increase of the light, young RBC fraction. On the contrary, hereditary spherocytosis, autoimmune hemolytic anemia, aplastic anemia and erythroleukemia were associated with an increase of the dense, old RBC fractions. These features, which are linked to the circulatory stress, or to the actual life span of the RBC, or both, may theoretically provide an in vivo model to study the relationship between the RBC aging process and the tissue oxygenation, but the severity of the underlying disease poses a serious problem because it may induce responses not directly linked to the RBC age distribution.

To explore whether there are subgroups in the healthy human populations suitable to correlate the RBC age distribution to the O_2 transport mechanisms, we have determined the average RBC density in 59 males and females subdivided into the following age classes: 2-12, 25-35, and 62-85. In addition, samples were obtained from 10 pregnant women in the age range 23-34. For the measurements, two isotonic solutions were obtained mixing an hyperosmotic (2.66M NaCl + 0.09 M KCl) solution, water and Percoll (Sigma Chemicals, St. Louis, Mo) in variable ratios to yield densities of 1.099 and 1.102 g/ml. These densities were selected because they appeared the most indicative after running several complete distribution vs density curves. Small tubes were loaded with 0.15 ml of this solution and 0.05 ml of freshly drawn heparinized blood. The tubes were centrifuged at 12 000 rpm for 2 min at 0°C, and the blood layers above and below the Percoll were removed by a syringe and diluted into 5 ml Dabkin's reagent. The amount of RBCs lighter than the Percoll was finally calculated from the absorbance readings.

The adult females in the age range 25-35 exhibited the youngest RBCs with respect to all the other classes of subjects (p<0.0005), including the age-matched males and pregnant women (figure 1),leading to the following considerations: 1) the factor(s) shifting the average RBC distribution towards the young RBC is (are) to be searched in the events inherent to menstruating women, possibility the anemia-induced increase of erythropoiesis; 2) either the oxidative factors are less, or the RBC life span is shorter in the blood of adult females than in the other subjects; and 3) the age of the subject per se does not affect the RBC age distribution in healthy subjects.

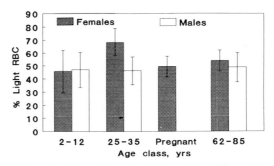

Fig. 1. Percent of RBCs lighter than 1.099 and 1.102 g/ml in various classes of subjects (average between duplicate measurements at the two densities, mean±S.D., n=69).

Some hematologic findings (Fig. 2) obtained in healthy male and female subjects of the same age group (25-35) are highly correlated to this result and to the previously discussed observation that the young RBCs have higher DPG and lower O_2 -affinity than the old RBCs. Whatever the factor for the younger or less stressed RBCs in females' blood, this is an unique opportunity to investigate in healthy well-doing populations whether the altered RBC age distribution has an effect on the RBC-O_2 transport. However, it is mandatory first to understand whether a lowered RBC-O_2 affinity is really advantageous for the oxygenation of the tissues.

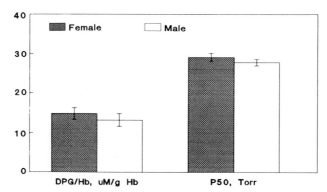

Fig. 2. Some hematologic findings in blood from 9 male and 16 female subjects of the same age (25-35). The $[DPG]/[Hb]$ and P_{50} differences are significant at the $p < 0.005$ level.

The RBC-O_2 affinity and the O_2 transport to the tissues

Theory

Whether a decreased RBC-O_2 affinity, whatever the cause, is advantageous for tissue oxygenation, is a still debated question. Theoretical considerations, sketched in figure 3, indicate that when the P_{50}

increases from 28 to 32 Torr as a consequence of an increase of the [DPG]/[Hb] ratio from 12.9 to 17.7 μM/g Hb, or 0.8 to 1.1 M/M, (Samaja et al., 1981), the amount of O_2 that can be delivered to the tissues at constant arterial and venous PO_2 is augmented, in the case of figure 3, from 43% to 49% of the total O_2 carried by blood. Although it is expected that the organism may adjust the venous PO_2, it is likely that a decreased RBC-O_2 affinity is favourable for the organism because the same amount of O_2 can be delivered to the tissues either at a higher PO_2 or at a lower flow through a specific organ. In the former case, the higher venous PO_2 prevents tissue hypoxemia, and in the latter the lower blood flow reduces the circulatory load. It was calculated (Samaja, 1988) that the blood flow through the major organs may decrease about 10% following an increase of the P_{50} of 2 Torr, without compromising the oxygenation of the tissues at the physiological metabolic level.

The observations that the average RBC age distribution is younger in female than in male blood, and that the young RBC has a lower O_2 affinity than the old RBC, provide an opportunity to test in vivo the theoretical indications that a lowered RBC-O_2 affinity favours the oxygenation of the tissues. As expected, [Hb] is lower in females with respect to males (13.7 and 15.5 g/dl, respectively, on the average). Consequently, the blood O_2 capacity in females is lower and the lower O_2 delivery to the tissues must be compensated with a higher cardiac output. The required increase of the cardiac output when [Hb] decreases from 15.5 to 13.7 g/dl was calculated (Samaja et al., 1986) at constant value of the other factors, such as arterial and venous PO_2's (95 and 40 Torr, respectively), arterial PCO_2 (40 Torr), and O_2 uptake (0.35 1/min): the cardiac output in females should increase from 6.34 to 7.12 1/min to compensate their lower Hb and to support an adequate oxygenation to the tissues. However, since the [DPG] / [Hb] ratio and the P_{50} are higher in females than in males, the enhanced O_2 carrying properties tend to compensate the lower [Hb] in females, and it was calculated that owing to the increase of the [DPG]/[Hb] ratio only, the cardiac output in females may decrease from 7.12 to 6.80 1/min without comprimising the oxygenation of the tissues. This implies a considerable saving of the energy cost of blood pumping.

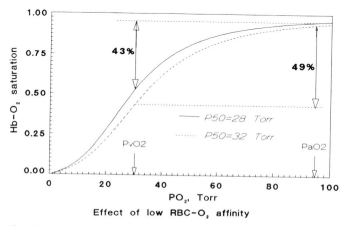

Effect of low RBC-O$_2$ affinity

Fig. 3. When the P_{50} shifts from 28 to 32 Torr, the fraction of O_2 delivered to tissues increase from 43% to 49% at constant $P_a O_2$ and $P_v O_2$

Isolated heart perfusion

The direct experimental confirmation of the above theories is not univocal despite the very large body of experimental studies using in vivo and in vitro (or ex-vivo) models. The in vivo studies aimed to show an advantage of a lowered RBC-O_2 affinity necessarily have the drawback that the studied organisms can compensate the experimentally induced changes of the RBC-O_2 affinity with both circulatory (blood flow vs venous PO_2 adjustment, organ to organ compensation, capillary recruitment) and metabolic (erythropoiesis long-term adaptation, acid/base adjustment) adaptations. On the other hand, the major problem of the in vitro and ex-vivo studies is that the often drastic changes of the induced RBC-O_2 affinity have little physiological relevance. Thus, it is still uncertain whether the relatively smaller fluctuations of the O_2 affinity are physiologically significant in in vivo models. However, the perfused isolated heart is perhaps ideal to test the physiological significance of the RBC-O_2 affinity changes which, under appropriate experimental conditions, may be correlated to a variety of myocardium responses without confounding effects. Nevertheless, the data obtained from these studies still remain quite contradictory. Part of them showed a high degree of insensitivity of the heart O_2 consumption to a decreased P_{50} : Martin et al. (1979) attributed this result obtained in working rat hearts to a 50% increase of the capillary density; and Ross and Hlastala (1981) hypothesized that the drop of the O_2 consumption observed in isolated skeletal muscles was associated with the presence of humoral factors rather than with the RBC O_2 affinity. On the other hand, in experiments in paced isolated rabbit hearts, the myocardial function increased when switching from RBCs at P_{50} =16.9 to RBCs at P_{50} =32.6 Torr (Apstein et al., 1985) showing an evident advantage associated with a decreased RBC-O_2 affinity. In other studies, inositol hexaphosphate (IHP) was used to drastically increase the P_{50} from 18-20 to 42-47 Torr. In perfused isolated rat and rabbit hearts (Stucker et al., 1985; Baron et al., 1987), low O_2 affinity RBCs decreased the coronary blood flow and increased the O_2 consumption again showing an advantage with respect to high-O_2 affinity RBCs, but the allowance for flow compensation, the low arterial PO_2 (184-188 Torr) and a possible direct effect of the IHP-loaded RBCs (Samaja et al., 1990b) do not clarify the question.

In our Langendorff-perfused isolated hearts, the aorta of excised hearts from male Sprague Dawley rats was mounted onto a stainless-steel cannula and the heart perfused retrogradely with an oxygenated Krebs-Henseleit buffer (PO_2=670 Torr, PCO_2=42 Torr, pH 7.4). A ballon connected to a pressure transducer (Harvard Apparatus mod. 52-9966, Natick, MA) was inserted into the left ventricle and filled with saline to monitor the developed pressure. A teflon cannula was inserted in the pulmonary artery to collect the perfusate for the measurement of the venous PO_2 and of the O_2 uptake. An additional pressure transducer above the aortic cannula monitored the coronary pressure. The age-induced O_2 affinity changes were simulated treating human stored blood (3 to 10 days old) from a local blood bank in two different ways. To obtain RBCs with low O_2 affinity, the blood was filtered through a Pall RC-100 Leukocyte Removal Filter (Pall Biomedical Products Corp., East Hills, NY) and incubated 60 min at 37°C in the presence of 50 ml rejuvenating solution (103 mM sodium phosphate, 100 mM inosine, 100 mM pyruvate, 5 mM adenine, pH 7.0 at 37°C). The RBCs were then washed (IBM 2991 blood cell processor, Hopewell Junction, NY) with 2 1 isotonic NaCl. RBCs with high O_2 affinity were obtained using 10 days old blood filtered and washed as above. The resulting P_{50} values were 18 and 30 Torr at pH 7.4, $PCO_2 = 40$ Torr, 37°C.

Table 1. Myocardial function change (mean ±SEM) when switching from buffer to RBC perfusion. The change in the developed pressure only is significant at the $p < 0.005$ level.

| | RBC-O_2 affinity | |
	high	Low
Coronary pressure, Torr	8.1±3.5	14.1±9.5
Developed pressure, Torr	-10.6±5.1	-0.8±4.1
Oxygen uptake, µM/min	0.7±0.4	-0.1±0.2

Table 1 shows the change of the heart performance when the perfusion was switched from the buffer to the RBCs with high or low O_2 affinity at 10% hematocrit. All the other parameters were unchanged, including the arterial gasses, the flow through the heart (15 ml/min) and the heart preload (10 Torr). The RBCs with low O_2 affinity have a positive effect on the perfused hearts, because of the better preserved developed pressure when switching from the buffer perfusion to the low O_2 affinity RBCs. The coronary pressure, index of the vascular state, did not differ appreciably between the two groups as well as the O_2 uptake. It can thus be inferred that an increased RBC-O_2 affinity is not favourable for the oxygenation of the tissues in this experimental model.

CONCLUSIONS

It is possible that under anemic or hypoxic conditions which increase the erythropoiesis, the average RBC population become relatively younger. The size of the alteration of the consequent whole blood affinity for O_2 is sufficient to directly tune up the O_2 transport and adapt it to the new situation. This hypothesis has had three independent lines of confirmation: 1) a theoretical inference based on the calculation of the amount of the O released at constant values of the other factors that regulate the O_2 delivery to the tissues; 2) an experimental approach based on the perfusion of isolated hearts with RBCs of different O_2 affinities; and 3) comparing the relevant hematologic data in two healthy populations that have different RBC age distributions. We conclude that the RBC aging process appears biologically relevant for the adjustment of the average blood O_2 affinity and transport.

ACKNOWLEDGEMENTS

We thank Mr. M. Beretta, Drs. F. Rossi and L. Sabbioneda, and Prof. A. Zanella for great help provided in performing part of the work with the perfused hearts. RM and MT are recipients of grants from the Istituto Scientifico San Raffaele. Supported in part by the C.N.R. Target project Biotechnology and Bioinstrumentations and by the NATO grant n. 900530.

REFERENCES

1. C. S. Apstein, R. C. Dennis, L. Briggs, W. M. Vogel, J. Frazer and C. R. Valeri, Effect of erythrocyte storage and oxyhemoglobin affinity changes on cardiac function, Am. J. Physiol. 248:H508 (1985).
2. J. F. Baron, E. Vicaut, O. Stucker, M. C. Villereal, C. Ropars, B. Teisserie and M. Duvelleroy, Isolated heart as a model to study the effects of the decrease in oxygen hemoglobin affinity, Adv. Biosciences 67:73 (1987).
3. R. Benesch and R.E. Benesch, The effect of organic phosphates from the human erythrocyte on the allosteric properties of hemoglobin, Biochem. Biophys. Res. Comm. 26:162 (1967).
4. M. R. Clark, Senescence of red blood cells: progress and problems, Physiol. Rev. 68:503 (1988).
5. G. S. Eadie and I. W. Brown, Red blood cell survival studies, Blood 8:1110 (1953).
6. D. Haidas, D. Labie and J. C. Kaplan, 2,3-diphosphoglycerate content and oxygen affinity as a function of red cell age in normal individuals, Blood 38:463 (1971).
7. J. H. Jandl and R. A. Cooper, Hereditary spherocytosis, in:"Hematology", W. J. Williams, E. Beutler, A. J. Erslev, M. A. Lichtman ed., McGraw-Hill, New York (1983).
8. M. Kadlubovski and P. S. Agutter, Changes in the activities of some membrane-associated enzymes during in vivo ageing of the normal human erythrocyte, Br. J. Hematol. 37:111 (1977).
9. O. Linderkamp and H. J. Meiselman, Geometric osmotic and membrane chemical properties of density-separated human red cells, Blood 59:1121 (1982).
10. H. Mairbaurl, W. Schobersberger, O. Oelz, P. Bartsch, K. U. Eckardt and C. Bauer, Unchanged in vivo P at high altitude despite decreased erythrocyte age and elevated 2,3-diphosphoglycerate, J. Appl. Physiol. 68:1186 (1990).
11. J. L. Martin, M. Duvelleroy, B. Teisserie and M. Duruble, Effect of an increase in HbO_2 affinity on the calculated capillary recruitment of an isolated rat heart, Pflug. Arch. 382:57 (1979).
12. W. C. Mentzer and M. R. Clark, Disorders of erythrocyte cation permeability and water content associated with hemolytic anemia, in: "Pathological Membranes", A. Nowotny ed, Plenum, New York (1983).
13. K. Nakashima, O. Susuma and S. Miwa, Red cell density in various blood

disorders, J. Lab. Clin. Med. 82:297 (1973).

14. B. K. Ross and M. P. Hlastala, Increased hemoglobin oxygen affinity does not decrease skeletal muscle oxygen consumption, J. Appl. Physiol. 72:211 (1981).

15. M. Samaja, A. Mosca, M. Luzzana, L. Rossi-Bernardi and R. M. Winslow, Equations and nomogram for the relationship of human blood p_{50} to 2,3-diphosphoglycerate, CO_2, and H+, Clin. Chem. 27:1856 (1981).

16. M. Samaja, P. E. diPrampero, P. Cerretelli, The role of 2,3-DPG in the oxygen transport at altitude, Resp. Physiol. 64:191 (1988).

17. M. Samaja, Prediction of the oxygenation of human organs at varying blood oxygen carrying properties, Resp. Physiol. 72:211 (1988).

18. M. Samaja, E. Rovida, R. Motterlini, M. Tarantola, A. Rubinacci and P. E. diPrampero, Human red cell age, oxygen affinity and oxygen transport, Resp. Physiol. 79:69 (1990).

19. M. Samaja, R. Motterlini, M. Tarantola, M. Beretta, F. Rossi, L. Sabbioneda, M. Porcellati and A. Zanella, Viability of the IHP-loaded red cell in the hypoperfused isolated rat heart, Adv. Biosciences in press (1990).

20. W. Schmidt, D. Boning, and K. M. Brauman, Red cell age effects on metabolism and oxygen affinity in humans, Resp. Physiol. 68:215 (1987).

21. C. Seaman, S. Wyss and S. Piomelli, The decline in energetic metabolism with ageing of the erythocyte and its relationship to cell death, Am. J. Haematol. 8:31 (1980).

22. O. Stucker, E. Vicaut, M. C. Villereal, C. Ropars, B. P. Teisserie and M. A. Duvelleroy, Coronary response to a large decrease of hemoglobin O_2 affinity in isolated rat heart, Am. J. Physiol. 249:H12244 (1985).

RED CELL METABOLISM, NORMAL AND ABNORMAL

IMPLICATIONS FOR RED CELL AGING

William N. Valentine and Donald E. Paglia

Departments of Medicine and Pathology
and Laboratory Medicine
University of California
Center for Health Science
Los Angeles, California, U.S.A.

INTRODUCTION

Aging of the human erythrocyte consists of changes in its structure and function beginning at birth and culminating in its demise. This is commonly regarded as a chronic process, progressing inexorably in small increments to senescence and ultimate death and its time frame for the normal red cell averages about 120 days. However, this transition can be greatly accelerated by intrinsic cellular defects or extrinsic environmental influences. Among many others, the latter include trauma to the red cell as in the microangiopathic hemolytic syndromes, autoimmune disorders where the erythrocyte may be an innocent victim, and the individually varying effects of infection, diet, medications and exposure to hazardous substances in the environment. Red cell life span also may be dramatically shortened by genetically determined birth defects whose impact in some instances can be substantially ameliorated by calculated interventions. Witness the near-normal survival of the hereditary spherocyte after splenectomy, in contrast to that of the same cell circulating in a subject with an intact spleen. Witness the shortened red cell life span accompanying primaquine medication in some glucose-6-phosphate dehydrogenase (G-6-PD) deficient individuals.

THE NORMAL RED CELL AND MAINTENANCE OF THE ADENINE NUCLEOTIDE POOL

The red cell maintains its high energy phosphates and its energy charge (1) (ATP+1/2ADP/ATP+ADP+AMP) at an effective operating level exclusively through the conversion of hexose to lactate. The loss of metabolic integrity with severe glucose deprivation is clearly demonstrable in vitro and is the basis of the standard autohemolysis and incubated osmotic fragility tests (2).

Red Blood Cell Aging, Edited by M. Magnani and
A. De Flora, Plenum Press, New York, 1991

When normal fresh blood is incubated at 37°C, the adenine nucleotide pool is dramatically depleted within a few hours. ATP–ADP–AMP ratios are initially shifted toward accumulation of AMP. However, within 12–24 hours, only hypoxanthine (Hx), a degradation product irretrievable as a source of adenine ribonucleotides, remains. Despite total inability to fuel energy needs, hemolysis does not ensue for some forty-eight hours. Even then only some 2 per cent of cells have hemolysed. Initially the erythrocyte volume increases as its Na^+-K^+ pump fails and intracellular cations and water content are augmented. But long before there is overt hemolysis, cell shrinkage supervenes, K^+ and H_2O loss (the Gardos effect) more than compensating for Na influx. Intracellular Ca^{++} increases as the Ca^{++} pump also fails. There is budding and loss of bilayer lipids and integral proteins from the increasingly less deformable plasma membrane. Membrane surface area/volume ratio decreases while the mean corpuscular Hb concentration (MCHC) is increased. Subtle and complex changes occur in the cytoskeleton proteins and finally, hemolysis becomes evident. In the case of the hereditary spherocyte, the same spectrum of events transpires, but in a foreshortened time frame (2-5). It is remarkable that total loss of energy-dependent metabolic capabilities precedes by so long the extravasation of Hb. Long before the latter is evident, the erythrocyte is no longer viable and there is no prospect of resuscitation.

Under physiologic conditions, while glucose availability varies, such progressive, prolonged and total metabolic deprivation should not occur. The adenine nucleotide pool is, nonetheless, subject to losses which must be compensated by salvage mechanisms and a delicate balance maintained between degradation and replenishment of crucial adenylates. While ATP may be totally dephosphorylated to adenosine as a result of a few enzymatic reactions such as those resulting in the synthesis of adenosylmethionine and adenosylcobalamin, these are quantitatively minor and relatively insignificant. For practical purposes, essentially only AMP is at risk of being lost from the pool. It can undergo dephosphorylation to adenosine (Ado) or be deaminated to inosine monophosphate (IMP). In the human red cell the latter reaction is irreversible.

Adenosine (Ado) generated by dephosphorylation of AMP may suffer several fates. It may rejoin the nucleotide pool by rephosphorylation via red cell Ado Kinase (AK) as long as ATP is available (the adenosine cycle (6). It may diffuse out of the red cell or it may be enzymatically deaminated to inosine and, ultimately, converted to Hx.

Since losses of adenine ribonucleotides are inevitable, maintenance of a stable pool must be dependent on counterbalancing salvage pathway. These potentially consist of the salvage of plasma adenine (Ad) through incorporation into AMP, or, alternately, the regeneration of AMP by the phosphorylation of plasma Ado. The human red cell possesses the enzymatic machinery to accomplish both. The salvage of Ad is dependent upon two sequential enzymatic reactions. The first synthesizes 5-phosphoribosyl-1-pyrophosphate (PRPP) from ribose-5-phosphate (R-5'-P) and ATP; in the second, adenine phosphoribosyl transferase (APRT) catalyses the formation of AMP from PRPP and adenine.

PRPP partecipates in a number of competing reactions, and its formation in highly regulated (7). Ad is present in plasma and erythrocytes, and as far as is known, is derived virtually entirely from methylthioadenosine, a metabolite in polyamine synthesis (8-10).

The second salvage pathway involves the phosphorylation of Ado via AK (6). Ado has been shown to be released by fatty tissue, hepatocytes, cardiac muscle and other tissues (11). Its main source is ubiquitous AMP, a cytosol constituent of all body tissues. A second potentially substantial source is S- adenosylhomocysteine via its hydrolase (12). The S- adenosylhomocysteine is a by-product of methylations mediated by S- adenosylmethionine (12). In normal erythrocytes, Ado is a substrate not only for AK but also for competing Ado deaminase which converts it to inosine.

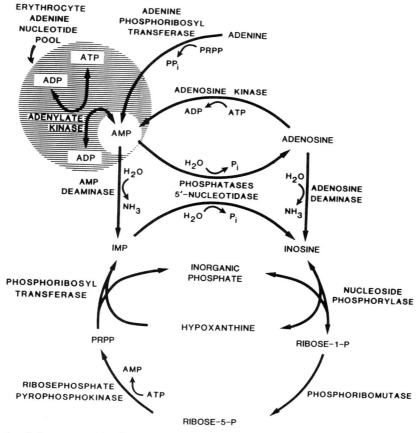

Fig. 1. Salvage and degradative reactions of erythrocyte adenine nucleotides. Reproduced with permission of <u>Blood</u> 74:2161-2165, 1989, Grune & Stratton, Inc.

While plasma concentrations of both Ad and Ado are measurable and have been widely cited, their accuracy is clouded by the rapid equilibria of the extracellular metabolites with body cells, their active catabolism, and lack of pertinent information concerning their transient and potentially highly variable concentrations at tissue sites of liberation. Indeed the half-life of Ado incubated with undiluted fresh whole blood has been reported as less than 10 seconds (13). It is a long voyage from the myocardium and the liver to the antecubital vein where measurements are often made. Every centimeter of the voyage is inhabited by cellular sponges and cannibals, including a myriad of red cells. It seems fair to assume that the spectrum of Ado concentration to which the circulating erythrocyte is exposed during a given time frame in its travels on the vascular highways and alleys is at best uncertain.

Both PRPP synthetase and AMP deaminase are allosteric enzymes. P_i is an absolute requirement of the former, whose activity increases markedly over a wide range of P_i concentrations. In contrast, increasing P_i inhibits AMP deaminase. PRPP synthetase is inhibited by ATP, ADP, and 2,3-DPG while AMP deaminase is activated allosterically by ATP and dATP. These and other complex regulatory factors render fresh intact cells in native serum more reliable in terms of reflecting in vivo metabolism than are artificial buffered systems, particularly those employing phosphate, hemolysates, and stored blood. Still, difficulties to reproduce variables remain.

RELATIVE ROLE OF DEPHOSPHORYLATION AND DEAMINATION IN ERYTHROCYTE CATABOLISM OF AMP

While under certain conditions dephosphorylation of AMP can be a significant contributor to its catabolism, deamination appears to be the dominant mechanism of loss from the adenine nucleotide pool under physiologic conditions. Utilizing specific inhibitors of Ado deaminase and AK as aids, in defining metabolic pathways of AMP catabolism in vitro, Bontemps et al (14) have concluded that with glucose present deamination of AMP followed by dephosphorylation of IMP accounts for virtually 100% of AMP losses. Production of Ado and its conversion to Hx occurred only when glucose was nearly completely utilized and AMP levels are increased 30 fold. In the presence of glucose, but under conditions of alkalinization, intact red cell AMP increased 15-20 fold as a result of selective augmentation of reactions of the first half of glycolysis, during which ATP degradation was enhanced and not fully balanced by ATP-generating reactions of terminal glycolysis. All Hx generated was nonetheless formed via intial deamination of AMP. Some dephosphorylation of AMP to Ado occurred, but the latter was 100% rephosphorylated by uninhibited AK as long as small amounts of ATP remained. Net losses of red cell adenylates via dephosphorylation of AMP were minor at most except in the near absence of glucose and under conditions where red cell AMP was increased by many fold.

Dephosphorylation via 5-nucleotidase is strongly inhibited in mammalian cells by several purine and pyrimidine nucleoside triphosphates (15). For example, the K_i for ATP in Ehrlich ascites tumor cells is 12 μM (15), whereas normal ATP concentrations in human red cells exceed 1000 μM.

We have come to conclusions similar to those of Bontemps et al., (14) utilizing both intact red cell systems and hemolysate incubations with multiple purine substrates over a wide range of media and in the presence and absence of genetically-determined pyrimidine-5'-nucleotidase deficiency (P-5'-Nase) (16,17).The red cell also contains a nucleotidase with special affinity for deoxyribonucleotides which is clearly distinct from P-5'-nase (18-20). Evidence presented by Bagnora and Hershfield (21) for lymphoblastoid cell lines confirms that their catabolism of adenine ribonucleotides precedes by initial deamination of AMP. Similar conclusions by others suggest that this is the dominant catabolic pathway in mammalian cells in general (22). This dominance is further supported by actual increases in the red cell adenine nucleotide pool observed under conditions where AMP deaminase activity is diminished by aging (23,24), by inherited deficiency (25), and possibly by an unfavorable regulatory environment.

RELATIVE ROLES OF ADENINE AND ADENOSINE SALVAGE IN REPLENISHING THE RED CELL ADENINE NUCLEOTIDE POOL

While salvage of Ad has been proposed as a viable normal mechanism of replenishing the adenylate pool (26,27), there is contrary evidence that this pathway contributes little to adenylate replenishment under physiologic conditions. The salvage of Ad requires availability of PRPP and the catalytic activity of APRT. The red cell concentration of ATP and the total adenine nucleotides have been documented in a substantial number of patients with an essentially complete, genetically-determined deficiency of APRT(9) and, hence, total inability to utilize adenine salvage for regeneration of AMP. Such subjects have entirely normal concentrations of red cell ATP and adenine ribonucleotides (9,28,29). Further evidence resides in a single kindred well studied by Wada et al (30) in which the proband had very marked PRPP synthetase deficiency and his parents had substantially subnormal activities compatible with heterozygosity for the same abnormality. Convincingly, the proband had all the expected features of deficient PRPP - megaloblastic anemia, hypouricemia etc. The proband and all studied members of the kindred had entirely normal red cell concentration of ATP.

Bartlett (31) found no incorporation of ^{14}C-labeled Ad into adenylates of the human red blood cell provided freshly collected blood was employed in the incubation. This was despite the fact that Ad concentrations employed were some 1000x greater than normal plasma levels, and that incubations were for a period of one hour, PRPP was present in such blood in amounts too small to be detected by the methods employed. In contrast, additions of P_i to the incubation permitted substantial incorporation of the labelled Ad. Hershko et al (32) had earlier documented the large increase in PRPP when fresh blood was incubated with phosphate buffers. Inorganic phosphate concentrations clearly play a crucial role in regulating PRPP synthesis, and studies relative to red cell incorporation of Ad under conditions where P_i concentrations are increased unphysiologically create a serious departure from normal in vivo metabolism.

In contrast, diminished or augmented salvage of Ado appear respectively to be associated with diminished or augmented red cell adenylates. Ado is a substrate for two red cell enzymes, a kinase returning it to the pool, and a deaminase eliminating it from potential salvage. In four well documented kindreds (33-36) a dominantly inherited syndrome has been associated with 50-100 fold increases in Ado deaminase activity, hemolytic anemia, and marked reduction in adenine nucleotides. The concentration of the latter in the index kindred was less than one-half that expected in comparably reticulocyte-rich blood (33). The massive increase in deaminating activity creates a metabolic imbalance, diminishing Ado available for phosphorylation. It is believed that this diminution in salvage results in depletion of the adenine nucleotide pool, and ultimately leads to metabolic deprivation as the probable cause of hemolysis (33).

In a converse manner, severe inherited Ado deaminase deficiency creates the opposite imbalance. Not only are plasma levels of dAdo and Ado increased, but also AK acts essentially unopposed and at its maximum capacity and efficiency. As a result of salvage, red cells accumulate large amounts of ribonucleotides which, interestingly, are largely deoxy in type. Adenine nucleotides may actually decrease, and in inherited and 2'-deoxycoformycin-induced Ado deaminase deficiency there appears to be a reciprocal relation between levels of ATP and dATP in red cells (22). The explanation of this apparent paradox is believed to lie in certain properties of deoxynucleotides (21,37). Once phosphorylated to dAMP, the latter, unlike AMP, is an extremely poor substrate for AMP deaminase, and shares minimally in AMP's susceptibility to deamination. Further, dATP is a much less active participant in the sum total of reactions requiring high energy phosphate than its preferred ATP counterpart. The net effect is retention of the deoxynucleotide with at least partially selective loss of the non-deoxy compounds (21,37). This in turn provides further evidence of the importance of AMP deamination to losses from the normal red cell adenine nucleotide pool.

The above discussion has centered on metabolic features of maintenance of the red cell adenylate pool under normal in vivo conditions. The relative contributions of salvage pathways may vary in other circumstances. For example, when Bartlett employed blood stored in ACD solution under blood bank conditions (31), PRPP concentration was too low to be measured on day zero, but seven days later was substantial. The allosteric activator P_i of PRPP synthetase also increases as part of the storage lesion (38). While the importance of adenine salvage under physiologic conditions remains controversial (26), it is our view that the weight of evidence currently indicates a minor role, and this may be due in part to relative inavailability of adequate PRPP. Under conditions of blood storage, the adenine salvage pathway is clearly operative if the purine base is made available (38).

Higher than normal ATP concentrations may characterize the red cells in uremia and these are accompanied by increased concentrations of P_i (39,40). In one study (39) ATP concentrations averaged some 70 per cent greater than normal and there was strong correlation with levels of P_i. Increased P_i in uremia thus could by a special circumstance favoring adenine salvage and simultaneously inhibiting adenylate losses via deamination. The matter requires further investigation before firm conclusions can be drawn.

TRANSIENT ERYTHROBLASTOPENIA OF CHILDHOOD AS A MODEL FOR STUDIES IN RED CELL
AGING

While in the past it has been widely assumed that metabolic deprivation
secondary to deteriorating enzymatic machinery of glycolysis played a
significant role in limiting normal erythrocyte life span, this is now open
to serious question. The characteristics of the oldest red cells have
commonly been derived from metabolic evaluations of the most dense fractions
of erythrocytes separated by density gradients or by a variety of other
techniques (41). The assumption has been that these were reasonably uniform
representatives of the oldest circulating red cells. Data are emerging
strongly indicating that this is not the case (42-44), and, indeed, that
diminished ATP concentrations may not be present in old erythrocytes. In
point of fact, under several circumstances increased ATP has been observed
in this cohort of cells.

Clark (45) has pointed out that somewhat lower than normal ATP levels
in the densest erythrocytes does not necessarily indicate global
deterioration in ATP levels. The most dense cells have high MCHC and low
membrane surface area/volume ratios and hence diminished cell water.
Although functions requiring ATP are dependent on ATP concentration, the
latter have usually not been appropriately expressed in terms of cell
water (45). The work of Dale and his colleagues (23,24,46) has demonstrated
that ATP levels in the oldest rabbit erythrocytes are actually increased.
Increased red cell ATP has also been demonstrated in subjects with
genetically-determined complete deficiency of AMP deaminase (25).

Transient erythroblastopenia of childhood present the opportunity of
studying red cells aging in vivo without the necessity of in vitro
separatory manipulations of cell cohorts of differing ages. This entity
presents as a self-limited cessation of erythropoiesis lasting a few
weeks to a few months, during which time the pool of circulating
erythrocytes progressively ages. Reticulocytes are absent, anemia worsens
progressively, and as the period of selective aplasia is prolonged, mean
cell age approaches that of normal red cell death. Erythrocytes from 11 such
patients were evaluated, in some of which cessation of erythropoiesis
had endured to the point that Hb concentrations of blood were as low as
1.5-2.0 g/dl (47). Activities of AMP deminase were consistently decreased
to 5-70% of normal control means. ATP and total adenine nucleotides
increased in concentration up to three fold. In the index case, which
unfortunately was not available for repeated studies, AMP deaminase activity
was virtually absent, there was partial deficiency of the enzyme in both
parents, and total adenine nucleotide concentrations were three times
normal. Inability to carry out repeated studies precluded determination
with certainty as to whether in this instance transient erythroblastopenia
had fortuitously intervened in a subject with an inherited deficiency of
AMP deaminase. In our experience, AK and Ado deaminase do not change in
activity appreciably as the red cell ages. The association between
diminished AMP deaminase activity and increased red cell ATP is probably
not coincidental. It has now been observed in fractionated aged rabbit
erythrocytes (23,24,46), in inherited human AMP deaminase deficiency (25),
and in the aged cohort of red cells of children with transient

erythroblastopenia (47). Further, it is a reasonable relationship, given the fact that AMP deaminase deficiency also diminishes the drain on the adenine nucleotide pool imposed by irreversible deamination of AMP. In a personal communication, Dale and Beutler (48) have reported similar observations in cases of transient erythroblastopenia. On a speculative level, these observations suggest that certain other "high ATP" syndromes should be reinvestigated in terms of possible imbalance in degradative and salvage pathways of adenine nucleotide metabolism.

While many factors contributing to normal red cell senescence have been documented or hypothesized (2) the process of aging is still incompletely understood. The rate of catalytic decay of a variety of erythrocyte enzyme activities has been reported (41,49-52), but without a universal consensus. It is now clear that a biphasic loss of activity is common, a rapid decay accompanying the brief period of reticulocyte maturation, and a variably slow phase characterizing the remainder of the cellular life span (43). Initial activity losses have been most profound in the case of hexokinase (53-55). However, events of the postreticulocyte stage are more poorly defined. In the case of some enzymatic proteins catalytic activity has been little changed with age. In others, such as P-5'-Nase (43) and AMP deaminase (23,24,47), major activity losses are the rule. Demonstration of normal to substantially elevated concentrations of ATP in old erythrocytes supports the adequancy of glycolytic machinery responsible for maintaining the energy change of the cell. However, this does not imply that other ATP-dependent and non-dependent processes are universally intact since these rely on apoenzymes as well as co-factors. Whether or not enzymatic decay plays any role in the myriad of irreversible membrane abnormalities heralding ultimate cell death requires further investigation.

THE ERYTHROENZYMOPATHIES

The above does not deny metabolic deprivation a major contribution in the markedly shortened life span of red cells with severe, inherited impairment of anaerobic glycolysis. While the issue remains controversial (56), we believe inability to sustain an effective energy charge and concentrations of ATP is the primary event resulting in the premature red cell destruction characterizing these hemolytic erythroenzymopathies (57).

There is ample evidence that deficient glycolysis inevitably results in depletion of red cell adenylates and hemolysis. Such metabolic deprivation may be the result of lack of glucose substrate as in the autohemolysis test, or of inability to metabolize available glucose. A common feature of the glycolytic erythroenzymopathies is the accumulation of metabolites behind the blocking molecular lesion. An exception is hexokinase since it catalyses the initial step in glycolysis. Additionally, there is inappropriately low production of lactate and frequently low concentrations of ATP. The latter were noted in the first descriptions of the "nonspherocytic" hemolytic anemias by Dacie and his collegues (58,59) and later confirmed by workers in Australia (60,61). Further the true extent of ATP depletion in the subset of the measured red cell population on the verge of destruction is deceptively obscured by the fact that all measurements are means. The latter reflect in part the substantial numbers of reticulocytes capable of deriving energy from oxidative phosphorylation and non-glucose substrates and an additional population of very young erythrocytes still marginally capable of compensating for severe glycolytic deficiencies. Unfortunately, it is virtually impossible to asses ATP in the crucial, small subset of cells whose failing metabolism at the moment of measurement has brought them to the brink of death. Further, it is likely that when a threshold level of

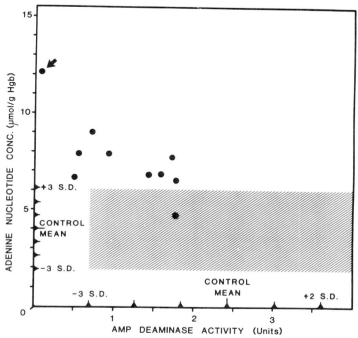

Fig. 2. Total adenine nucleotide concentrations as a function of AMP deaminase activity in erythrocytes from patients with transient erythroblastopenia of childhood. The index patient, who may have hereditary deficiency of AMP deaminase, is indicated by the arrow. The shaded area encompasses ± 3 SD around each mean as derived from 44 (ordinate) and 142 (abscissa) normal control determinations. Reproduced with permission of Blood 74:2161–2165, 1989. Grune & Stratton, Inc.

deprivation is reached, terminal changes, fueled by diminished energy change, increased AMP, and enhanced from the adenylate pool, rapidly cascade. Such changes would be compounded by an unstable variant enzyme with progressevely diminishing catalytic capacities. The common denominator of impaired glycolysis, regardless of its origin or the location of the molecular lesion, is that at some point of severity energy metabolism must become deficient.

This view does not imply that all hemolytic syndromes associated with erythroenzymopathies have a common primary etiology characterized by severe diminution in the red cell energy change. Certainly inability to protect hemoglobin and other red cell constituents from oxidative denaturation characterizes various enzymopathies of the hexosemonophosphate shunt and glutathione metabolism and is a major contributor to premature erythrocyte destruction. Certainly enzymatic as well as structural defects pertinent to the integrity and function of the plasma membrane play a different role than do the anaerobic glycolytic enzymopathies, and the mechanism of hemolysis in P-5'-Nase deficiency with its large accumulation of pyrimidine nucleotides remains a matter of speculation, though a number of possible contributory factors have been postulated. Nor does it imply that events secondary to metabolic deprivation may not play crucial roles in the demise of the erythrocyte. Such events include among many others loss of membrane deformability, a more spheroidal shape, an array of potential alterations of the membrane cytoskeleton, the lipid bilayer and its integral proteins, inability to pump cations against electrochemical gradients, and deleterious effects of accumulation of metabolites or, conversely, of their impaired synthesis. It addresses only the primary and initiating event, and not the littany of consequences.

REFERENCES

1. D. E. Atkinson, The energy charge of the adenylate pool as a regulatory parameter: Interaction with feedback modifiers, Biochemistry 7:4030 (1968).
2. J. V. Dacie, The Haemolytic Anemias, 3rd edition, Vol. 1. The hereditary haemolytic anaemias, Churchill Livingstone, New York (1985).
3. R. I. Weed, Hereditary spherocytosis – A review, Arch Int Med 135:1316 (1969).
4. W. N. Valentine, Hereditary spherocytosis revisited, West J Med. 128:35 (1978).
5. S. E Lux and P. S. Burke, Disorders of the red cell membrane skeloton: hereditary spherocytosis and hereditary elliptocytosis, in: "The Metabolic Basis of Inherited Disease", CR Scriver, AL Beaudet, WS Sly, D Valle, eds., 6th edition, Chap 95, McGraw-Hill, New York (1989).
6. I. H. Fox and W. N. Kelley, The role of adenosine and deoxyadenosine in mammalian, cells, Annu Rev Biochem 47:655 (1978).
7. T. D. Palella and I. H. Fox, Hyperuricemia and gout, in: The Metabolic Basis of Inherited Disease., eds C R Scriver, AL Beaudet, WS Sly, D Valle 6th edition, Vol. 1, Chap 37, Mc Graw-Hill, New York (1989).
8. H. G. Williams-Ashman, J. Seudenfekd and P. Galetti, Trends in the biochemical pharmacology of 5'-deoxy-5'-methylthioadenosine, Biochem Pharmacol 31: 277 (1982).
9. A. Simmonds, A. S. Sahota and K. J. VanAcker, Phosphoribosyltransfe-rase deficiency and 2,8-dihydroxyadenine lithiasis, in: The Metabolic Basis of Inherited Disease, eds CR Scriver, AL Beaudet, MS Sly, D Valle, 6th edition, Vol. 1, Chap 39, Mc Graw Hill, New York (1989).
10. A. Sahota, D. R. Webster, C. F. Potter, H. A. Simmons, A.V Rogers and T. Gobson, Methylthioadenosine phosphorylase activity in human erythrocytes, Clin Chim Acta 128:283 (1983).

11. Physiological and Regulatory Functions of Adenosine and Adenine Nucleotides. eds HP Baer, GI Drummond, Raven Press, New York (1977).
12. H. S. Mudd and J. R. Poole, Labile methyl balances for normal humans on various dietary regimes, Metabolism 24:721 (1975).
13. R. E. Klabunde Dipyridamole inhibition of adenosine metabolism in human blood, Eur J Pharmacol 93: 21 (1983).
14. F. Bontemps, G. Van den Berge and H. G. Hers, Pathways of adenine nucleotide metabolism in erythrocytes, J Clin Invest 77:824 (1986).
15. J. F. Henderson, Regulation of adenosine, in: Physiological and Regulatory Function of Adenosine nucleotides, eds HP Baer, GI Drumond, Raven Press, New York (1979).
16. D. E. Paglia, W. N. Brockway and M. Nakatani, Substrate specificity and pH sensitivity of deoxyribonucleotidase and pyrimidine nucleotidase in human hemolysates, Exp Hematol 15:1041 (1987).
17. D. E. Paglia W. N. Valentine, M. Nakatani and R.A. Brockway. Mechanisms of adenosine 5'-monophosphate catabolism in human erythrocytes, Blood 67:988 (1986).
18. D. E. Paglia, W. N. Valentine, A. S. Keitt, R. A. Brockway and M. Nakatani, Pyrimidine nucleotidase deficiency with active dephosphorylation of dTMP: evidence for existence of thymidine nucleotidase in human erythrocytes, Blood 62:1147 (1983).
19. D. E. Paglia W. N. Valentine, and R. A. Brockway, Identification of thymidine nucleotidase and deoxyribonucleotidase activities among normal isozymes of 5'-nucleotidase in human erythrocytes, Proc Natl Acad Sci USA 81:588 (1984).
20. D. M. Swallow, I. Aziz, D. A. Hopkinson and S. Miwa, Analysis of human erythrocyte 5'nucleotides in healthy individuals and a patient deficient in pyrimidine 5'-nucleotidase, Ann Hum Genet 47:19 (1983).
21. A S. Bagnara, and M. S. Hershfield, Mechanism of deoxyadenosine-induced catabolism of adenine ribonucleotides in adenosine deaminase-inhibited human T lymphoblastoid cells, Proc Natl Acad Sci USA 79: 2673 (1982).
22. H. M. Kredich, and M. S. Hershfield, Immunodeficiency diseases caused by adenosine deaminase deficiency and purine nucleoside phosphorylase deficiency, in: The Metabolic Basis of Inherited Disease, eds CR Scriver, AL Beaudet, WS Sly, D. Valle, 6th Edition Vol. 1, Chap 40, McGraw Hill, New York (1989).
23. T. Suzuki, and G. L. Dale, Senescent erythrocytes; isolation of in vivo aged cells and their biochemical characteristics, Proc Natl Acad Sci USA 85:1647 (1988).
24. G. L. Dale and S. L. Norenberg, Time-dependent loss of adenosine 5'-monophosphate deaminase activity may explain elevated adenosine 5'-triphosphate levels in senescent erythrocytes, Blood 74:2157 (1989).
25. N. Ogasawara, H. Goto, Y. Yamada, I. Nishigaki, T. Itoh and I. Hasegava, Complete deficiency of AMP deaminase in human erythrocytes, Biochem Biophys Res Comm 122:1344 (1984).
26. K. R. Tanaka and C. R. Zerez, Red cell enzymopathies of the glycolytc pathway, Semin Hem 27:165 (1990).
27. C. R. Zeren, N. A. Lachant and K.R. Tanaka, Decreased erythrocyte phosphoribosylpyrophosphate synthetase activity and impaired formation in thalassemia minor: A mechanism for decreased adenine nucleotide content, J Lab Clin Med 114:43 (1989).

28. B. M. Dean, D. Perrett, H. A. Simmonds, and K. J. Van Acker, Adenine and adenosine metabolism in intact erythrocytes deficient in adenosine monophosphate-pyrophosphate phosphoribosyltransferase; a study of two families, Clin Sci Mol Med 55:407 (1978).

29. I. H Fox, S. La Croix, G. Planet, and M. Moore, Partial deficiency of phosphoribosyltransferase in man, Medicine (Baltimore) 56: 515 (1977).

30. Y. Wada, Y. Nishimura, M. Tanabu, Y. Yoskimura, K. Linuma. T. Linuma, T. Yoshida and T. Arakawa, Hypouricemic mentally retarded infant with a defect of 5-phosphoribosyl-1-pyrophosphate synthesis of erythrocytes, Tokoku J. Exp Med 113:149 (1974).

31. G. R. Bartlett, In vitro age-related alterations in adenine nucleotide metabolism in human red cells, in: Progress in Clinical and Biological Research Vol. 195. Cellular and Molecular Aspects of Aging. The Red Cell as a Model, eds JW Eaton, DK Konzen, J. G. White, Alan R. Less, Inc., New York (1985).

32. A. Hershko, A. Razin, T. Shoshani and J. Mager, Turnover of purine nucleotides in rabbit erythrocytes. II Studies in vitro, Biochim Biophys Acta 149:59 (1967).

33. W. N. Valentine, D. E. Paglia, D. E. Tartaglia and F. Gilsanz Hereditary hemolytic anemia with increased red cell adenosine deaminase (45-70 fold) and decreased adenosine triphosphate, Sci 195:783 (1977).

34. S. Miwa, H. Fujii, N. Matsumoto, T. Nakutsuji, S. Oda, H. Asano, S. Asano, and Y. Miura, A case of red-cell adenosine deaminase overproduction associated with hereditary hemolytic anemia found in Japan, Am J Hematol 5:107 (1978).

35. J. L. Perigon, M. Hamet, H. A. Buc, P. Cartier and M. Derycke, Biochemical study of a case of hemolytic anemia with increased (85-fold) red cell adenosine deaminase, Clin Chim Acta 124:205 (1982).

36. H. Kanno, K. Tani, H. Fujii, S. M. M. Iguchi-ariga, H. Ariga, T. Kozaki and S. Miwa. Adenosine deaminase (ADA) over production associated with congenital hemolytic anemia: Case report and molecular analysis Japan J Exp Med 58:1 (1988).

37. W. N. Valentine, D. E. Paglia, S. Clarke, B. H. Morimoto, M. Nakatani, and R. A. Brockway, Adenine ribo- and deoxyribonucleotide metabolism in human erythrocytes, B- and T- lymphocyte lines, and monocyte-macrophages, Proc Natl Acad Sci USA 82:6682 (1985).

38. S. P. Masouredis, Preservation and clinical use of erythrocytes and whole blood, in: Hematology, 4th edition, eds W. J Williams, E. Beutler, A. J. Erslev, M. A. Lichtman, Chap 169, McGraw-Hill, New York (1990).

39. M. A. Lichtman and D. R. Miller, Erythrocyte glycolysis, 2-3 diphosphoglycerate and adenosine triphosphate concentration in uremic subjects: relationship to extracellular phosphate concentration, J. Lab Clin Med 76: 267 (1970).

40. P. J. Crowter, and M. N. Cauchi, Red cell nucleotide levels in patients with thalassemia, iron deficiency and renal failure, Brit J Haematol. 53:113 (1983).

41. M. R. Clark and S. B. Shohet, Red cell senescense. Clin Haematol 14:223 (1985).

42. M. Morrison, C. W. Jackson, T. J. Mueller, T. Huang, M. E. Doktor, W. S. Walker, J.A. Singer and H. H. Edwards, Does red cell density correlate with red cell age?, Biomed Biochim Acta 42:107 (1983).
43. E. Beutler, How do red cell enzymes age?,: A new perspective. Br J Haematol 61:377 (1985).
44. E. Beutler, Isolation of the aged, Blood Cells 14:1 (1988).
45. M. R.Clark, Selected ionic and metabolic characteristics of human red cell populations separated on stractan density gradients., in: Progress in Clinical and Biological Research Cellular and Molecular Aspects of Aging. The Red Cell as a Model Vol. 195 eds JW Eaton, DK Konzen, JG White, Alan R Liss, Inc, New York (1985).
46. T. Suzuki, and G. L. Dale, Biotinylated erythrocytes; in vivo surivival and in vitro recovery, Blood 70:791 (1987).
47. D. E. Paglia, W. N. Valentine, M. Nakatani, and R. A. Brockway, AMP deaminase as a cell-age marker in transient erythroblastopenia of childhood and its role in the adenylate economy of erythrocytes, Blood 74:2161 (1989).
48. G. L. Dale and E. Beutler, Personal communication (1989).
49. F. Brok, B. Ramot, E. Zwang and D. Danon, Enzyme activities in human red blood cells of different age groups, Isrl J Med Sci 2:291 (1966).
50. M. A. Lichtman, Does ATP decrease exponentially during red cell aging?, Nouv Rev Fr Hematol 15:625 (1975).
51. N. S. Cohen, J. E. Ekhelm, M. G. Luthr and D. L. Hanahan, Biochemical characteristcs of density separated human erythrocytes, Biochim Biophys Acta 419:229 (1976).
52. G. Bartosz, E. Grzelinska and J. Wagner, Aging of the erythrocyte. XIV. ATP content does decrease, Experientia 38-575 (1976).
53. W. N. Valentine, F. A. Oski, D. E. Paglia, M. A. Baughan, A. S. Schneider and J. L. Naiman, Hereditary hemolytic anemia with hexokinase deficiency. Role of hexokinase in aging, N Engl J Med 276:1 (1967).
54. G. Jansen, L. Kuenderman, G. Rijksen, B. P. Cato and E. J. Staal, Characteristics of hexokinase, pyruvate kinase, and glucose-6-phosphate dehydrogenase during adult and neonatal reticulocyte maturation, Am J Hematol 20:203 (1985).
55. M. Magnani, V. Stocchi, L. Chiarantini, G. Serafini, M. Dachà and G. Fornaini, Rabbit red blood cell hexokinase. Decay mechanism during reticulocyte maturation, J Biol Chem 261:8327 (1986).
56. E. Beutler, "The primary cause of hemolysis in enzymopathies of anearobic glycolysis.", A commentary. Blood Cells 6:827 (1980).
57. W. N. Valentine and D. E. Paglia, The primary cause of hemolysis in enzymopathies of anaerobic glycolysis: A viewpoint, Blood Cells 6:819 (1980).
58. J. V. Dacie, P. L. Mollison, N. Richardson, J. G. Selwyn, and L. Shapiro, Atypical congenital haemolytic anemia, O J Med 22:79 (1953).
59. J. G. Selwyn, and J. V. Dacie, Autohemolysis and other changes resulting from the incubation in vitro of red cells from patients with congenital hemolytic anemia, Blood 9:414 (1954).
60. G. C. DeGruchy, J. N. Santamaria, I. C. Parsons, and H. Crawford, Nonspherocytic congenital hemolytic anemia, Blood 16:1371 (1960).
61. M. A. Robinson, P. B. Loder, and G. C. DeGruchy, Red-cell metabolism in nonspherocytic congenital haemolytic anemia, Br J Haematol 7:327 (1961).

SIGNIFICANCE AND RELEVANCE OF NAD SYNTHESIS IN HUMAN

ERYTHROCYTE LIFE SPAN

Vanna Micheli, Silvia Sestini, Marina Rocchigiani,
Monica Pescaglini and Carlo Ricci

Istituto di Chimica Biologica
Università degli Studi di Siena, Italy

INTRODUCTION

NAD and NADP are known to play a key role in the general metabolism of human erythrocytes, owing to their involvement in all oxidative-reductive reactions occuring both in the energetic catabolism and in the processes preventing the cells from oxidative damage.

The general pathways for the synthesis and breakdown of pyridine coenzymes have long been outlined in humans, using the red blood cells as a tool (1) for many of the investigations. Nevertheless many questions have remained unsolved concerning the specific pathway followed by these particular cells and its regulation. Questions are also arising about the relevance and actual extent of NAD synthesis in the general economy of the erythrocyte, and on the importance of maintaining steady endocellular levels of NAD and NADP, also in connection with possible pathological involvement in specific disorders (2,3).

Fig. 1 illustrates the pathways of NAD metabolism in human erythrocytes. Preformed pyridine precursors, namely nicotinamide (NAm) or nicotinic acid (NA), are required, since the de novo synthesis from tryptophan is not active in mature erythrocyte. NA is usually believed to be a better precursor than NAm for NAD synthesis in human erythrocytes, on the basis of studies conducted in vivo or in vitro at high and unphysiological concentrations of substrates and very long incubation times (3,4). Using very sensitive HPLC methods coupled with radioactivity, we have recently reported (6) that NAm is probably the best precursor when short times of incubation and very low concentrations of the substrates, closer to the physiological levels, are used. Thus both the amidated pathway (from Nam) and the deamidated one (from NA) are possibly active in the human erythrocyte.

Only few enzymes of pyridine metabolism have been characterised and the occurence of some reactions of both synthesis and breakdown in human red blood cells is still uncertain. Methods to assay the activities of some enzymes involved in the synthesis (NAPRT, NAmPRT, NMN-AT, NAMN-AT) have been developed by us in recent years (7-9), and some apparent kinetic constants have been determined.

NA and NAm seem to be metabolised through distinct routes, by means of distinct phosphoribosyltranferases and possibly of distinct adenylyl-transferases acting on NMN or NAMN. Possible connections between the two

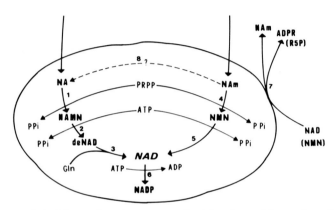

Fig. 1. NAD metabolism in human red blood cells. 1: NA phosphoribosyltransferase; 2: NAMN adenylyltransferase; 3: NAD synthetase; 4: NAm phosphoribosyltransferase; 5: NMN adenylyltransferase; 6: NAD kinase; 7: NAD glycohydrolase; 8: NAm deamidase. Abbreviations: NA: nicotinic acid; NAm: nicotinamide; NAMN: NA mononucleotide; NMN: NAm mononucleotide; DeNAD: deamido-NAD (NA-adenine-dinucleotide); ADPR: ADP-ribose; PPribP: 5-phosphoribosyl-1-pyrophosphate.

routes have not been identified yet, as no deamidating activity has been determined in human erythrocytes, though the formation of some deamidated compounds after incubation of cells in the presence of NAm has been reported (10).

The present study has been carried out on human erythrocytes of different mean age, separated according to density, in order to investigate the behaviour of the process of NAD synthesis through the two pathways during the erythrocyte life-span.

METHODS

Erythrocytes of different density were separated by centrifugation on discontinuous density gradients of Percoll-bovine serum albumine as described elsewhere (11). Three fractions of increasing density indicated as 1, 2 and 3 were obtained. Reticulocytes were almost completely confined in fraction 1 (30-40% of the fraction cell content) and only occasionally a minimal amount was detectable by routine means in the other fractions.

Intact washed erythrocytes were used for the determination of the endogenous nucleotide content and to study the uptake of extracellular precursors and the rate of synthesis of pyridine nucleotides.

Protein-free extracts were obtained by adding two volumes of 0.6 N perchloric acid (PCA) to one volume of erythrocytes. Precipitated proteins were discarded by centrifugation and the supernatants, neutralized with K_2CO_3, were processed by HPLC to quantify the endocellular nucleotides, monitoring both 260 and 280 nm absorption.

Pyridine nucleotide synthesis was studied incubating the erythrocytes with three volumes of 30 mM potassium phosphate, pH 7.4, containing 107.8 mM NaCl, 19 mM glucose, 0.5 μM ^{14}C NA or NAm (53 mCi/mMole), at 37°C for 1h. The same incubation medium was used to assess the uptake of NA or NAm by intact cells, as previously described (12). Protein-free PCA extracts of incubation mixtures, neutralized with K_2CO_3, were processed by HPLC monitoring both 260 nm absorbance and radioactivity counts.

The HPLC system consisted of a System Gold Programmable Solvent module 126 supplied with Scanning Detector module 167 and with in line Radioisotope Detector 171 (Beckman). A Beckman Ultrasphere XL ODS Column (3 μm; 4.6x7.0 cm) was used. The elution was carried out as previously described (13) with methanol (A) and 0.1 M potassium dihydrogen phosphate buffer containing 6mM TBA, pH 6 (B) as eluents.

Enzyme activities were assayed on crude lysates, using HPLC methods for NAPRT, APRT, NAmPRT (7-9), NMN-AT and NAMN-AT (14). A spectrophotometric method was used to assay G6PD activity (15).
Hemoglobin concentration was assayed by Drabkin's method (16).

RESULTS

The activities of G-6-PD and APRT were assayed on the lysates from erythrocytes separated into three fractions of increasing density, as a tool to ascertain the age-dependent loss of activity expected (Table 1). The mean percentage distribution of the erythrocytes throughout the density gradient over 20 healthy donors is also shown.

Endogenous purine and pyridine nucleotide patterns have been determined (Table 2). A moderate decay in NAD and NADP level (23 and 27%, respectively) is shown by the heavier older cells in comparison with the lighter younger ones. This contrasts with the more marked decrease shown by adenine nucleotides (30-50%) and by guanine nucleotides (60-80%).

Table 1. Percent Distribution of Erythrocytes on Percoll
Discontinuous Density Gradient and Activity of G-6-PD
and APRT (nmoles/h/mg Hb) in Crude Lysate (mean ± SD).

Fractions	% total Hb	G-6-PD	APRT
1	1.8± 1	358.6±30.8	24.3±10.6
2	26.1±15	262.7±72.7	20.5± 3.2
3	67.3±30	202.5±28.4	11.8± 2.1
Total		260.7±63	18.2± 2.5

Table 2. Levels of Some Purine and Pyridine Nucleotides (nmoles/ml cells) in
Total Human Erythrocytes and in Fractions of Increasing Density.

Fractions	AMP	ADP	ATP	GDP	GTP	NAD	NADP
1	36.1	266.2	2341.6	42.1	330.6	74.1	52.4
2	26.2	210.1	1494.1	20.7	92.4	60.1	41.5
3	20.0	190.4	1399.9	16.4	53.7	57.3	38.3
Total	11.6	160.1	1469.1	16.5	61.6	60.7	38.3

In spite of the decay of the adenine nucleotides on the whole, the
energy charge (calculated as: $[ATP] + 0.5 [ADP] / [ATP] + [ADP] + [AMP]$) is not
significantly modified in the three populations of different mean age.

The first steps of NAD synthesis have been studied on the density
separated erythrocytes in order to find out whether the observed decreased
levels of pyridine coenzymes in aged cells might be due to impairments in
the synthesis rate. Increased breakdown seemed to be unlikely, as no
breakdown products, such as ADPR or the mononucleotides NMN and NAMN, could
be found in any of the observed samples.

The study was undertaken following the incorporation of ^{14}C labeled NA
and Nam into the nucleotides by intact cells. Moreover the activities of
the enzymes catalysing the first reactions of the pathway were determined on
crude lysates.

Table 3 shows the results obtained when the amidated pathway, starting
from nicotinamide, is followed. In the condition used for the incubation
of the intact cells, NMN and NAMN were the main products whereas minimal
amounts of NAD were found. NMN production rate was impaired in aged cells;
interestingly, some deamidated mononucleotide was found, with increasing
production in old cells, leading to the conclusion that still unknown
deamidating activities might be unmasked during the life span, acting on

Table 3. Nucleotide Production by Intact Cells Incubated for 1 Hour at 37°C With 0.5 μM ^{14}C NAm (53 mCi/mMol) and Activities of NAmPRT and NMN-AT in Crude Lysate. Erythrocytes of Increasing Age are Indicated with Numbers 1, 2 and 3. Results are Expressed as nmoles/h/g Hemoglobin.

	Intact cells			Lysates	
	NMN	NAMN	NAD	NAmPRT	NMN-AT
Fractions					
1	15.2	0.3	0.06	81	532
2	10.8	1.8	trace	54	504
3	4.7	4.2	0.04	50	300
total	1.4	2.2	0.1	77	508

NAm or on its mononucleotide. The activities of the two enzymes, NAmPRT and NMNAT, were also impaired in old cells compared with the young ones.

The production of nucleotides from (^{14}C)NA by intact cells from the fractions and the total erythrocyte population was measured, and the activities of NAPRT and NAMN-AT were determined on crude lysates, as shown in Table 4. In the condition used for the incubation of intact cells with NA, NAMN was the only product and its rate of production was decreased in aged cells. The activities of the two enzymes, NAPRT and NAMN-AT showed similar decays.

Table 4. Nucleotide Production by Intact Cells Incubated for 1 Hour at 37°C With 0.5 μM (^{14}C)NA (53 mCi/mMol) and Activities of NAPRT and NAMN-AT in Crude Lysates. Erythrocytes of Increasing Age Are Indicated With Numbers 1,2 and 3. Results Expressed As nmoles/h/g Hemoglobin.

	Intact cells	Lysates	
	NAMN	NAPRT	NAMN-AT
Fractions			
1	5.6	2,262	478
2	4.6	1,680	446
3	2.7	854	400
total	5.0	1,407	468

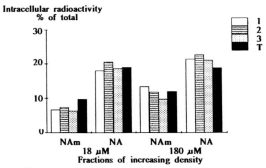

Fig. 2. Uptake of ^{14}C NA or NAm by intact red blood cells separated by
density (1, 2 and 3: fractions of increasing age). Data are
expressed as percent of the total radioactivity found in the
cells incubated for 1 min as described under methods.

The first step of nucleotide production by intact cells includes two
processes: uptake of the precursors from the extracellular medium and
following metabolism. We have previously reported that the rate of uptake
of NAm by intact human erythrocytes is significantly lower than that
of NA (12). Uptake of NA and NAm by the density separated cells was then
investigated. Fig. 2 shows that the rate of NA uptake was not altered with
age at low (20 µM) nor at high (200 µM) extracellular concentration. By
contrast, the uptake of NAm was slightly impaired with age at high but not
at low concentration. This finding enlights a new aspect of the different
behavior of the two precursors of NAD, depending not only on the activity of
the committed enzyme, but also on their rate of entrance into the cell.

DISCUSSION

Present results show that both the amidated and the deamidated pathway
of NAD synthesis are impaired during erythrocyte maturation and aging. This
may account for the lowered level of pyridine coenzymes found in aged red
cells separated by density. The decay observed in the synthesis rate by
intact cells and in the enzyme activities shows different impairments
between the route via NA and the one via NAm, which may indicate different
regulations.

On NAm pathway, NAmPRT and NMN-AT activities show a similar decay in
the oldest cells. On NA pathway, the impairment of NAPRT activity is much
more marked than the one of NAMN-AT.

In intact cells incubated with NAm the rate of production of NMN in the
oldest cells is more markedly decreased than the production of NAMN from NA.
Moreover, during NAm incubation, aged erythrocytes show an increasing
production of NAMN, which seems to represent an alternative pathway implying
the presence of some deamidase activity. Different experimental conditions,
however, might strongly influence this process in intact cells.

The lower uptake of NAm compared with NA and its decrease in senescent cells might be a further barrier for the utilization of the amidated pyridine precursor at high unphysiological concentrations. This finding is consistent with our previous reports (6,17).

The decay in the effectiveness of the routes of synthesis of NAD is comparable, when not less marked, with that observed for other metabolic pathways in the senescent erythrocytes. Both amidated and deamidated pathways seem to be active in mature erythrocytes, but it is conceivable that the two different routes play different roles during aging.

Nevertheless the actual significance and the occurrence of a real turnover of NAD during erythrocyte life span is still an open question. The hypothesis should be taken into account that cellular NAD turnover might be a minor metabolic fate for the extracellular pyridine precursors and that the mature circulating erythrocytes might also have a role in their transport among different tissues.

ACKNOWLEDEGEMENTS

This work was supported in part by the Italian Ministero della Università e della Ricerca Scientifica e Tecnologica (60% and 40% funds).

REFERENCES

1. J. Preiss, P. Handler, Biosynthesis of DPN. I. Identification of intermediates, J. Biol. Chem. 233:759 (1957).
2. H. A. Simmonds, L. D. Fairbanks, G. S. Morris, D. R. Webster, E. H. Harley, Altered erythrocyte patterns are characteristic of inherited disorders of purine and pyrimidine metabolism, Clin. Chim. Acta 171:197 (1988).
3. C. R. Zerez, K. R. Tanaka, Impaired NAD synthesis in Pyruvate-Kinase-deficient human erythrocytes: a mechanism for decreased total NAD content and a possible secondary cause of hemolysis, Blood 69:999 (1987).
4. J. Preiss, P. Handler, Synthesis of DPN from NA by human erythrocytes in vitro, J. Amer. Chem. Soc. 79:1514 (1957).
5. V. Micheli, S. Sestini, C. Ricci, Purine and pyridine nucleotide production in human erythrocytes, Arch. Biochem. Biophys. 244: 54 (1986).
6. V. Micheli, H. A. Simmonds, S. Sestini, C. Ricci, Importance of nicotinamide as an NAD precursor in the human erythrocyte, Arch. Biochem. Biophys. 283:40 (1990).
7. V. Micheli S. Sestini, M. Rocchigiani, M. Pescaglini, C. Ricci, Nucleotide synthesis in the human erythrocyte: correlation between purines and pyridines, Biomed. Biochim. Acta 46:268 (1987).
8. S. Sestini, P. Lusini, V. Micheli, M. L. Ceccuzzi, C. Ricci, Evidence for NMN adenylyltransferase in human erythrocytes. Preliminary observations, Ital. J. Biochem. 37:63A (1988).

9. M. Rocchigiani, H. A. Simmonds, M. Bari, C. Ricci, NAmPRT assay in human red blood cells, Atti 35° Congresso Nazionale S.I.B., 1990.

10. M. Pescaglini, M. Rocchigiani, S. Sestini, V. Micheli, C. Ricci, Sintesi di NAD da Nicotinamide nei globuli rossi umani, Atti Accad. Fisiocritici Siena XV-Tomo VII:255 (1989).

11. V. Micheli, C. Ricci, A. Taddeo, R. Gili, Centrifugal fractionation of human erythrocytes according to age: comparison between Ficoll and Percoll density gradients, Quad. Sclavo Diagn. 22:236 (1985).

12. M. Rocchigiani, M. Pescaglini, S. Sestini, V. Micheli, C. Ricci, Nicotinic acid and Nicotinamide uptake by human erythrocytes, Ital. J. Biochem.38:159 (1989).

13. M. Rocchigiani, M. Bari, V. Micheli, M. Pescaglini, S. Sestini, Pyridine and purine nucleotide metabolism in human erythrocytes: HPLC determination of the intermediates, Ann. Nutr. Metab. 33:225 (1989).

14. C. Ricci, S. Sestini, V. Micheli, Role of NMN adenylyltransferase in the regulation of NAD metabolism of human erythrocytes, "Macromolecules in the functioning cell", Proceedings of the 6th Soviet-Italian Symposium (Kiev, 1988), in press.

15. E. Beutler, "Red Cell metabolism. A manual of biochemical methods" eds. Grune & Stratton (1975).

16. D. L. Drabkin, Spectrophotometric studies. XV. Hydration of macro-sized crystals of human hemoglobin and osmotic concentration in red cells, J. Biol. Chem. 185:231 (1950).

17. V. Micheli, S. Sestini, P. Lusini, M. L. Ceccuzzi, C. Ricci, Reflections on the relevance of pyridine nucleotide synthesis in human circulating erythrocytes. "Macromolecules in the functioning cell" Proceedings of the 5th Soviet-Italian Symposium (Roma, 1988).

PROTEIN MODIFICATIONS

HYPOTHESES ON THE PHYSIOLOGICAL ROLE OF ENZYMATIC PROTEIN METHYL

ESTERIFICATION USING HUMAN ERYTHROCYTES AS A MODEL SYSTEM

Patrizia Galletti, Caterina Manna, Diego Ingrosso,
Patrizia Iardino and Vincenzo Zappia

Department of Biochemistry of Macromolecules
1st Medical School, University of Napoli, Italy

INTRODUCTION

Among the different theories of aging, the "error catastrophe" hypothesis, first postulated by Orgel in 1963 (1), attracted the interest of scientists by proposing that errors in protein biosynthesis might accumulate with age and reach a catastrophic level at which several vital components of an organism could no longer function to sustain life. Despite the overwhelming evidence against this theory, an important prediction of Orgel's hypothesis was that abnormal proteins would accumulate during cell aging. As a matter of fact, abnormal forms of several proteins were found in aged tissues (2); however, they do not arise from errors in protein biosynthesis, as originally predicted, but as a consequence of several spontaneous modifications which occur post-translationally. There are, indeed, a number of time-dependent spontaneous reactions (3), i.e. deamidation, racemization, glycation and oxidation, that may alter the covalent structure of proteins, which occur at such rates that significant damage to long-lived proteins accumulates during the life time of the cells. In this respect, it is conceivable that systems have evolved to avoid the accumulation of aberrant proteins, either by their selective proteolysis or through the action of protective enzymes, and a failure of both these systems could be one of the mechanisms responsible for the cell aging. The protective enzymes should be particularly efficient in those cells where an impairment or a loss of protein biosynthesis had occurred, thus preventing the protein turnover. Therefore human erythrocytes represent an excellent model system to study either the relationship between post-translational chemical modifications of proteins and cell aging, or the presence of enzymatic systems that deal with this time-dependent protein damage.

This article will describe hypotheses on the physiological role of the enzyme S-adenosylmethionine: protein carboxyl-O-methyl transferase, type II (or D-aspartyl/L-isoaspartyl protein methyl transferase, EC 2.1.1.77) which

Red Blood Cell Aging, Edited by M. Magnani and
A. De Flora, Plenum Press, New York, 1991

149

selectively recognizes and modifies proteins containing altered aspartyls, such as D-aspartyl and L-isoaspatyl residues, in the light of the possible involvement of this enzyme in the processing of these age-damaged proteins.

ORIGIN OF ALTERED ASPARTYLS IN RED BLOOD CELLS

The quantitatively most important covalent modification which accounts for the accumulation of L-isoAsp and D-Asp residues in aged proteins is the deamidation of intrinsically labile asparagine residues, such as those followed by glycine or serine (4,5). The deamidation occurs via a cyclic succinimidyl intermediate, whose opening on either side of the imide nitrogen generates both a normal peptide bond and an atypical isopeptide bond where the aspartyl is linked with its ß-carbonyl to the following residue in the peptide chain (4); the ratio of the two peptides is always in favor of the isoform and is generally 7:3 (Fig. 1).
Furthermore, it has been suggested that also L-Asp residues, present in flexible regions of a protein, can form the cyclic imide, even though this process is extremely slow (5,6).

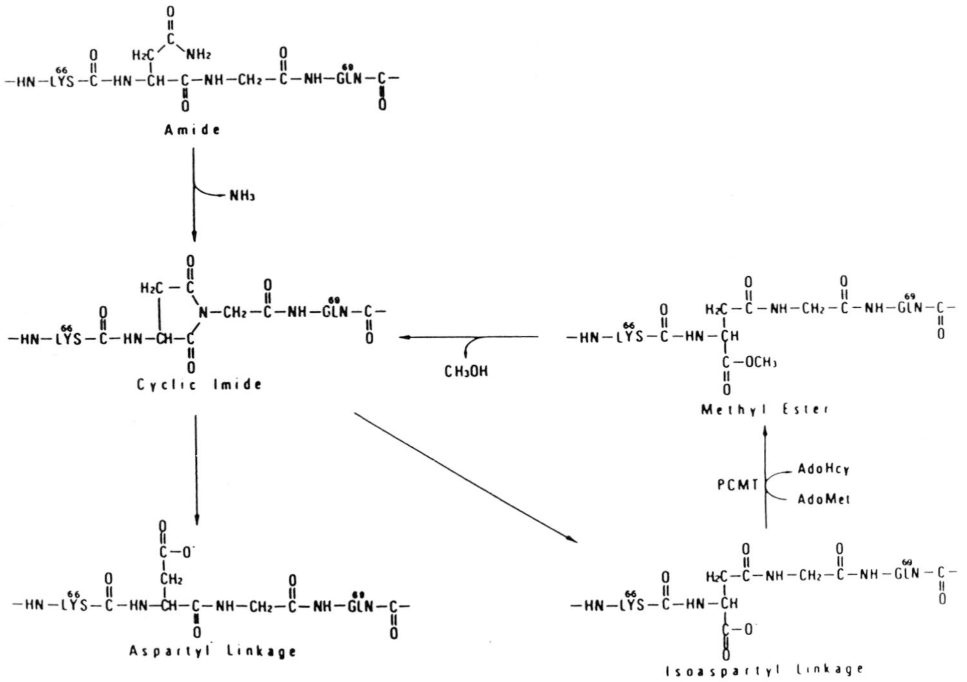

Fig. 1. Proposed reaction mechanisms for deamidation of Asn residues and PCMT-mediated repair of the resulting isoAsp residues. PCMT= Protein carboxyl methyl transferase, type II; AdoMet= S-adenosylmethionine; AdoHcy= S-adenosylhomocysteine.

The relationship between protein deamidation and aging has been demonstrated in different model systems and it has been proposed by Robinson (7) that deamidation may represent a biological timer for the aging of protein molecules. The presence of L-isoAsp residues can affect both structure and function of the deamidated protein. Recently Di Donato et al. (8) clearly demonstrated the selective deamidation of Asn-67 in bovine seminal ribonuclease (BS-RNase). The enzymatic activity of the fully deamidated form of BS-RNase is significantly lower than that of the amidated protein (9). In calmodulin, the deamidation of Asn-60 and/or Asn-97 generates an isoAsp-containing protein which shows an increased molecular radius and exhibits only 20% of the activity of the native calmodulin in the activation of calmodulin-dependent protein kinase (10). Finally, in the triose phosphate isomerase the deamidation of Asn-15 and Asn-71 in both subunits causes the dissociation of this dimeric protein and increases the rate of its proteolytic degradation (11).

The cyclic succinimide, intermediate of the deamidation and isomerization processes, is also particularly prone to racemization, leading to the formation of both D-aspartyl and D-isoaspartyl residues. The accumulation of racemized Asp residues has been extensively reported to occur in long-lived proteins such as human tooth enamel (12) and dentine (13) and eye lens proteins (14). The measure of the extent of racemization has been proposed as a tool for evaluating protein age (13).

The rate of formation of D-aspartyl residues in human erythrocytes has been evaluated by Clarke and coworkers (15): intrinsic membrane proteins accumulate D-residues at such a rate that 1 in every 28 Asx residues in these proteins could be in the D configuration after 120 days. The racemization rate is slower for extrinsic membrane proteins and even more for the cytosolic ones. In this respect, the reduction of anion transport activity of human erythrocyte band 3 observed in the oldest cells could be also partially due to the presence of D-aspartyls in this integral membrane protein (6).

Finally, it should be stressed that the propensity of a protein to form succinimide, the intermediate of deamination and racemization processes, is likely to be dependent on the structure and the environment around the Asx residue, such as the type of side chain on the adjacent amino acid residues and the constraints imposed by the secondary and tertiary structures around the residue (16,17).

In conclusion, although there are several structural restrictions, the formation of altered asparyl residues is one of the greatest protein damages occurring in vivo during cell aging. Evidence that these damaged proteins may be repaired comes from several studies by our and other laboratories on the enzyme S-adenosylmethionine: protein carboxyl-O-methyltransferase, type II (PCMT).

THE ENZYMATIC METHYL ESTERIFICATION OF ABNORMAL ASPARTYL RESIDUES: HUMAN ERYTHROCYTE AS A MODEL SYSTEM.

Among the different protein carboxyl methyl transferases identified so far, the enzyme PCMT selectively transfers the methyl group from 5-adenosylmethionine (AdoMet) to both the beta-carbonyl group of D-aspartic residues and the alfa carbonyl group of L-isoAsp, yielding D-aspartyl-beta-methyl ester and L-isoaspartyl-alfa-methyl ester, respectively (18). No methyl accepting activity has been observed with this enzyme on substrate containing normal L-aspartyl or L-glutamyl residues (19,20). The unusual substrate specificity of PCMT can explain the observed increase in the methyl esterification of erythrocyte membrane proteins which occurs during cell aging (21,22).

In human erythrocytes an active cytosolic PCMT is present in two isoenzymatic forms, whose sequence and structural features have been recently elucidated (23). The two species, which differ in their pI values, can not be distinguished on the basis of substrate specificity and kinetic properties <u>in vitro</u>; the physiological meaning of the presence of the two isoenzymes is still under investigation.

Preliminary studies performed <u>in vitro</u> with purified PCMT showed that most of the red cell proteins can be methylated (24-26), even though a highly substoichiometric extent, the membrane proteins being the major methyl accepting substrates (24). In order to identify the natural substrates of the erythrocyte PCMT, a number of experimental approaches were carried out using the intact cells.

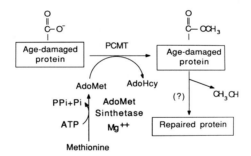

Fig. 2. Enzymatic protein methyl esterification in intact human erythrocytes. PCMT= Protein carboxyl methyl transferase, type II; AdoMet= S-adenosylmethionine; AdoHcy= S-adenosylhomocysteine

Figure 2 shows the methyl esterification reaction as it occurs in intact human red blood cells. Because the cells are impermeable to the methyl donor AdoMet (25), this compound has to be continuously biosynthesized through the action of the enzyme S-adenosylmethionine synthetase. This enzyme has been characterized and purified in red blood cells by Oden and Clarke (27). The cellular concentration of AdoMet in the erythrocytes was evaluated to be 3.5 μM (28), thus making the PCMT, whose Km value for AdoMet is 2 μM (29), saturated <u>in vivo</u> by the methyl donor substrate. Since no other major methyl transfer enzymes are active in the red blood cells it appears that the bulk of AdoMet utilization is in the protein methyl esterification reactions (27).

When intact human erythrocytes were incubated in presence of methyl labelled methionine, almost all the radioactivity appeared associated with membrane proteins (30). The 80% of the total membrane methyl incorporation was found to be hydrolyzed upon treatment in mild alkaline conditions and the product was identified as methanol (30); this result confirms that no other major methyl transfer reaction occurs at the level of erythrocyte membrane. The analysis of the labelled molecular species, performed by SDS-polyacrylamide gel electrophoresis, is reported in Fig. 3. The methyl esterified proteins, identified as radioactive species, include the cytoskeletal components bands 2.1 and 4.1 as well as the integral membrane proteins bands 3 and 4.5. Thus major methyl accepting substrates in red blood cells are proteins that accumulate altered aspartyl residues (6,15).

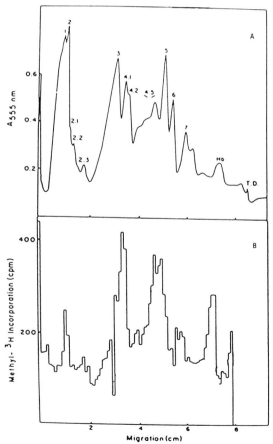

Fig. 3. SDS-polyacrylamide gel electrophoresis of human erythrocyte membrane proteins methyl esterified in intact cells. A: Scanning of Coomassie blue stained gel at 555 nm. B: Radioactivity profile determined from 1mm slices. For experimental details see reference 30.

Therefore, in order to study the methyl esterification process during cell aging, human erythrocytes were separated in three different fractions according to their age-dependent changes in density, as shown in Table 1: the effectiveness of separation was checked by testing their glucose-6-phosphate dehydrogenase activity, creatine content and MCHC value. The steady-state level of the overall methyl esterification of membrane proteins increases 4-5 fold in the oldest cells compared with the youngest ones. In all cell fractions the pattern of the methylated species is similar to that observed in the unfractionated erythrocytes. However in the membrane of the oldest fraction, the cytoskeletal protein bands 2.1 and 4.1 show an increase in methyl incorporation significantly more marked than that observed for other methylated species. Indeed the methyl esterification of bands 2.1 and 4.1 increases 7.8-fold and 9.8-fold respectively, while the average increase for the other species is 4.3-fold.

Table 1. Enzymatic methyl esterification of membrane proteins in different age-related fractions of human erythrocytes

Erythrocyte fraction	Cumulative distance from top	Relative density (d)	Methyl esterification (incorporation of [^{3}H] methyl groups)	
	%		cpm/mg protein	cpm/10^6 cells
A	5-10	1.050	31,000± 880	11.5 ±0.30
B	40-50	1.085	61,590±1950	21.54±0.68
C	85-95	1.123	133,638±4100	44.68±1.37

Intact red blood cells were incubated in presence of L-[methyl-^{3}H] methionine as described in reference 21. Fractions A-C correspond to erythrocyte populations of increasing age.

It should be pointed out that direct evidence for the enzymatic methyl esterification of altered aspartyl residues comes from the isolation of labelled D-aspartyl-beta-methyl ester by proteolytic digestion of membrane proteins, methyl esterified both in vitro (31) and in the intact cells (32). However, since this derivative represents only 0.1-10% of the total radioactivity associated with the membrane proteins, the presence of other methylated residues, such as L-isoAsp - alpha-methyl ester cannot be ruled out. As a matter of fact, the methyl esterification of this residue has been unequivocally demonstrated using several isoAsp-containing proteins and peptides. Surprisingly, in vitro none of the peptides so far tested containing D-Asp residues was found to be a substrate for PCMT. At present no definitive explanation has been proposed for the behavioral discrepancy between peptides and proteins containing D-aspartyl residues.

HYPOTHESES ON THE BIOLOGICAL ROLE OF PCMT TYPE II

The red blood cell has been an excellent model to demonstrate the relationship between the age-dependent accumulation of altered aspartyl residues and the appearance of methyl accepting sites for PCMT. Two possible roles can be hypothesized regarding the physiological meaning of this methylating enzyme: 1) the methyl esterification reaction could be part of a mechanism for the repair of age-damaged proteins; 2) the methyl esterified proteins could be selectively recognized and committed to catabolism. The complexity of cellular systems makes it difficult to approach this problem, therefore many data have been accumulated, in vitro using isoaspartyl-containing peptides and proteins as model systems.

It has been proposed by Aswad (33,34) that PCMT could participate in the repair of isopeptide bonds through the steps indicated in Fig. 1: 1) enzymatic methylation of the alfa-carboxyl group of the isoAsp residue; 2) hydrolysis of the methyl ester, yielding the succinimidyl intermediate; 3) cleavage of the cyclic intermediate yielding both the normal peptide and the isopeptide, generally in the ratio of 7:3. Therefore, since only 30% of original isopeptide is converted into the normal one, an effective repair would imply several consecutive cycles of methylation and demethylation reactions.

The validity of this hypothesis has been tested using a tryptic isopeptide from the naturally deamidated protein BS-RNase (36,37). The isopeptide was incubated in the presence of purified PCMT and AdoMet, and the assay mixture was analyzed at different times. As shown in Fig. 4, after 12 h incubation a significant conversion of the isopeptide into the normal one occurred. The formation, during the process, of both the methyl ester and the succinimidyl derivatives of the peptide confirms the validity of the proposed repair mechanism. At a longer incubation time the isopeptide/normal-peptide ratio was found to be 4.5:3, suggesting that an effective repair of the isopeptide had occurred through repeated methylation cycles. Similar results were obtained using synthetic isopeptides as model systems (38,39).

An elegant demonstration of the utility of this repair function was performed with the calcium binding protein calmodulin (10). When this protein is incubated for about 30 days under physiological conditions it is partially converted into two deamidated forms can be easily separated by electrophoresis under non-denaturing conditions. The deamidation occurs at Asn-60 and/or Asn-97 in two of the calcium-binding sites, and the two forms contain IsoAsp and Asp residues, respectively. The isoAsp-form has only 10% of the activity of native calmodulin in the activation of calmodulin-dependent protein kinase and represents an excellent substrate for the methylating enzyme. The Asp-form more closely resembles native calmodulin, since it exhibits 40% of the original activity and is a very poor methyl accepting substrate. Incubation of the isoAsp-form in the presence of both PCMT and AdoMet results in a significant repair of the isopeptide bond, restoring about 50% of the activity of the native protein. The physiological meaning of this result is further stressed by the evidence that in intact human erythrocytes, calmodulin is effectively methyl esterified by the endogenous methyl transfer enzyme.

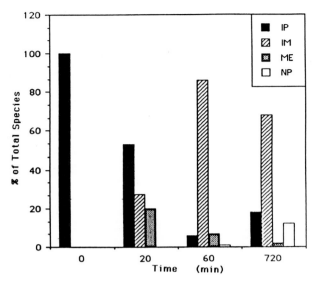

Fig. 4. PCMT-mediated repair of the isopeptide (62-76) from the deamidated
subunit of BS-RNAase. The isopeptide was incubated in the presence
of both PCMT and AdoMet and the assay mixture was analyzed by
reverse-phase HPLC at different incubation times, as described in
reference (36). IP= isopeptide; IM= succinimidyl derivative of
isopeptide; ME=isopeptide; IM=succinimidyl derivative of isopeptide;
ME=methyl ester derivative of isopeptide; NP= normal peptide.

Some limitations in the effectiveness of the repair pathway can be
imposed by the tri-dimensional structure of the protein substrates. The
deamidated forms of several proteins, such as BS-RNase (36,37), EGF (17,40),
RNase A, cytochrome C, aldolase and triosephosphate isomerase (41), indeed,
are not modified in vitro by PCMT. At least in some cases (17,36) the
unfolding of the protein results in a dramatic increase in the methyl
accepting capability to a near stoichiometric extent. It is conceivable that
the repair mechanism may be selective in vivo toward a subset of deamidated/
isomerized proteins either because these proteins are particularly prone to
molecular damage, or because the damage itself affects their function more
dramatically than in the average of protein molecules. It is worth noting
that EGF is resistant to deamidation in its native form, while it undergoes
this damage when treated under conditions which cause the complete unfolding
of the protein (17). Regarding the second possibility, it should be pointed
out that the effects of the deamidation/isomerization processes on protein
function are greatly dependent on the location of the molecular damage in
the protein structure (6).

All together these observations support the hypothesis of the repairing
role of PCMT toward altered isoaspartyl residues, provided that the tri-
dimensional organization of the individual protein substrates allows the
recognition of the methyl accepting site by the methyltransferase.
Regarding the question whether or not a similar mechanism could be operative
in the case of proteins and peptides containing the racemized D-aspartyl
residues, it should be pointed out that preliminary experiments using highly

racemized eye lens proteins speak against the exisitence of such a repair pathway. However further studies on different model systems are needed to definitively rule out this possibility.

The second hypothesis on the physiological role of PCMT postulates that this enzyme could be involved in the selective proteolysis of deamidated proteins, in order to prevent the accumulation of these defective molecules. The methylated isoAsp residue and/or its succinimide derivative could act as signals for the selective proteolysis of damaged proteins by specific proteolytic enzymes. In this respect, in human erythrocytes a post-proline endopeptidase has been detected, which rapidly cleaves succinimidyl-containing peptides (42). On the other hand no relationship seems to exist between the methyl esterification of the cytoskeletal protein band 4.1 (21) and its selective proteolysis by the calcium-dependent neutral proteinases (43,44). It has been demonstrated, indeed, that the exposure of intact red blood cells to high calcium concentrations leads to a significant proteolytic degradation of several membrane proteins, including band 4.1. Under similar calcium-loading conditions we reported that membrane protein methyl esterification is significantly reduced (Table 2), as a result of the impaired AdoMet biosynthesis, due to the ATP depletion (45).

Table 2. Effect of the Ca^{2+} loading on enzymatic methyl esterification of membrane proteins in intact human erythrocytes.

Cell sample	Membrane protein methyl esterification		
	cpm/mg protein	cpm/10^6 cells	%
No Addition	13,760±400	17.10±0.51	100
+10 μM A 23187 +200 μM CaCl$_2$	4,370±130	4.68±0.14	31
+ 10 μM A23187 + 200 μM CaCl$_2$ + 50 mM EGTA	19,137±574	20.53±0.61	139

Intact red blood cells were treated as reported in reference 45, in the presence of L-[methyl-^{3}H] methionine and the methyl incorporation of membrane proteins was evaluated as described by Galletti et al. (21).

A final consideration on the biological role of PCMT arises from our observation that in vitro the preferential substrates of this enzyme are short peptides that can easily adopt the right alignment of the enzyme-binding sites. Therefore, taking into account that the isopeptide bond is generally resistant to the proteolytic attack, it is possible to speculate that the methyl esterification of short isopeptides the products of the partial proteolytic degradation of deamidated proteins, would allow their complete catabolism through the PCMT-mediated conversion of the isopeptide bond into a normal one.

REFERENCES

1. L. E. Orgel, The maintenance of the accurancy of protein synthesis and
   its relevance to ageing, Proc. Natl. Acad. Sci. U.S.A., 49:517
   (1963).
2. M. Rothstein, Detection of altered proteins, in: Altered Proteins and
   Aging", R. C. Adelman and G. S. Roth, ed., CRS Pres, Boca Raton, FL
   (1983).
3. J. McKerrow, Nonenzymatic post-translational amino acid modifications in
   aging, A brief review. Mech. Ageing Dev. 10:371 (1979).
4. P. Bornstein and G. Balian, Cleavage at Asn-Gly bonds with
   hydroxylamine, Methods Enzymol. 47:132 (1977).
5. R. C. Stephenson and S. Clarke, Succinimide formation from aspartyl and
   asparaginyl peptides as a model for the spontaneous degradation of
   proteins, J. Biol. Chem. 264:6164 (1989).
6. J. Lowenson and S. Clarke, Does the chemical instability of aspartyl and
   asparaginyl residues in proteins contribute to erythrocyte aging?,
   Blood Cells 14:103 (1988).
7. A. B. Robinson and C. J. Rudd, Deamidation of glutaminyl and asparaginyl
   residues in peptides and proteins, Curr. Top. Cell. Regul. 8:247
   (1974).
8. A. Di Donato, P. Galletti and G. D'Alessio, Selective deamidation and
   enzymatic methylation of seminal ribonuclease, Biochemistry, 25:8361
   (1981).
9. A. Di Donato, A. Ciardiello, R. Piccoli and G. D'Alessio, Protein
   deamidation and ageing in a model system: bovine seminal and
   pancreatic ribonucleases, in "Protein Metabolism in Aging",
   Wiley-Liss, Inc. (1990).
10. B. A. Johnson, E. L. Langmack and D. W. Aswad, Partial repair of
    deamidation-damaged calmodulin by protein carboxyl methyltransferase,
    J. Biol. Chem., 262:12283 (1987).
11. P. M. Yuan, J. M. Talent, R. W. Gracy, Molecular basis for the
    accumulation of acidic isozyme of triose phosphate isomerase on
    aging, Mech. Ageing Develop. 17:151 (1981).
12. P. M. Helfman, J. L. Bada, Aspartic acid racemization in tooth enamel
    from living humans, Proc. Natl. Acad. Sci. U. S. A., 72:2891 (1975).
13. P. M. Helfman, J. L. Bada, Aspartic acid racemization in dentine as a
    measure of ageing, Nature, 262:279 (1976).
14. P. M. Masters, J. L. Bada, J. S. Zigler Jr., Aspartic acid racemization
    in the human lens during ageing and cataract formation, Nature,
    268:71 (1977).
15. L. S.Brunauer, and S. Clarke, Age-dependent accumulation of protein
    residues which can be hydrolyzed to D-aspartic acid in human
    erythrocytes, J. Biol. Chem. 261:12538 (1986).
16. S. Clarke, Propensity for spontaneous succinimide formation from
    aspartyl and asparaginyl residues in proteins Int. J. Pept. Protein
    Res. 30:308 (1987).
17. P. Galletti, P. Iardino, D. Ingrosso, C. Manna, and V. Zappia, Enzymatic
    methyl esterification of a deamidated form of mouse epidermal growth
    factor, Int. J. Pept. Protein Res. 33:397 (1989).
18. S. Clarke, Protein carboxyl methyltransferases: Two distinct classes of
    enzymes, Ann. Rev. Biochem. 54:479 (1985).

19. P. Galletti, D. Ingrosso, C. Manna, F. Sica, S. Capasso, P. Pucci and G. Marino, Enzymatic methyl esterification of synthetic tripeptides: structural requirements of the peptide substrate: detection of the reaction products by fast-atom-bombardment mass spectrometry, Eur. J. Biochem., 177:233 (1988).
20. D. W. Aswad and B. A. Johnson, The unusual substrate specificity of eukaryotic protein carboxyl methyltransferases, Trends biochem. Sci. 12:155 (1987).
21. P. Galletti, D. Ingrosso, A. Nappi, V. Gragnaniello, A. Iolascon and L. Pinto, Increased methyl esterification of membrane proteins in aged red blood cells: preferential esterification of ankyrin and band 4.1 cytoskeletal proteins, Eur. J. Biochem., 135:25 (1983).
22. J. R. Barber and S. Clarke, Membrane protein carboxyl methylation increases with human erythrocytes age, J. Biol. Chem. 258:1189 (1983).
23. D. Ingrosso, A. W. Fowler, J. Bleibaum and S. Clarke, Sequence of the D-aspartyl/L-isoaspartyl protein methyltransferase from human erythrocytes: common sequence motifs for protein, DNA, RNA, and small molecule S-adenosylmethionine-dependent methyltransferases, J. Biol. Chem. 264:20131 (1989).
24. P. Galletti, W. K. Paik and S. Kim, Methyl acceptors for Protein Methylase II from human erythrocyte membrane, Eur. J. Biochem. 97:221 (1979).
25. S. Kim and P. Galletti, Enzymatic methyl esterification of proteins: studies on erythrocyte membrane as model system, in "Transmethylation", E. Usdin, R. T. Borchardt and C. R. Creveling eds., Elsevier/North Holland (1979).
26. C. M. O' Connor and S. Clarke, Carboxyl methylation of cytosolic proteins in intact human erythrocytes: identification of numerous methyl-accepting proteins including hemoglobin and carbonic anhydrase, J. Biol. Chem. 259:2570 (1984).
27. K. L. Oden and S. Clarke, S-adenosyl-L-methionine synthetase from human erythrocytes: role in the regulation of cellular S-adenosylmethionine levels, Biochemistry, 22:2978 (1983).
28. J. R. Barber, B. H. Morimoto, L. S. Brunauer and S. Clarke, Metabolism of S-adenosyl-L-methionine in intact human erythrocytes, Biochim. Biophys. Acta 887:361 (1986).
29. S. Kim, Purification and properties of protein methylase II, Arch. Biochem. Biophys. 157:476 (1973).
30. S. Kim, and P. Galletti, In vivo carboxyl methylation of human erythrocyte membrane proteins, J. Biol. Chem. 255:338 (1980).
31. C. M. O' Connor and S. Clarke, Methylation of erythrocyte membrane proteins at extracellular and intracellular D-aspartyl sites in vitro: saturation of intracellular sites in vivo, J. Biol. Chem., 258:8485 (1983).
32. P. N. McFadden and S. Clarke, Methylation at D-aspartyl residues in red cells: a possible step in the repair of aged membrane proteins, Proc. Natl. Acad. Sci. U. S. A. 79:2460 (1982).
33. D. W. Aswad, Stoichiometric methylation of porcine adrenocorticotropin by protein carboxyl methyltransferase requires deamidation of asparagine 25: evidence for methylation at alpha-carboxyl group of atypical L-isospartyl residues, J. Biol. Chem. 259:10714 (1984).

34. B. A.Johnson and D. W. Aswad, Enzymatic protein carboxyl methylation at physiological pH: cyclic imide formation explains rapid methyl turnover, Biochemistry 24:2581 (1985).

35. I. M. Ota and S. Clarke, Multiple sites of methyl esterification of calmodulin in intact human erythrocytes, Arch. Biochem. Biophys. 279:320 (1990).

36. P. Galletti, A. Ciardiello, D. Ingrosso, A. Di Donato and G. D'Alessio, Repair of isopeptide bands by protein carboxyl-O-methyltransferase: seminal ribonuclease as a model system, Biochemistry 27:1752 (1988).

37. P. Galletti, P. Ingrosso, G. Pontoni, A. Oliva and V. Zappia, Mechanism of protein carboxyl methyl transfer reactions: structural requirements of methyl accepting substrates, in "Advances in post-translational modifications of proteins and aging", V. Zappia, P. Galletti, R. Porta and F. Wold eds., Plenum Publishing Co. New York (1988).

38. P. N. McFadden and S. Slarke, Conversion of isoaspartyl peptides to normal peptides: implications for the cellular repair of damaged proteins, Proc. Natl. Acad. Sci. U. S. A. 84:2595 (1987).

39. B. A. Johnson, E. D. Murray Jr., S. Clarke, D. B. Glass and D. W. Aswad, Protein carboxyl methyl transferase facilitates conversion of atypical L-isoaspartyl peptides to normal L-aspartyl peptides, J. Biol. Chem. 262:5622 (1987).

40. P. Galletti, D. Ingrosso, C. Manna, P. Iardino and V. Zappia, Enzymatic methyl esterification in the processing of age-damaged proteins, in "Protein metabolism in aging", H. L. Segal, M. Rothstein and E. Bergami eds., Wiley-Liss, New York, 1990.

41. B. A. Johnson, J. M. Shirokawa and D. W. Aswad, Deamidation of calmodulin at neutral and alkaline pH: quantitative relationships between ammonia loss and the susceptibility of calmodulin to modification by protein carboxyl methyltransferase, Arch. Biochem. Biophys. 268:276 (1989).

42. J. Momand and S. Clarke, Rapid degradation of D- and L-succinimide-containing peptides by a post-proline endopeptidase from human erythrocytes, Biochemistry, 26:7798 (1987).

43. S. Pontremoli, E. Melloni, B. Sparatore, M. Michetti and L. Horecker, A dual role for the calcium-requiring proteinase in the degradation of hemoglobin by erythrocyte membrane proteinases, Proc. Natl. Acad. Sci. U. S. A. 81:6714 (1984).

44. D. E. Croall, J. S. Morrow and G. N. De Martino, Limited proteolysis of the erythrocyte membrane skeleton by calcium-dependent proteinases, Biochim. Biophys. Acta, 882:287 (1986).

45. P. Galletti, D. Ingrosso, P. Iardino, C. Manna, G. Pontoni and V. Zappia, Enzymatic basis for the calcium-induced decrease of membrane protein methyl esterification in intact erythrocytes, Eur. J. Biochem. 154:488 (1986).

HETEROGENEITY OF GUANINE NUCLEOTIDE BINDING PROTEINS IN HUMAN RED BLOOD CELL MEMBRANES

Antonio De Flora, Gianluca Damonte, Adina Sdraffa,
Luisa Franco and Umberto Benatti

Institute of Biochemistry
University of Genoa
Genoa, Italy

SUMMARY

Membranes from human erythrocytes bind radioactive GTP and GTP analogs according to apparently homogeneous patterns. In spite of this uniform type of association, multiple guanine nucleotide binding proteins have been identified both by SDS-PAGE analysis of native and of variously ADP-ribosylated membrane preparations and by FPLC chromatography of solubilised erythrocyte membranes preliminarily incubated with [α-^{32}P] GTP in the presence of 5 mM $MgCl_2$. From eight to nine peak fractions of pronase-digestible GTP-binding activity were separated on a MA7Q anion exchange column, this pattern being highly reproducible with different membrane preparations. Prior incubation of membranes with [α-^{32}P] GTP in the presence of excess unlabeled GDP resulted in displacement of bound labeled nucleotide from all FPLC fractions. The patterns of GTP binding were also markedly modified by preliminary treatment of membranes with N-ethylmaleimide. Detectable GTPase activity was present in each of the FPLC peak fractions. This wide heterogeneity of guanine nucleotide binding proteins raises so far unanswered questions as to their physiological significance in the mature erythrocyte.

INTRODUCTION

All eukaryotic cells contain a number of guanine nucleotide binding regulatory proteins (G proteins) in their membranes which play a fundamental role in the transduction of receptor-linked extracellular signals to intracellular effector systems (1-3). In addition to the classical and better characterized heterotrimeric G proteins involved in several signal-transducing pathways, the wide family of monomeric, small molecular size, G proteins is receiving increasing attention and interest in different cell types (4,5).

Red Blood Cell Aging, Edited by M. Magnani and
A. De Flora, Plenum Press, New York, 1991

Erythrocytes too have been reported to have a complement of different G proteins, some of which have been purified and characterized in their structural properties (6-12). However, a systematic study of erythrocyte G proteins has not been undertaken yet and this may represent a serious limitation to knowledge of erythrocyte biochemistry and physiology, notably of the functional changes undergone by the circulating red cell during aging and of the onset of alterations that take place in several red cell diseases.

The _in vivo_ occurring chemical signals that can be identified and specifically bound by receptors on the erythrocyte membrane are certainly limited in number, as compared to excitable and secretory cells. Moreover, the limited information so far available on red cell G proteins cannot discriminate between the hypothesis of a vestigial role played at earlier times during the erythropoietic stage and the alternative possibility of regulation of specific functions in the mature erythrocyte. In the latter case, by analogy with other cell types, a reasonable target function of erythrocyte G proteins might be control of electrolyte homeostasis and accordingly of cell shape and deformability (11). Specifically, we were prompted to screen the panel of G proteins present in the erythrocyte by the fact that the intraerythrocytic content of calcium undergoes a progressive increase during red cell aging (13) and a sharp increase in genetic erythropathies including sickle disease (14), thalassemias (15) and favism (16,17). Recent reports of correlations between G proteins and calcium influx in a number of cell types (18-26) raise the possibility that in the erythrocyte too calcium homeostasis is under the control of so far unidentified, G protein-related mechanisms.

MATERIALS AND METHODS

Reagents and detergents were the highest grade commercially available. Unlabeled nucleotides were further purified on an FPLC system (Pharmacia, Uppsala, Sweden) using a Mono-Q HR 5/5 column. All labeled nucleotides were purchased from New England Nuclear, Dreierich, FRG.

Unsealed, calmodulin-free membranes were prepared from human erythrocytes according to Niggli et al.(27) in the presence of 2 mM PMSF, 1 mM leupeptin and 0.1 μM pepstatin and stored at -80°C. Protein concentration was determined according to Lowry et al. (28). Metabolic depletion and repletion of human erythrocytes were performed as described previously (29).

Binding of labeled guanine nucleotide to erythrocyte membranes was measured according to Northup et al. (30) with some modifications as reported previously (29).

ADP-ribosylation of human erythrocyte membranes by pertussis toxin (PTX), cholera toxin (CTX) and exoenzyme C3 from C. botulinum was carried out as described previously (31).

Western blotting was performed following the method of Towbin et al.(32); the nitrocellulose sheets were incubated with either $[\alpha-^{32}P]$ GTP, as described (31), or the pan-reactive monoclonal antibody Y13-259 according to Hsu et al.(33).

FPLC analysis of GTP-binding activity was carried out as follows. After incubation of human erythrocyte membranes with $[\alpha-^{32}P]$GTP and solubilisation with 3% CHAPS in the same conditions as reported by Wolfman et al. (34), membrane proteins were separated by FPLC on an anion exchange column (MA7Q, Bio-Rad). A linear gradient of NaCl from zero to 0.09 M in 18 min was used first, then isocratic 0.09 M NaCl was applied for 10 min and again a linear 0.09-0.18 M gradient of NaCl in 20 min was used in 10 mM Tris, 1 mM MgCl$_2$, 1 mM dithiothreitol, 0.5% CHAPS, pH 6.8. Each fraction was assayed for $[\alpha-^{32}P]$ GTP binding by rapid filtration using HA 0.45 µM nitrocellulose filters (Millipore, Bedford, MA). The filters were washed three times with 3 ml of 50 mM Tris, 10 mM MgCl, 1 mM dithiothreitol, 0.1 mM ATP, pH 6.8, then added to 5 ml of scintillation liquid and the retained radioactivity was determined in a ß-counter. For the competition experiments, GDP (or ATP) at the final concentration of 0.25 mM was added to the incubation mixture. To estimate the effect to the thiol reagent N-ethylmaleimide (NEM) on GTP binding, the membranes were pre-incubated with 20 mM Hepes, pH 7.4, and 20 mM NEM for 20 min at 30°C prior to incubation with $[\alpha-^{32}P]$ GTP as above.

In the experiments designed to estimate GTPase activity, after incubation of membranes with $[\alpha-^{32}P]$GTP, solubilisation with 3% CHAPS and elution of individual peaks from the MA7Q column, a time zero point was taken by spotting samples onto PEI cellulose plates for thin layer chromatography (34). Each peak fraction was then supplemented with both GTP and GDP, each to a final concentration of 300 µM. The fractions were finally incubated at 30°C for 60 min and spotted again onto PEI cellulose. The changes in the GTP/GDP ratio were detected by autoradiography.

RESULTS

Guanine nucleotide-binding properties of erythrocyte membranes

Previous results obtained in our laboratory had shown that binding of GTP, GTPγS and GDP to unsealed, right side out erythrocyte membranes, obeys first order kinetics. A single rate constant of association of 0.15±0.02 min was calculated for GTPγS, with a half-life of 4.8±0.5 min (29). Equilibrium binding experiments revealed a homogeneous type of association, whose parameters are summarized in Table 1. The stoichiometries of equilibrium binding of guanine nucleotides to membranes proved to undergo marked variations according to the metabolic state of erythrocytes from which membranes had been prepared. Thus, metabolic depletion caused the total number of binding sites to decrease to 35 picomoles/mg protein, while repletion of preliminarily depleted erythrocytes resulted in a stoichiometry of 50 picomoles/mg protein (29).

Table 1. Stoichiometries and Dissociation Constants of Equlibrium
Binding of GTP, GTPγS and GDP to Human Erythrocyte
Membranes[a].

Labeled nucleotides	$[\alpha-{}^{32}P]$ GTP	$[\alpha-{}^{32}P]$ GDP	$[\gamma-{}^{35}S]$ GTPγS
n=picomoles ligand/mg protein	65±6.0	60±4.2	68±10
Kd (nM)	18±4.3	29±3.5	10± 2.3

[a] Data redrawn from ref. 29.

Sodium dodecyl sulfate-PAGE analysis of guanine nucleotide-binding proteins

A subsequent approach to the study of G proteins was made by SDS-PAGE separation of erythrocyte membrane proteins. Electrophoretic separation was carried out, a) either following pre-incubation of membranes with $[\alpha-{}^{32}P]$ NAD and various bacterial toxins (pertussis toxin, cholera toxin and exoenzyme C3 from Clostridium botulinum), or b) prior to western blotting and subsequent exposure of the nitrocellulose sheets to $[\alpha-{}^{32}P]$ GTP or c) to the rat Y13-259 MoAb which is panreactive with ras proteins.

The results of these experiments allowed to establish the presence of four GTP-binding proteins (b) with Mr values of 27, 26, 22 and 21 kDa, respectively (31).

In addition to $G_{i\alpha}$ and $G_{0\alpha}$ (substrates of pertussis toxin) and to $G_{s\alpha}$ (substrate of cholera toxin), three membrane proteins that were identified by exoenzyme C3-catalyzed ADP-ribosylation (a) were shown to comigrate on the SDS-PAGE slabs with three of the four above mentioned GTP-binding proteins at apparent Mr values of 27, 26 and 22 kDA, respectively (31). Finally, immunostaining revealed a ras p21 protein whose Mr was 22 kDa (Table 2).

In spite of an indentical migration at an apparent Mr value of 22 kDa, at least two (a rho and a ras protein, respectively) or even three (possibly a GTP-binding protein unrelated to both rho and ras) different small molecular size G proteins are therefore consistently found in erythrocyte membranes in addition to those of 27, 26 and 21 kDa revealed by GTP binding on the transblots.

FPLC analysis of guanine nucleotide-binding proteins

A number of small molecular size G proteins present in various cell types have been shown to retain firmly bound guanine nucleotides upon increasing Mg^{2+} up to 5-10 mM (34). This property was exploited by first incubating erythrocyte membranes with $[\alpha-^{32}P]$ GTP, then solubilising with 3% CHAPS detergent and finally submitting the solubilised supernatant to FPLC on an anion exchange MA7Q column. Fractions were analysed with a filter binding assay which allowed us to investigate the individual $[\alpha-^{32}P]$ GTP-binding peaks separated from both native and variously treated membranes.

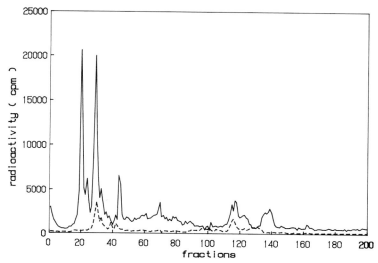

Fig. 1. FPLC analysis of $[\alpha-^{32}P]$ GTP-binding activity from solubilised erythrocyte membranes (34). Membranes were labeled $[\alpha-^{32}P]$ GTP in the absence (continuous line) or in the presence (broken line), of 0.25 mM GDP as described under "Materials and Methods". Ten mM Mg^{2+} was then added and after solubilisation with 3% CHAPS the sample was centrifuged and the supernatant applied onto the MA7Q column.

Table 2. Mr Values of Erythrocyte Membrane Proteins Identified as GTP-Binding Proteins, Bacterial Toxin Substrates and Ras-related Proteins[a]

Mr (kDa)	$[\alpha-^{32}P]$ GTP binding	ADP ribosylation	Y13-259 MoAb
44 (G$_s$)	–	CTX	–
42 (G$_i$)	–	PTX	–
40 (G$_0$)	–	PTX	–
27	+	C3	–
26	+	C3	–
22	+	C3	+
21	+		–

[a]Data redrawn from ref. 31

Fig. 1 shows a typical pattern of $[\alpha-^{32}P]$ GTP-binding activity as obtained from solubilised membranes from native, metabolically normal erythrocytes. Eight major GTP-binding peak fractions, showing remarkably constant relative proportions, were identified by this procedure. The only variability was due to an additional GTP-binding peak eluted at fraction no. 7 and present in some membrane preparations. Exposure of $[\alpha-^{32}P]$ GTP-pre-loaded membranes to pronase resulted in the complete abolition of all GTP-binding FPLC peaks.

Competition experiments performed with an excess of unlabeled GDP over the $[\alpha-^{32}P]$ GTP incubated with membranes demonstrated almost complete displacement of the radioactive GTP from all FPLC peaks (Fig. 1). Lower concentrations of GDP indicated quantitative differences of $[\alpha-^{32}P]$ GTP binding among the individual peaks, probably as a result of different affinities of these fractions towards GDP. ATP was ineffective in displacing $[\alpha-^{32}P]$ GTP from the FPLC peak fractions (not shown).

Treatment of membranes with 20 mM NEM prior to their incubation with $[\alpha-^{32}P]$ GTP at high Mg^{2+} concentrations resulted in distinctively different patterns of GTP binding as compared with untreated membrane preparations (not shown). Overall, the most striking difference was consistent reduction, to almost complete disappearance, of peaks 1-6 of $[\alpha-^{32}P]$ GTP-binding activity and especially of peaks 1, 2, 5 and 6. This pattern contrasted with that of peaks 7 and 8 which were virtually unaffected by preliminary exposure of membranes to 20 mM NEM.

Since low GTPase activity is known to be one of the strict requirements of G proteins, no matter what their molecular sizes, preliminary experiments were carried out in order to establish its presence in the individual FPLC peaks. Although the procedure followed, which is essentially the same as described by Wolfman et al. (34), is not yet a quantitative one because of intrinsically low enzyme activity, all peaks appear to hydrolyze GTP to GDP (Fig. 2), with the highest GTPase activity being in peaks no. 1-5.

Table 3 summarizes the properties so far explored of the eight (nine) peaks of GTP-binding activity separated by FPLC.

Table 3. Properties of the FPLC-separated $[\alpha-^{32}P]$ GTP Binding Peaks

1. Reproducible FPLC patterns, i.e. eight-nine peaks of GTP-binding activity
2. Concentration-dependent competition of GDP for $[\alpha-^{32}P]$ GTP-binding in the individual peaks
3. NEM alters the $[\alpha-^{32}P]$ GTP binding profile significantly
4. All peaks have intrinsic GTPase activity

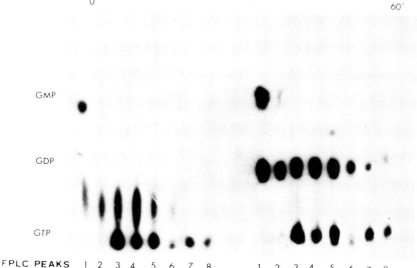

Fig. 2. Detection of GTPase activity in the $[\alpha-^{32}P]$ GTP-binding peaks obtained by fractionation on MA7Q column. Immediately after elution a time zero point was taken by spotting 20 ul of each of the peak fractions on a PEI cellulose sheet. Both GTP and GDP at the final concentration of 300 uM each were added to the fractions and after one hour at 30°C a second sample was taken and spotted. The distribution of guanine nucleotides was evaluated by thin layer chromatography (34) and the dried sheets were exposed to autoradiography, as described under "Materials and Methods".

DISCUSSION

The observed multiplicity of G proteins in human erythrocyte membranes is to some extent surprising in view of the limited range of functions that may be under the control of signal-transducing systems in the circulating red cell.

Although this may be in a sense a facilitating condition for a correlation between G proteins and erythrocyte functions to be identified, the opposite may hold since these G protein-related functions might be present at earlier stages of differentiation of erythroid cells only. However, the occurrence of largely unexplored processes in the control of electrolyte homeostasis within mature erythrocytes seems to support the working hypothesis of involvment of G proteins in the regulation of ionic channels, as clearly demonstrated to occur in a number of cell types. Accordingly, characterization of the G proteins present in the erythrocyte membrane seems to be desirable in order to shed light on still undefined aspects of the red cell physiology and pathophysiology.

The numerous G proteins identified by FPLC analysis of solubilised membranes still escape characterization of their Mr values because of their

167

apparently limited amount within red cell membrane (not exceeding a total number of 18,000 copies per cell - see ref. 29). This makes direct estimation on the SDS-PAGE slabs impossible even by sensitized silver staining procedures (35). On the other hand, if the total content of G proteins were equal to or lower than 0.1% of the proteins present in the erythrocyte membrane, they would be individually undetectable following FPLC separation and direct SDS-PAGE analysis. Therefore we are currently attempting to circumvent the problem by using alternative precedures such as SDS-PAGE analysis of lyophilized FPLC peak fractions combined with either ADP-ribosylation by exoenzyme C3 (rho proteins) or Western blotting and subsequent immunostaining (ras proteins).

Detectable, although not yet measured, GTPase activity in the FPLC peaks opens interesting questions concerning modulators of this activity and therefore of functions of G proteins. Specifically it becomes now feasible to explore whether control systems such as GAP (GTP-ase activating protein) activity (36) and GDP-dissociating protein factors (37) are present in the red cell like in other cell types. Moreover, possible regulation of GTPase intrinsinc activity of individual FPLC peak fractions by specific ADP-ribosylation or chemical reagents such as NEM seems to deserve attention. It is interesting to mention that pre- treatment of intact, normal erythrocytes with 20 mM NEM (i.e., the same concentrations resulting in remarkable changes of patterns of $[\alpha-{}^{32}P]$ GTP-binding activity - see Results) has been recently found to enhance the influx of ${}^{45}Ca$ in the red cells considerably (Guida, L., and De Flora, A., unpublished data). Furthermore, erythrocytes from favic patients during acute hemolysis were reported to have abnormally high levels of intracellular Calcium (16,17). Although the mechanisms underlying this event are still undefined, it is tempting to speculate that the red-ox imbalance of erythrocytes following ingestion of fava beans could affect some critical sulfhydryl groups of cellular proteins, thereby opening Calcium channels in the membrane. The result of an enhanced influx of ${}^{45}Ca$ in erythrocytes from Glucose 6-phosphate dehydrogenase (G6PD) deficient, but not from normal subjects, as obtained with minute concentrations of the favism-related aglycone, divicine (38), is consonant with this hypothesis. The involvement of specific G proteins in the opening of Calcium channels through perturbation of selected sulfhydryl groups affecting GTP binding, such as reported to occur for ras p21 (39) and observed in our laboratory following reaction with NEM, is an attractive possibility requiring further investigation.

ACKNOWLEDGMENTS

This study was supported by Consiglio Nazionale delle Ricerche, Rome (Target Project "Genetic Engineering" and Special Project "Mechanisms of Signal Transduction") and by the Italian Ministry of University and Research.

REFERENCES

1. A. G. Gilman, G proteins: transducers of receptor-generated signals, Ann. Rev. Biochem. 56:615 (1987).

2. L. Stryer, Cyclic GMP cascade of vision, Ann. rev. Neurosci. 9:87 (1986).

3. L. Birnbaumer, J. Abramowitz and A. M. Brown, Receptor-effector coupling by G proteins, Biochim. Biophys. Acta 1031:163 (1990).

4. M. Barbacid, ras genes, Ann. Rev. Biochem. 56:779 (1987).

5. R. D. Burgoyne, Small GTP-binding proteins, Trends Biochem. Sci., 14:394 (1989).

6. J. Codina, J. D. Hildebrandt, R. D. Sekura, M. Birnbaumer, J. Bryan, C. R. Manclark, R. Iyengar and L. Birnbaumer, N_s and N_i, the stimulatory and inhibitory regulatory components of adenylyl cyclase, J. Biol. Chem. 259:5871 (1984).

7. J. Codina, J. Hildebrandt, R. Iyengar, L. Birnbaumer, R. D. Sekura and C. R. Manclark, Pertussis toxin substrate, the putative N component of adenylyl cyclases, is an heterodimer regulated by guanine nucleotide and magnesium, Proc. Natl. Acad. Sci. U.S.A. 80:4276 (1983).

8. A. Yatani, J. Codina, A. M. Brown and L. Birnbaumer, Direct activation of mammalian atrial muscarinic potassium channels by GTP regulatory protein G ,Science 235:207 (1987).

9. J. Codina, A. Yatani, D. Grenet, A. M. Brown and L. Birnbaumer, The subunit of the GTP binding protein G_k opens atrial potassium channels, Science 236:442 (1987).

10. K. Ikeda, A. Kikuchi and Y. Takai, Small molecular weight GTP-binding proteins in human erythrocyte ghosts, Biochem. Biophys. Res. Commun. 156:889 (1988).

11. T. Tanimoto, M. Hoshijima, M. Kawata, K. Yamamoto, T. Ohmori, H. Shiku, H. Nakano and Y. Takai, Binding of ras p21 to bands 4.2 and 6 of human erythrocyte membranes, FEBS Lett. 226:291 (1988).

12. D. J. Carty and R. Iyengar, A 43 kDa form of the GTP-binding protein G_{i3} in human erythrocytes, FEBS Lett. 262:101 (1990).

13. P. Arese and A. De Flora, Pathophysiology of hemolysis in glucose 6-phosphate dehydrogenase deficiency, in: "Seminars Hematol.", E. A. Rachmilewitz, ed., 27:1 (1990).

14. J. W. Eaton, T. D. Skelton, H. S. Swofford, C. E. Kolpin and H. S. Jacob, Elevated erythrocyte calcium in sickle cell disease, Nature 246:105, 1973.

15. O. Shalev, S. Mogilner, E. Shinar, E. A. Rachmilewitz and S. L. Schrier, Impaired erythrocyte calcium homeostasis in ß-thalassemia, Blood 64:564 (1984).

16. A. De Flora, U. Benatti, L. Guida, G. Forteleoni and T. Meloni, Favism: disordered erythrocyte calcium homeostasis, Blood 66:294 (1985).

17. F. Turrini, A. Naitana, L. Mannuzzu, G. Pescarmona and P. Arese, Increased red cell calcium, decreased calcium adenosine triphosphatase, and altered membrane proteins during fava bean hemolysis in glucose-6-phosphate dehydrogenase-deficient (Mediterranean variant) individuals, Blood 66:302 (1985).

18. J. Heschler, W. Rosenthal, W. Trautwein and G. Schultz, The GTP-binding protein, G_0, regulates neuronal calcium channels, Nature 325:445 (1987).

19. D. L. Ghosh, J. M. Mullaney, F. I. Tarazi and D. L. Gill, GTP-activated communication between distinct inositol 1,4,5-triphosphate-sensitive and -insensitive calcium pools, Nature 340:236 (1989).

20. D. L. Gill, J. M. Mullaney, T. K. Ghosh, and S. H. Chueh, Cell calcium metabolism. Physiology, biochemistry, pharmacology and clinical implications, G. Fiskum, ed., Plenum Press, New York, pp. 157–168 (1989).
21. R. Fulceri, A. Romani, G. Bellomo and A. Benedetti, Liver cytosolic non-dialysable factor(s) can counteract GTP-dependent Ca^{2+} relase in rat liver microsomal fractions, Biochem. Biophys. Res. Commun. 163:823 (1989).
22. T. Kitazawa, S. Kobayashi, K. Horiuti, A. V. Somlyo and A. P. Somlyo, Receptor-coupled, permeabilized smooth muscle. Role of the phosphotidylinositol cascade, G. proteins and modulation of the contractile responde to Ca^{2+}, J. Biol. Chem. 264:5339 (1989).
23. I. Kojima, M. Kitaoka and E. Ogata, Guanine nucleotides modify calcium entry induced by insulin-like growth factor-I, FEBS Lett. 258:150 (1989).
24. B. P. Hughes and G. J. Barritt, Evidence that guanosine 5'- -thio triphosphate stimulates plasma membrane Ca^{2+} inflow when introduced into hepatocytes, Biochem. J. 257:591 (1989).
25. S. Muallem and T. G. Beeker, Relationship between hormonal, GTP and $Ins(1,4,5)P_3$-stimulated Ca^{2+} uptake and release in pancreatic acinar cells, Biochem. J. 263:333 (1989).
26. R. B. Moore, G. A. Plishker and S. K. Shriver, Purification and measurement of calpromotin, the cytoplasmic protein which activates calcium-dependent potassium transport, Biochem. Biophys. Res. Commpun. 166:146 (1990).
27. V. Niggli, E. S. Adunyah, J. T. Penniston and E. Carafoli, Purified (Ca -Mg)-ATPase of the erythrocyte membrane, J. Biol. Chem. 256:395 (1981).
28. O. H. Lowry, N. J. Rosebrough, A. L. Farr and R. J. Randall, Protein measurement with the folin phenol reagents, J. Biol. Chem. 193:265 (1951).
29. G. Damonte, A. Morelli, M. Piu, P. Longone and A. De Flora, "In situ" characterization of guanine nucleotide-binding properties of erythrocyte membranes, Biochem. Biophys. Res. Commun. 159:41 (1989).
30. J. K. Northup, M. D. Smigel and A. G. Gilman, The guanine nucleotide activating site of the regulatory component of adenylate cyclase, J. Biol. Chem. 257:11416 (1982).
31. G. Damonte, A. Sdraffa, E. Zocchi, L. Guida, C. Polvani, M. Tonetti, U. Benatti, P. Boquet and A. De Flora, Multiple small molecular weight guanine nucleotide-binding proteins in human erythrocyte membranes, Biochem. Biophys. Res. Commun. 166:1398 (1990).
32. H. Towbin, T. Staehelin and J. Gordon, Electrophoretic transfer of proteins from polyacrylamide gels to nitrocellulose sheets: procedure and some applications, Proc. Natl. Acad. Sci. U.S.A. 76:4350 (1959).
33. S. M. Hsu, L. Raine and H. Fanger, A Comparative Study of the Peroxidase-Antiperoxidase Method and an Avidin-Biotin Complex Method for Studying Polypeptide Hormones with Radioimmunoassay Antibodies, Am. J. Clin. Pathol. 75:734 (1981).
34. A. Wolfman, A. Moscucci and I. G. Macara, Evidence for multiple, ras-like, guanine nucleotide-binding proteins in Swiss 3T3 plasma membranes, J. Biol. Chem. 264:10820 (1989).

35. J. H. Morrissey, Silver stain for proteins in polyacrylamide gels: a modified procedure with enhanced uniform sensitivity, _Anal. Biochem._ 117:307 (1981).

36. M. Trahey and F. McCormick, A cytoplasmic protein stimulates normal N-ras p21 GTPase, but does not affect oncogenic mutants, _Science_ 258:542 (1987).

37. C. J. Der, B. T. Pan and G. M. Cooper, ras mutants deficient in GTP binding, _Molec. Cell. Biol._ 6:3291 (1986).

38. P. Arese, L. Mannuzzu, F. Turrini, S. Galiano and G. F. Gaetani, Etiological aspects of favism, _in_: "Glucose 6-phosphate dehydrogenase", A. Yoshida and E. Beutler, eds. Academic Press, San Diego, pp. 45-75 (1986).

39. S. Hattori, T. Yamashita, T. D. Copeland, S. Oroszlan and T. Y. Shih, Reactivity of a Sulphydryl Group of the ras Oncogene Product p21 Modulated by GTP Binding, _J. Biol. Chem._ 261:14582 (1986).

ROLE OF HEMOGLOBIN DENATURATION AND BAND 3 CLUSTERING IN

INITIATING RED CELL REMOVAL

Philip S. Low

Department of Chemistry
Purdue University
West Lafayette, Indiana

While multiple mechanisms likely exist to assure that a defective erythrocyte does not escape removal by macrophages, we believe that the more heavily used clearance pathways will have certain characteristics in common. First, the pathway should involve a change in components already present in the circulating erythrocytes, since de novo protein synthesis will have terminated before the erythrocyte reaches maturity. Second, the changes initiating the removal sequence must eventually be manifested on the exofacial surface of the cell, since a macrophage has little means of detecting an intracellular biochemical lesion. And finally, the exofacial changes recognized by the macrophage must be inducible by a change in the biochemistry of the cytoplasm, since cells that develop intracellular defects early in their lifespans are also removed early (e.g., sickle cells, (1) ß-thalassemic cells, (2) cells with enzyme deficiencies, (3) cells treated with oxidants, (4) etc). That is, a linkage of some sort must exist between the functional state of components in the cytoplasm and markers at the cell surface recognized by macrophages. The hypothesis outlined below describes how hemoglobin, the most abundant protein in the cytoplasm, and band 3, the most prominent protein in the membrane cooperate to establish this linkage, transducing information regarding the biochemical integrity of the cell to the reticuloendothelial system which is responsible for aged/abnormal cell clearance.

OUR HYPOTHESIS

We propose that the molecular marker that reports the physiological health of a circulating erythrocyte to the reticuloendothelial system lies in the lateral distribution of membrane-spanning proteins, predominantly band 3. In the healthy erythrocyte, band 3 is evenly distributed within the plane of the membrane. However, as hemoglobin denatures, or oxidative species increase, or cytoskeletal components dissociate, one or more microscopic clusters of band 3 can form at a localized site on the membrane.

Red Blood Cell Aging, Edited by M. Magnani and
A. De Flora, Plenum Press, New York, 1991

Because this clustered site is not normally present in a healthy cell, it is viewed as foreign or "nonself" by the immune system and rapidly opsonized with autologous IgG. Once a threshold size or density of IgG has accumulated on the band 3 aggregate, the site is further opsonized with complement. This IgG/complement cluster then serves as the recognition marker to trigger cell removal by macrophages. While the role of autologous IgG and complement are specifically emphasized as mediators of senescent cell recognition, the hypothesis does not exclude the possibility that clusters of integral membrane proteins might be directly recognized by an IgG-independent mechanism (5,6). Furthermore, since band 3 clustering can be initiated by numerous biochemical lesions, (7-14) the pathway ensures that virtually any cell in biochemical distress will be removed, regardless of its chronological age. Documentation of this pathway (Fig. 1), which is thought to be driven mainly by hemoglobin denaturation in vivo, is provided below.

RED CELLS DEVELOP BIOCHEMICAL LESIONS AS THEY AGE

An extraordinary variety of biochemical changes are reported to accompany red cell aging (see review by Clark (15)). These changes include carboxymethylation of proteins (16), activation of the proteases (17,18), glycosylation of proteins (19), loss of membrane area (20,21), decline in enzyme activities (22,23), binding of hemichromes to the membrane (24-26), changes in band 4.1a: 4.1b ratio (27,28), increases in oxidized lipids (29) and proteins (30), changes in cell rheology (31) and fragility (32), changes in exposure of cell surface sugars (33,34), and gradual accumulation of Ca^{++} (35). There is unfortunately little information regarding how most of these changes impact on red cell lifespan. However, it is noteworthy that band 3 clustering, an event that will later be shown to trigger IgG opsonization, can be caused by hemichrome binding (7-10,36,37), ATP depletion (11), malondialdehyde formation (12), Ca^{++} accumulation (13), oxidative cross-linking (14), or membrane skeletal weakening (38). For this reason, we have felt that band 3 clustering might warrant closer scrutiny as a possible transducer of distress signals from the cytoplasm to the external surface of the cell. If antibodies were to opsonize these microscopic aggregates of band 3, then a variety of biochemical lesions could be detected by a single molecular marker, allowing for the processing of multiple cellular defects via a common clearance pathway.

HEMOGLOBIN DENATURATION CAUSES BAND 3 CLUSTERING

We discovered the avid interaction between denatured hemoglobin (hemichromes) and band 3 when we mixed a clear solution of hemoglobin fortuitously contaminated with small amounts of hemichromes with the purified, water-soluble cytoplasmic domain of band 3 (cdb3). Immediately upon mixing, a precipitate formed between cdb3 and the hemichromes, leaving the supernatant enriched in native hemoglobin. Further investigation of the precipitate revealed that it was polymeric in nature with a defined composition of 5 globin chains per band 3 chain, regardless of the ratio of components in the initial reaction medium (36,37). The oligomeric hemichromes were found to bind specifically at the acidic N-terminus of band 3 and at a site off the dyad axis of the hemichrome (36,37). Consistent with this localization, the copolymerization was shown to be inhibited by

174

phosphorylation of band 3 on Tyr 8 (7), enhanced by disulfide cross-linking of the hemichromes (37), and strongly stabilized by ionic interactions (37). Following its nucleation on band 3, the copolymer was found to solidify by formation of disulfide bonds between separate globin chains and also between globin and band 3. Only at extremely high native Hb concentrations was any reduction in copolymer formation observed (36), and further analysis of the stability of the copolymer under intracellular solution conditions indicated that it should form in vivo as soon as hemichromes were generated (36,37).

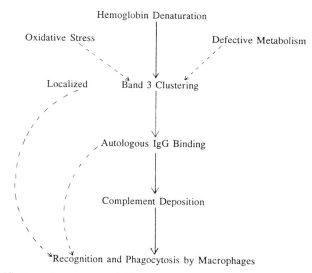

Fig. 1. Pathway of senescent/damaged red cell clearance
dashed arrows represent steps that are less well documented

Because band 3 is constrained to remain within the membrane in vivo, it was quickly realized that the 3-dimensional copolymerization of band 3 with hemichromes described above was not possible. However, to satisfy the strong affinity of the two components for each other, a two-dimensional cluster or aggregate of hemichromes with band 3 in the membrane was hypothesized. Four lines of evidence now support this cluster model. First, in isolated membranes, endogenous hemichromes have been shown to bind predominantly to the N-terminus of the cytoplasmic domain of band 3 (36), i.e. near Tyr 8 (7). Second, addition to whole cells of the Hb denaturant phenylhydrazine, when added sufficiently slowly, aggregates band 3 into small membrane clusters (7,10). Third, in cells with naturally occurring membrane bound hemichromes or Heinz bodies (e.g. sickle cells), clusters of band 3 can be seen by immunofluorescence microscopy and electron microscopy precisely at the membrane sites of Heinz body binding (2,9,39). And finally, isolation of hemichrome-membrane protein aggregates from sickle cells (41), ß-thalassemic cells (42), and the densest fraction of normal cells (43) yields a disulfide-bonded complex whose major membrane component is band 3 (41-43). We believe these aggregates, which are totally absent from younger healthy cells but constitute up to 1% of the membrane protein in aged/abnormal cells, represent the hemichrome-band 3 clusters which form when Hb begins to denature in vivo.

If it is assumed that hemichrome formation and band 3 clustering can trigger red cell removal, the question naturally arises as to when in an erythrocyte's lifespan do hemichromes bind and cluster membrane proteins? The answer to this question depends on the nature of the erythrocytes under scrutiny. In cells with unstable hemoglobins or inadequate reducing power, hemichromes may precipitate on the membrane soon after the cell is released into circulation. Certainly, a large fraction of circulating sickle cells contain one or more visible Heinz bodies (9). In contrast, normal cells develop Heinz bodies only late in their lifespans. Thus, only the densest fractions of HbA erythrocytes exhibit membrane-associated hemichromes (24-26), and Heinz body-containing normal cells only occur regularly in splenectomized individuals (26). In a hypertransfusion study conducted in mice, Mueller and colleagues (ref. 44, personal communications) observed that no hemichromes bound to the erythrocyte membrane until directly prior to red cell removal, at which point a rapid and massive deposition of globin chains was observed. We consider the timing of this globin deposition to be highly significant, since in all of the above examples the erythrocytes are removed soon after globin binds to their membranes. This correlation supports our contention that Hb denaturation followed by band 3 clustering initiates a major pathway of red cell clearance.

CLUSTERS OF BAND 3 BIND AUTOLOGOUS ANTIBODY

Initial experiments to determine whether simple clustering of band 3 might induce autologous IgG binding were conducted using exogenous clustering agents. For example, fresh erythrocytes treated with phenylhydrazine to induce intracellular hemichrome formation were found to bind tenfold more IgG upon reincubation in their own serum than untreated controls (10). Similarly, cells treated with the band 3 clustering agent, acridine orange (which has no effect on Hb stability, but instead directly clusters band 3), were observed to accumulate 30-fold more autologous IgG than controls (10). These experiments provided the first indication that the proposed causal relationship between band 3 clustering and autologous IgG binding indeed existed.

To more rigorously test the above correlation, a series of new clustering studies aimed at testing each of the putative events leading from band 3 clustering to phagocytosis was conducted. In collaboration with Professors Paolo Arese and Franco Turrini (40), the absolute requirement for a clustered distribution of band 3 was examined using reversible clustering agents to control band 3 distribution and chemical cross-linking agents to covalently stabilize any desired distribution of band 3. It was found that 1 mM Zn^{+2}, 1 mM acridine orange, and 0.35 µM melittin, i.e. treatments which cluster integral proteins in erythrocytes (45-47), all induced autologous IgG binding, complement fixation and phagocytosis by human macrophages (40). The essential role of integral membrane protein clustering was confirmed by demonstrating that none of the above steps occurred if clustering was prohibited. Thus, washing away of the clustering agent eliminated all 3 response, while stabilization of the clustered distribution with chemical cross-linkers (BS3 or DSP, Pierce Chemical, Co) preserved the responses even after removal of the clustering agent. Furthermore, subsequent cleavage of the cross-links maintaining the clustered distribution also reversed induction of IgG binding, complement fixation, and phagocytosis, and blockade of clustering by cross-linking the cells prior to addition of any clustering agents also blocked the subsequent responses, even in the presence of the clustering agents. Because the results were similar regardless of which clustering or cross-linking agent was used, it was clear that band 3 clustering per se was sufficient to initiate the entire red cell phagocytosis pathway. Furthermore, any treatment that inhibited band 3 clustering also prohibited expression of the same pathway.

Data gathered on untreated erythrocytes freshly collected from circulation also confirmed the connection between band 3 clustering and autologous IgG binding (41-43). When membranes were prepared from healthy erythrocytes, treated with 0.2 mM EDTA, pH 8 to dissociate the membrane skeleton, and then dissolved in a clean, nondenaturating detergent (e.g. 1% $C_{12}E_8$, Nikko Chemical Co), a solution of membrane components was generated which contained no pelletable material, even if centrifuged at 80,000g. However, when sickle cells, ß-thalassemic cells, or the densest 1% of normal cells (which survive only a few additional hours in circulation (48)) were treated similarly, a hemichrome and band 3-rich membrane protein pellet was readily harvested. Although this insoluble membrane protein aggregate constituted less than 1.3%, 1.6% and 0.09% of the total membrane protein in sickle, thalassemic and normal dense cells, respectively, it was still found to contain 75%, 27% and 55% of the total cell surface IgG (41-43). This enrichment represents a relative elevation in IgG binding at clustered sites on the membrane over nonclustered sites of 253-fold, 160-fold, and 640-fold, for sickle cells, ß-thalassemic cells, and dense normal cells, respectively. We were, therefore, able to conclude that naturally occurring band 3 clusters are seen as antigenic by the immune system and are consequently aggressively opsonized with autologous IgG.

Although several control experiments were conducted in the aforementioned studies, perhaps one particularly illuminating study should by mentioned here. To assure that the pelletable aggregates contained IgG, band 3 and hemichromes in the same insoluble particles, the detergent extract containing the insoluble particles was incubated with agarose beads coated with either anti-human IgG or the nonspecific protein, avidin. After sedimenting and washing, the former beads were found to contain large amounts of hemichromes (spectral data), while the latter beads attracted no material from the same suspension. This control study thus suggested that a physical link exists between autologous IgG in the particle suspension and the hemichrome aggregate in the same detergent extract. Since band 3 was also found in the same particle, we presume the protein linking the autologous IgG to the hemichromes is band 3.

One intriguing question that still remains concerns why clusters of band 3 and other integral proteins should be more antigenic than an even distribution of the same polypeptides. Three hypotheses, but unfortunately no concrete answers, can be offered at this point. First, aggregation could distort the normal conformation of band 3 or glycophorin so that the aggregated population would appear foreign or denatured to the immune system. Second, clustering could simply remove native epitopes from a protected or occluded location, for example, by simply clearing away overlying regions of the glycocalyx. The newly exposed native epitopes would then appear foreign to the immune system because they would not be normally encountered on a healthy erythrocyte. The hypothesis we currently favor is that senescent cell antibodies have evolved a low affinity for their antigens such that a single arm attachment cannot be maintained in vivo (10). In this scenario, clustering of band 3, because it would bring the antigenic sites together, would permit the exponentially more avid bivalent interaction. The potential increase in affinity (E) due to clustering of band 3 can actually be calculated from the valence (V) of the antibody (i.e. 2 in the case of IgG) and the two-dimensional mole fraction of the antigen in the membrane (X_0), where:

$$E = \frac{1}{V} \left[\frac{2}{X_0} \right]^{v-1}$$

Assuming band 3 is the antigen and that it is present at 600,000 dimers per cell, the enhancement factor for IgG binding upon clustering calculates to 10^3 and for IgM binding to 10^{12}. Clearly, these increases in affinity are sufficient to cause clustering-induced antibody binding. Consistent with this interpretation, it should be noted that anti-band 3 antibodies are normal components of human serum (11, 49-51) and the cross-linking of band 3 enhances their binding to red cells (49,50).

AUTOLOGOUS IgG BINDING PROMOTES COMPLEMENT DEPOSITION

That antibody opsonization of an antigen promotes complement fixation in general is well established in the immunology literature. However, that complement fixation plays an important role in senescent erythrocyte removal has largely been ignored until recently. Thus, an early report by Ehlenberger and Nussenzweig (52) that complement deposition greatly enhanced red cell phagocytosis received little attention until Lutz et al. (53) demonstrated that inactivation of the complement present in autologous serum largely abolished phagocytosis of diamide-treated cells opsonized with anti-band 3 IgG. Since this time, other reports have appeared supporting the strong enhancing effect of complement C3 fragments on phagocytosis, especially at low levels of bound autologous IgG (54,55). However, Lutz et al. (53) were the first to show that bound anti-band 3 IgG promoted fixation of complement on the surface of erythrocytes.

We have also examined the involvement of complement fixation in our model studies (40) using fresh human erythrocytes in autologous serum and Zn^{+2}, melittin, or acridine orange to generate the clustered distribution of band 3. In the absence of clustering, only basal le els of complement could be detected on the erythrocyte surfaces. However, when the clustered distribution was promoted, the basal level rose up to 8-fold. Since diisopropylfluorophosphate largely inhibited the enhanced C3b deposition, the observed deposition must have been catalyzed by a convertase activated by a clustered deposit of autologous IgG. Further,

178

since complement deposition invariably followed IgG binding in these model studies, regardless of the reagents used to generate or stabilize the clustered distribution of integral membrane proteins, we conclude that complement fixation must be an important intermediate in the red cell senescence pathway. In this respect it should be noted that the hemichrome-band 3-IgG aggregates isolated from the densest fraction of normal erythrocytes were also found to contain C3b (unpublished observations).

LOCALIZED DENSITIES OF CELL SURFACE IgG AND C3b TRIGGER PHAGOCYTOSIS

The erythrocyte senescence pathway obviously culminates in phagocytosis. While antibody independent phagocytosis can be measured in vitro (5,6), it is questionable whether it contributes significantly to physiological cell removal in vivo. Thus, when autologous antibodies are allowed to opsonize an aged or altered erythrocyte, phagocytosis is invariably enhanced (40,49-56). Furthermore, low levels of bound autologous IgG, which by themselves are insufficient to induce phagocytosis, can aggressively promote phagocytosis if complement deposition is permitted at the same sites (40,52-55). In our studies using exogenous clustering agents we monitored the dependence on Zn^{+2} concentration of IgG binding, C3b deposition, and phagocytosis, and observed that IgG binding was hyperbolically dependent on Zn^{+2} concentration, while C3b and phagocytosis varied sigmoidally with the same parameter. Thus, a minimum cluster density of IgG appears to be required before the autocatalytic (i.e. cooperative) deposition of complement can begin. Phagocytosis was obviously more sensitive to complement deposition than autologous IgG binding since it followed the complement curve almost exactly (40).

In light of our clustering model, we feel several observations from other labs observe at least brief mention. First, Shaw et al. (57) report that a clustered distribution of IgG on a red cell surface is much more effective in promoting lymphocyte attack than an even distribution of the same number of IgG. Second, Hermanowski et al. (55) demonstrate that erythrocytes bearing clustered C3b on their surfaces are bound "far more avidly" by both macrophages and polymorphonuclear leukocytes than erythrocytes bearing the same number of complement in random array, and finally, in ultrastructural studies, Rifkind (58) has shown that the site of recognition of phenylhydrazine-treated erythrocytes by macrophages is directly over the site where hemichromes cluster on the membrane. Thus, these studies all point to the involvement of antigen clustering in mediating red cell removal.

In conclusion, each of the proposed steps in the red cell clearance pathway outlined in Fig. 1 can be supported by both in vitro and in vivo studies. Although hemoglobin denaturation is not only erythrocyte defect that can lead to membrane protein clustering, at present it would appear to constitute the major pathway. The fact that multiple defects can all lead to a common molecular marker recognized by macrophages allows an erythrocyte to be cleared from circulation as soon as it falters in any of its major biochemical tasks. Thus, in our model erythrocyte removal is not determined by an invariable clock but instead by cell viability.

ACKNOWLEDGEMENT

Supported by a grant from the National Institute of Health (GM24417)

REFERENCES

1. P. R. McCurdy and A. S. Sherman, Irreversibly sickled cells and red cell survival in sickle cell anemia, Amer. J. Med. 64:253 (1978).

2. E. A. Rachmilewitz, E. Shinar, O. Shalev, U. Galili and S. L. Schrier, Erythrocyte membrane alterations in ß-thalassemia, Clinics in Haematology 14:163 (1985).

3. J. J. Kaneko, Comparative erythrocyte metabolism, Adv. Vet. Sci. Comp. Med. 18:117 (1974).

4. D. A. Bates and C. C. Winterbourn, Hemoglobin denaturation, lipid peroxidation and hemolysis in phenylhydrazine-induced anemia, Biochim. Biophys. Acta 798:84 (1984).

5. S. Horn, J. Copas and N. Bashan, A lectin-like receptor on murine macrophage is involved in the recognition and phagocytosis of human red cells oxidized by phenylhydrazine, Biochem. Pharm. 39: 775 (1990).

6. M. Beppu, M. Ochiai and K. Kikugawa, Macrophage recognition of periodate-treated erythrocytes: involvement of disulfide formation of the erythrocyte membrane proteins, Biochim. Biophys. Acta 979:35 (1989).

7. P. S. Low, Interaction of native and denatured hemoglobins with band 3: consequences for erythrocytes structure and function, in "Red Blood Cell Membranes" P. Agre and J. C. Parker, ed., Marcel Dekker, Inc. NY, p. 237 (1989).

8. P. S.Low and R. Kannan, Effect of hemoglobin denaturation on membrane strucutre and IgG binding: role in red cell aging, in "The Red Cell: Seventh Ann Arbor Conference", Alan R. Liss, Inc. p. 525 (1989).

9. S. M. Waugh, B. M. Willardson, R. Kannan, R. J. Labotka and P. S. Low, Heinz bodies induce clustering of band 3, glycophorin, and ankyrin in sickle cell erythrocytes, J. Clin. Invest. 78:1155 (1986).

10. P. S. Low, S. M. Waugh, K. Zinke and D. Drenckhahn, The role of hemoglobin denaturation and band 3 clustering in red blood cell aging, Science 227: 531 (1985).

11. H. Muller and H. U. Lutz, Binding of autologous IgG to human red blood cells before and after ATP-depletion, Biochim. Biophys. Acta 729:249 (1983).

12. P. Hochstein and S. K. Jain, Association of lipid peroxidation and polymerization of membrane proteins with erythrocyte aging, FASEB 40:183 (1981).

13. A. Elgsaeter, D. M. Shotton and D. Branton, Intermembrane particle aggregation in erythrocyte ghosts, Biochim. Biophys. Acta 426: 101 (1976).

14. F. Turrini, A. Naitana, L. Mannuzzu, G. P. Pescarmona and P. Arese, Increased red cell calcium, decreased calcium adenosine triphosphatase, and altered membrane proteins during Fava bean hemolysis in glucose-6-phosphate dehydrogenase-deficient (Mediterranean variant) individuals, Blood 66: 302 (1985).

15. M. R. Clark, Senescence of red blood cells: progress and problems, Physiol. Rev. 68: 503 (1988).
16. J. R. Barber and S. Clarke, Membrane protein carboxyl methylation increase with human erythrocyte age, J. Biol. Chem. 258:1189 (1983).
17. K. Yamamoto, M. Yamada and Y. Kato, Age-related and phenylhydrazine-induced activation of the membrane-associated cathepsin E in human erythrocytes, J. Biochem. 105: 114 (1989).
18. M. M. B. Kay, K. Sorensen, P. Wong and P. Bolton, Antigenicity, storage and aging: physiologic autoantibodies to cell membrane and serum proteins and the senescent cell antigen, Molec. and Cell. Biochem. 49: 65 (1982).
19. H. Vlassara, J. Valinsky, M. Brownlee, C. Cerami, S. Nishimoto and A. Cerami, Advanced glycolysation endproducts on erythrocyte cell surface induce receptor-mediated phagocytosis by macrophages. A model for turnover of aging cells, J. Exp. Med. 166: 539 (1987).
20. K. Miyahara and M. J. Spiro, Nonuniform loss of membrane glycoconjugates during in vivo aging of human erythrocytes: studies of normal and diabetic red cell saccharides, Arch. Biochem. Biophys. 232: 310 (1984).
21. M. A. Zago, D. T. Covas, M. S. Figueiredo, C. Bottura, Red cell pits appear preferentially in old cells after splenectomy, Acta Haemat. 76: 54 (1986).
22. A. Fazi, E. Piatti, A. Accorsi and M. Magnani, Cell age dependent decay of human erythrocytes glucose-6-phosphate isomerase, Biochim. Biophys. Acta 998:286 (1989).
23. C. Seaman, S. Wyss and S. Piomelli, The decline in energetic metabolism with aging of the erythrocyte and its relationship to cell death, Amer. J. Hemat. 8: 31 (1980).
24. H. Q. Campwala and J. F. Desforges, Membrane-bound hemichrome in density-separated cohorts of normal (AA) and sickled (SS) cells, J. Lab. Clin. Med. 99: 25 (1982).
25. D. A. Sears, J. M. Friedman and D. R. White, Binding of intracellular protein to the erythrocyte membrane during incubation: the production of Heinz bodies, J. Lab. Clin. Med. 86: 722 (1975).
26. J. G.Selwyn, Heinz bodies in red cells after splenectomy and phenacetin administration, Brit. J. Haematol. 4: 173 (1955).
27. T. J. Mueller, C. W. Jackson, M. E. Dockter and M. Morrison, Membrane skeletal alterations during in vivo mouse red cell aging: increase in the band 4.1a:4.1b ratio, J. Clin. Invest. 79: 492 (1987).
28. T. Suzuki and G. Dale, membrane proteins in senescent erythrocytes, Biochem. J. 257: 37 (1989).
29. S. K. Jain, Evidence for membrane lipid peroxidation during the in vivo aging of human erythrocytes, Biochim. Biophys. Acta 937: 205 (1988).
30. A. Brovelli, C. Seppi and C. Balduini, Modification of membrane protein organization during in vitro aging of human erythrocytes, Int. J. Biochem. 16: 1115 (1984).
31. S. P. Sutera, R. A. Gardner, C. W. Boylan, G. L. Carroll, K. C. Chang, J. S. Marvel, C. Kilo, B. Gonen and J. R. Williamson, Age-related changes in deformability of human erythrocytes, Blood 65:275 (1985).
32. J. M. Rifkind, K. Araki and E. C. Hadley, The relationship between the osmotic fragility of human erythrocytes and cell age, Arch. Biochem. Biophys. 222: 582 (1983).

33. D. Aminoff, M. A. Ghalambor and C.J. Henrich, GOST, galactose oxidase and sialyl transferase, substrate and receptor sites in erythrocyte senescence, in "Erythrocyte Membranes 2. Recent Clinical and Experimental Advances", W. C. Kruckerberg, J. W. Eaton and G. J. Brewer, ed., Liss: New York, p. 269 (1981).

34. U. Galili, I. Flechner, A. Knyszynski, D. Danon and E.A. Rachmilewitz, The natural anti-α-galactosyl IgG on human normal senescent red blood cells, Br. J. Haematol. 62:317 (1986).

35. T. Shiga, M. Sekiya, N. Maeda, K. Kon and M. Okazaki, Cell age-dependent changes in deformability and calcium accumulation of human erythrocytes, Biochim. Biophys. Acta 814:289 (1985).

36. S. M. Waugh and P. S. Low, Hemichrome binding to band 3: nucleation of Heinz bodies on the erythrocyte membrane, Biochemistry 24:34 (1985).

37. S. M. Waugh, J. A. Walder and P. S. Low, Partial characterization of the copolymerization reaction of erythrocyte membrane band 3 with hemichromes, Biochemistry 26: 1777 (1987).

38. P. P. DaSilva, Translational mobility of the membrane intercalated particles of human erythrocyte ghosts, J. Cell. Biol. 53: 777 (1972).

39.K. Schlüter and D. Drenckhahn, Co-clustering of denatured hemoglobin with band 3: its role in binding of autoantibodies against band 3 to abnormal and aged erythrocytes, Proc. Natl. Acad. Sci.USA 83: 6137 (1986).

40. F. Turrini, P. Arese, J. Yuan and P. S. Low, Clustering of integral membrane proteins of the human erythrocyte membrane stimulates autologous IgG binding, complement deposition and phagocytosis, submitted for publication (1991).

41. R. Kannan, R. Labotka and P. S. Low, Isolation and characterization of the hemichrome-stabilized membrane protein aggregates from sickle erythrocytes, J. biol. Chem. 263: 13766 (1988).

42. R. Kannan, Mechanism of aging of human red cells, Ph. D. Dissertation, Purdue University, 71 (1990).

43. R. Kannan, J. Yuan and P. S. Low, Isolation and characterization of antibody-enriched complexes from membranes of density fractionated human erythrocytes, manuscript submitted (1991).

44. M. Morrison, C. W. Jackson, T. J. Mueller, T. Huang, M. E. Dockter, W. S. Walker, J. A. Singer and H. H. Edwards, Does cell density correlate with red cell age?, Biomed. Biochim. Acta 42: S107 (1983).

45. M. J. Clague and R. J. Cherry, A comparative study of band 3 aggregation in erythrocyte membranes by melittin and other cationic agents, Biochim. Biophys. Acta 980: 93 (1989).

46. S. W. Hui, C. M. Stewart and R.J. Cherry, Electron microscopic observation of the aggregation of membrane proteins in human erythrocyte by mellitin, Biochim. Biophys. Acta 1023: 335 (1990).

47. G. Lelkes, G. Lelkes, K. S. Merse and S. R. Hollan, Intense, reversible aggregation of intramembrane particles in non-haemolyzed human erythrocytes, Biochim. Biophys. Acta 732: 48 (1983).

48. M. TenBrinke and J. DeReget, Cr-half time of heavy and light human erythrocytes, Scand. J. Haematol. 7: 336 (1970).

49. H. U. Lutz, A naturally occurring autoantibody to band 3 protein of human red blood cells and its possible role in removal of senescent red cells, in "Red Cell Membrane Glycoconjugates and Related Genetic Markers", J-P Cartron, P. Rouger and C. Salmon, eds., p. 273 (1983).

50. M. Beppu, A. Mizukami, M. Nagoya and K. Kikugawa, Binding of anti-band 3 autoantibody to oxidatively damaged erythrocytes, J. Biol. Chem. 265: 3226 (1990).

51. M. M. B. Kay, Localization of senescent cell antigen on band 3, Proc. Natl. Acad. Sci. USA 81: 5753 (1984).

52. A. G. Ehlenberger and V. Nussenzweig, The role of membrane receptors for C3b and C3d in phagocytosis, J. Exp. Med. 145: 357 (1977).

53. H. U. Lutz, F. Bussolino, R. Flepp, S. Fasler, P. Stammler, M. D. Kazatchkine and P. Arese, Naturally occurring anti-band 3 antibodies and complement together mediate phagocytosis of oxidatively stressed human erythrocytes, Proc. Natl. Acad. Sci. USA 84:7368 (1987).

54. N. Yousaf, J. C. Howard and B. D. Williams, Studies in the rat of antibody-sensitized and N-ethylmaleimide-treated erythrocyte clearance by the liver: effects of immune complex infusion and complement activation, Immunolgy 64:193 (1988).

55. A. Hermanowski-Vosatka, P. A. Detmers, O. Götze, S. C. Silverstein and S. D. Wright, Clustering of ligand on the surface of a particle enhances adhesion to receptor-bearing cells, J. Biol. Chem. 263: 17822 (1988).

56. M. M. B. Kay, Role of physiologic autoantibody in the removal of senescent human red cells, J. Supramol. Struct. 9: 555 (1978).

57. G. M. Shaw, D. Aminoff, S. P. Balcerzak and A. F. LoBuglio, Clustered IgG on human red blood cell membranes may promote human lymphocyte antibody-dependent cell-mediated cytotoxicity, J. Immunol. 125: 501 (1980).

58. R. A. Rifkind, Heinz body anemia: an ultrastructural study. II. Red cell sequestration and destruction, Blood 26: 433 (1965).

A POTENTIAL, INTRACELLULAR TRIGGER FOR REMOVAL OF SENESCENT

ERYTHROCYTE: HEMOGLOBIN WITH METHIONINE BETA (55) D6

OXIDIZED TO THE SULFOXIDE DERIVATIVE

Gino Amiconi, Alberto Bertollini,
Rosa Maria Matarese and Donatella Barra

Department of Biochemical Sciences
University of Rome "La Sapienza"
and the CNR Center of Molecular Biology
Rome, Italy

INTRODUCTION

Most of the prevalent theories of aging are working hypotheses designed to measure and explain various age-related changes. The basic assumption of the theories commonly referred to as stochastic is that cellular aging may be due to cumulative and deteriorative effects of random events, such as the chemical damages due to free radicals (1). In recent years (since the discovery of superoxide dismutase), oxygen radicals have been invoked to favor debilitating processes, including aging of molecules and cells (1,2). Although oxidation of thioesther groups has long been known in biochemistry, the importance of methyonyl residues as possible targets of oxygen radicals has only recently been recognized (3). Methionyl residues in proteins and peptides can readily be oxidized also _in vitro_ to the sulfoxide derivative by suitable chemical agents, under mild conditions, which do not modify other residues except exposed sulfhydryl groups. Therefore, in view of the effect of the oxidation of methionine in several biological processes (such as aging), it seemed of interest to investigate the functional properties of human hemoglobin with some methionyl groups oxidized by chloramine T.

RESULTS

Human hemoglobin contains 3-methionines per alpha-beta dimer at positions alpha(32)B13, alpha(76)EF5, and beta(55)D6, the latter being located at the alpha1-beta1 subunit interface (4).

Red Blood Cell Aging, Edited by M. Magnani and
A. De Flora, Plenum Press, New York, 1991

Oxidation of methionyl groups was obtained by adding chloramine T (purchased from Fluka AG, Switzerland) in H_2O to a solution ($5x10^{-4}$ M in heme) of oxyhemoglobin or deoxyhemoglobin.

Chemical characterization of oxidized human hemoglobin clearly shows that mild treatment of this protein with chloramine T (5:1 molar ratio of reagent to methionyl residues for 3-5 min at 25°C) modifies almost exclusively methionyl and cysteinyl residues. Evidence to support this statement is from the following experiments: (i) amino acid analysis under standard conditions; (ii) tryptic peptide maps obtained from the separated chains of treated and untreated hemoglobin; (iii) analysis of amino- and carboxyl-terminal groups of the alpha and beta chains, that yields 100% recovery with respect to control; (iv) spectroscopic determination of tryptophan; (v) titration of accessible sulfhydryl groups in the treated protein shows (in accord with amino acid analysis) partial to complete absence of freely titrable cysteinyl residues (at position beta (93)F9).

In order to identify the methionyl residues oxidized after partial modification of the protein under mild experimental conditions, the product of the reaction was treated with CNBr, and the resulting peptide mixture was subjected to automated Edman degradation.

The results clearly indicate the extent of successful cleavage at the various methionyl residues in the reacted protein and, conversely, the oxidation state of some of the methionines. By comparing the yields of the expected residues following each methionine on each subunit with those of the aminoterminal sequences at the two chains (see Table 1), one can conclude that, after mild exposure to chloramine T, Met-beta(55)D6 is completely oxidized to sulfoxide, Met-alpha(32) B13 appears to be intact, and Met-alpha(76)EF5 can be modified to a different extent depending on the sample (30% in that reported in Table 1).

In 0.1M Tris pH 7.3, human hemoglobin with either 2 (i.e., Met-beta(55) D6) or 4 (i.e., Met-beta(55)D6 and Met-alpha(76)EF5) oxidized methionines has a very high oxygen affinity and is noncooperative. However, the value of oxygen affinity (in terms of P_{50}, the oxygen partial pressure at which one half of hemes is saturated with the gaseous ligand) is significantly different when the oxidized derivative is prepared starting from deoxyhemoglobin or from oxyhemoglobin, the latter derivative showing (see in figure 1) higher oxygen affinity (P_{50} =0.47 torr) than the former (P_{50} =1.1 torr). Moreover, the addition of inositol hexakisphosphate in 10-fold excess over tetramer results (see figure 1) in a decrease of oxygen affinity in both cases (P_{50} =1.8 torr), although the binding isotherm is always noncooperative (i.e., the value of the slope of the curves, reported in figure 1 and usually indicated by n, is equal 1). At any rate, the change in oxygen binding properties upon mild treatment with chloramine T is dramatic since, under the same experimental conditions (pH 7.3 with 2.5 mM inositol hexakisphosphate at 25°C), the P_{50} for normal human hemoglobin is 15.8 torr and n is equal to 2.4. Moreover, the experimental results indicate that the Bohr effect is completely abolished.

Table I. Automated Edman Degradation of CNBr-treated Human Hemoglobin (20 ug) After Mild Treatment With Chloramine T. Aminoterminal amino acid sequences of the peptides obtained after CNBr treatment of human hemoglobin: alpha1, Val-Leu-Ser-Pro; alpha33, Phe-Leu-Ser-Phe; alpha77, Pro-Asn-Ala-Leu; beta1, Val-His-Leu-Thr; and beta56, Gly-Asn-Pro-Lys.

Sequence cycle	Phenylthiohydantoin-derivative	Yield
		pmol
1	Val	521
	Phe	246
	Pro	84
	Gly	56
2	Leu	776
	Asn	40
	His	125
3	Ser	440
	Ala	53
	Leu	234
4	Pro	271
	Phe	235
	Leu	63
	Thr	251

In contrast to what is reported above, hemoglobin oxidized under more drastic conditions (e.g. 10:1 molar ratio of chloramine T to methionyl residues and 10 min incubation) shows (i) absence of unreacted thioether groups (i.e., all six methionines per tetramer are oxidized to sulfoxide) and (ii) optical spectra and molecular mass reminiscent of those typical of hemicromes.

Other experiments have been performed to control possible effects of the oxidation of cysteinyl groups at position beta93 on functional properties. Human oxyhemoglobin, previously reacted with iodoacetamide or p-hydroxymercuribenzoate (reagents which block thiols at position beta(93)F9 but do not significantly affect the functional properties of the protein) was oxidized with chloramine T. Oxygen equilibria of this doubly modified protein are superimposable to those determined starting with native hemoglobin (measured in 0.1 M Tris pH 7.3 at 25°C).

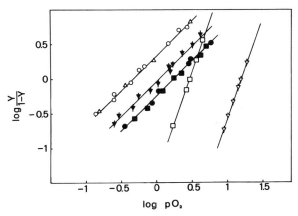

Fig. 1. Oxygen equilibrium (Hill plot) of human hemoglobin reacted with chloramine T.
Hemoglobin oxidized as oxy derivative: △, two oxidized methionyl groups/tetramer; O and ●, four oxidized methionyl groups/tetramer in the absence and presence of 2.5 mM inositol hexakisphosphate, respectively. Hemoglobin oxidized as deoxy derivative: ▼ and ■, four oxidized methionyl groups/tetramer in the absence and presence of 2.5 mM inositol hexakisphosphate, respectively. Normal human hemoglobin, in the absence (□) and presence (▽) of 2.5 mM inositol hexakisphosphate, is included for comparison. Experimental conditions were 0.1 M Tris (pH 7.3) + 0.1 M NaCl at 25°C; protein concentration was 5 mg/ml.

DISCUSSION

The results presented above clearly show that oxidation to sulfoxide of Met-beta(55)D6 (located at the alpha1-beta1 interface) in human hemoglobin leads to dramatic changes in functional and structural properties of the molecule. In particular, such hemoglobin derivative behaves as a high affinity, noncooperative molecule, reminiscent of the isolated chains (5) or carboxypeptidase A-digested hemoglobin (6). It may be recalled that, in the latter case, the addition of polyphosphate leads to a partial restoration of cooperativity, which is not apparent from the results on the oxidized derivative here reported. On the other hand, oxidized hemoglobin is more similar to the isolated beta chains than to the alpha subunits, the former showing somewhat different oxygen affinity depending on solvent composition (7). In spite of these considerations, it should not be forgotten that, in the presence of inositol hexakisphosphate, the oxygen affinity of the oxidized hemoglobin derivative is at least 10 times greater than that of native hemoglobin and the binding isotherm is noncooperative. Therefore, its functional properties appear to be due to stereochemical changes which completely destabilize the low affinity (T) conformation relative to the high affinity (R) structure, the starting point of such an effect being localized at the alpha1-beta1 subunit interface. Depending on the initial conditions (reaction with oxy or deoxy hemoglobin), somewhat different packing of alpha1 against beta1 may be achieved to account for the differences in oxygen affinity between the molecules obtained by treating oxy- or deoxy- derivatives with chloramine T.

The destabilization of the T-state implies that oxidation of Met-beta(55)D6 causes a partial disruption in the alpha1-beta1 contact and can therefore constitute the structural basis for a decreased solubility of the protein. In this connection, it is interesting to note that eleven out of the thirteen reported hemoglobins substituted in beta chain at the alpha1-beta1 interface (8) result in a distinct instability and, just as for the reported oxidized derivative, have high oxygen affinities, low cooperativities and low alkaline Bohr effect (in contrast, only one of the seven known hemoglobins with alpha chain substitutions at the same intersubunit contact is unstable). In particular, abnormal hemoglobin Matera beta (55)D6Met-Lys, showing a replacement at the same position as the present artificially-prepared derivative, has a decreased stability in heat and isopropanol test; in addition, numerous red blood cells of the carrier contained inclusion bodies (9).

In conclusion, within the framework of a general error theory of aging, one can speculate that (i) the amount of methionine sulfoxide in hemoglobin increases with the erythrocyte age (as observed for proteins of erythrocyte membrane, (10), since red blood cells appear to be lacking in methionine sulfoxide reductase; and consequently, (ii) due to its higher instability (which depends on the number of methionine sulfoxides in the tetramer and therefore on age), such hemoglobin derivative precipitates on band 3, thus possibly triggering the molecular mechanism (11) of erythrocyte removal in normal senescence.

ACKNOWLEDGEMENTS

This work was supported by grants of the Italian Ministry of Education (Ministero della Pubblica Istruzione) and the Italian Research Council (Consiglio Nazionale delle Ricerche).

REFERENCES

1. D. Harman, The aging process, Proc. Natl. Acad. Sci. USA 78:7124 (1981).
2. L. K. Obukhrova and H. M. Emanuel, The role of free radical oxidation reactions in the molecular mechanisms of the ageing of living organisms, Russian Chem. Rev. 52:201 (1983).
3. N. Brot and H. Weissbach, Biochemistry and physiological role of methionine sulfoxide residues in proteins Arch. Biochem. Biophys. 23:271 (1983).
4. G. Fermi and M. F. Perutz, Atlas of molecular structures in biology. 2. Haemoglobin and myoglobin, Clarendon Press, Oxford (1981).
5. E. C. De Renzo, C. Ioppolo, G. Amiconi, E. Antonini and J. Wyman, Properties of the α and ß chains of hemoglobin prepared from their mercuribenzoate derivatives by treatment with 3-dodecanethiol, J; Biol. Chem. 242:4850 (1967).
6. J. Bonaventura, C. Bonaventura, B. Giardina, E. Antonini, M. Brunori and J. Wyman, Partial restoration of normal functional properties in

carboxypeptidase-A- digested hemoglobin, Proc. Natl. Acad. Sci. USA 69:2174 (1972).

7. J. Bonaventura, C. Bonaventura, G. Amiconi, L. Tentori, M. Brunori and E. Antonini, Allosteric interactions in non-α chains isolated from normal human hemoglobin, and hemoglobin Abruzzo 143(H21)His Arg , J. Biol. Chem. 250:6278 (1975).

8. International Hemoglobin Information Center, IHIC Variant List, Hemoglobin 9:229 (1985).

9. G. V. Sciarratta and G. Ivaldi, Hb Matera 55(D6)Met-Lys : A new unstable hemoglobin variant in an Italian family, Hemoglobin 14:79 1990).

10. A. Brovelli, C. Seppi, G. Pallavicini and C. Balduini, Membrane processes during in vivo aging of human erythrocytes, Biomed. Biochim. Acta 42:S122.

11. P. S. Low, Structure and function of the cytoplasmic domain of band 3: consequences for erythrocyte membrane-peripheral protein interactions, Biochim. Biophys. Acta 864:145 (1986).

12. H. V. Lutz, F. Bussolino, R. Flepp, S. Fasler, P. Stammler, M. D. Kazatchkine and P. Arese, Naturally occurring anti-band 3 antibodies and complement together mediate phagocytosis of oxidatively stressed human red blood cells, Proc. Natl. Acad. Sci. USA 84:7368 (1987).

UBIQUITIN-MEDIATED PROCESSES IN ERYTHROID CELL MATURATION

Arthur L. Haas

Department of Biochemistry
The Medical College of Wisconsin
Milwaukee, Wisconsin U.S.A.

INTRODUCTION

The terminal differentiation of erythroblasts during erythrocyte maturation entails significant cellular remodeling. In avian cells this is accomplished in part by marked attenuation of nuclear transcription and in mammalian cells by the physical extrusion of the nucleus. Thus lacking the ability to replace all but those proteins required for maintenance of the mature erythrocyte, the normal complement of cellular constituents is subsequently modified by a highly active degradative mechanism(s) to yield a sub-population of proteins stable to such proteolysis. During this time many metabolic pathways are shunted by selective turnover of key enzymes. The enhanced degradation essential to erythroid cell maturation is assumed to involve the same ATP, ubiquitin-dependent multi-enzyme pathway responsible for cytosolic protein turnover within all eukaryotes. The mechanism(s) required to commit erythroblasts to enhanced degradation and to direct the resulting selective degradation of key enzymes provides a tractable model for the less acute regulation observed within nucleated cells. Characterization of the ATP, ubiquitin-dependent pathway in erythroid cells and recent observations in other cells and tissues subject to enhanced degradation in response to various experimental manipulations provides some understanding of the dynamics exhibited by this system during terminal differentiation.

DINAMICS OF UBIQUITIN POOLS WITHIN INTACT CELLS

Protein turnover via the ATP, ubiquitin-dependent pathway proceeds through degradative intermediates generated by the ATP-coupled covalent conjugation of ubiquitin to susceptible target proteins (1,2). First demonstrated in cell-free extracts of rabbit reticulocytes (3,4), this unique post-translational modification has subsequently been shown to be a general feature of all eukaryotic cells (5). During the decade since originally proposed, considerable evidence from diverse experimental

Red Blood Cell Aging, Edited by M. Magnani and
A. De Flora, Plenum Press, New York, 1991

approaches has supported the general validity of the <u>conjugate hypothesis</u> for energy-dependent degradation. In addition, mechanistic details obtained from reticulocyte extracts (6) and recognition of the acute susceptibility of ubiquitin to selective proteolytic inactivation (7) has allowed establishment of model ATP, ubiquitin-dependent degradative systems in cell-free extracts of rabbit liver and muscle (7,8).

If ubiquitin-protein conjugates represent degradative intermediates that flux through the ATP, ubiquitin-dependent pathway should be proportional to the concentration of these adducts (5,9). This prediction is generally supported by a variety of studies in different cells (1,2). However, direct measurements of free and conjugated ubiquitin pools within intact cells and ^{3}H/^{14}C pulse chase studies of ubiquitin adducts within cultured human fibroblasts have revealed a significant population of proteolytically stable ubiquitin-protein conjugates (5,9,10). The apparent discrepancy between proteolytically labile and stable conjugate pools has been reconciled by proposing that these adducts partition between alternate fates of degradation and disassembly, the latter process involving the action of isopeptidase(s) to specifically cleave the ubiquitin-protein linkage (5,11). The dynamics of intracellular ubiquitin pools can be formally described by a minimum kinetic model

$$\text{Protein} + \text{Ub} \underset{K_{dis}}{\overset{K_{form}}{\rightleftharpoons}} \text{Protein-Ub} \xrightarrow{K_{deg}} \text{Amino acids} + \text{Ub}$$

in which K_{form} represents the rate constant for ubiquitin-protein conjugate formation K_{dis} the rate constant for conjugate disassembly, and K_{deg} the rate constant for the limiting step of conjugate degradation (11).

Table 1. ATP, Ubiquitin-Dependent Proteolysis
Within Rabbit Reticulocyte Extracts[a]

	TCA soluble radioactivity (cpm)
−ATP	7,380
+ATP	6,790
+ATP,+Ub	111,000
+ATP,+rmUb	15,400
+ATP,+UbK48R	14,000

Degradation of ^{125}I-carboxy methylated bovine serum albumin (0.2 µM) was measured in duplicate for 1 hour at 37°C in incubations of 50 µl containing 50 mM TRIS-Cl (pH 7.5), 10 mM MgCl$_2$, 1 mM DTT, and 240 µg of Fraction II (14). Where indicated incubations also contained 2 mM ATP, 10 mM creatine phosphate, 1 IU/ml creatine phosphokinase, and 5 µM of either native ubiquitin (Ub), reductively methylated ubiquitin (rmUb), or a mutant ubiquitin for which lysine-48 has been changed to arginine (UbK48R).

The K_d for ATP in ubiquitin activation (0.5 μM) (12) requires the step of ligation to be satured with respect to the trinucleotide under all but the most extreme conditions so that ATP can be grouped into the term for K_{form}. Similarly, rates of conjugation are satured with respect to free ubiquitin since measurements of ubiquitin pools within intact cells and tissues are several fold above the K_d for free ubiquitin in the activation step (0.6 μM) (12).

The model does not preclude a role for target protein selectivity in the step of conjugate formation but emphasizes that proteolytic specificity reduces to a kinetic argument dependent on the relative magnitudes of K_{deg} versus K_{dis} for a given conjugate species. Therefore, the empirical ratio of K_{deg}/K_{dis} will be larger for proteolytically labile conjugates than for those adducts not subject to degradation. Recent studies demonstrate that multiubiquitination of target proteins, involving formation of extended homopolymer conjugates of ubiquitin specifically linked through lysine-48 of the polypeptide, manifests large ratios of K_{deg}/K_{dis} (13). The contribution of multi-ubiquitination to degradation of a model substrate by reticulocyte Fraction II can be seen in Table 1. Little stimulation in the initial rate of ^{125}I-cmBSA degradation over basal proteolysis (-ATP) is observed in the presence of ATP alone; however, addition of ubiquitin effects a 16-fold stimulation in degradation. Nearly all of the latter increase requires multi-ubiquitination at lysine-48 of the polypeptide since rates of degradation are greatly attenuated when supported by either rmUb of UbK48R, both of which are incapable of homopolymer formation (15). Separate rate studies have shown that all three forms of ubiquitin are kinetically indistinguishable in the steps of ubiquitin activation and mono-adduct formation (15).

Although ubiquitin conjugation appears to be relatively non-selective, based on the broad distribution of these adducts (3-5,10) certain features of specific target proteins have been identified that favor degradation. Abnormal protein conformation arising from amino acid modification and/or denaturation (16,17) as well as the identity of the amino terminal residue (the N-end rule) (13,18) have both been proposed to commit target proteins to degradation via multi-ubiquitination. Target protein specificity in both models assumes a central role for isopeptide ligase (E3), the terminal enzyme within the conjugation pathway, for the majority of basal ubiquitin-supported degradation. However, other observations support E3-independent mechanisms of multi-ubiquitination catalyzed by additional components of the conjugation cascade, notably $E2_{32K}$ of reticulocytes and the CDC34 gene product of yeast, that may have function-specific roles in initiating the degradation of particular proteins (15,19). Therefore, commitment to degradation via the ATP, ubiquitin-dependent system appears not to involve a single mechanism but rather represents the sum of contributions from multiple pathways.

MECHANISMS OF ENHANCED ATP, UBIQUITIN-DEPENDENT DEGRADATION

Measurements of ubiquitin pools within cultured cells and intact tissues subject to enhanced degradation have revealed well-conserved temporal dynamics between free and conjugated polypeptide (5). Increased rates of cellular protein degradation are consistently accompanied by a transient accumulation of ubiquitin conjugates (5), as predicted by the role of these adducts as degradative intermediates. In intact cellular systems examined to date, the accumulation of conjugates is paralleled by stimulation of ubiquitin synthesis, presumably to provide adequate polypeptide for the expanding adduct pool. The causal relationship between increased pools of free ubiquitin and protein conjugates is uncertain; that is, does accumulation of free ubiquitin drive the observed increase in the conjugate pool or does ubiquitin induction occur in response to an expanding conjugate pool. What is certain is that cells possess a homeostatic mechanism designed to maintain a constant ratio between the two pools, the set point, under normal conditions (5,11,20,21). Comparison of the dynamics of ubiquitin pools in cells subject to enhanced degradation provides insight into the general mechanism for up-regulation of flux through the ATP, ubiquitin-dependent pathways during erythrocyte maturation.

Cultured cells subjected to chemical or thermal stress exhibit increased pools of conjugated ubiquitin that are consistent with the expected increase in proteins of non-native structure following the environmental insult (5). This effect is equivalent to an increase for the protein term in the minimum kinetic model. Response of ubiquitin pools to stress has been most carefully examined in cultured chicken embryo fibroblasts for which temporal changes in free and conjugated ubiquitin levels are highly predictable (20). Within the first 30 minutes following shift up in temperature from 37 to 42 °C ubiquitin conjugates increase at the expense of the free pool without noticeable change in the total content of ubiquitin, leading to a net increase in the fractional level of conjugation within these cells relative to the homeostatic set point characteristic of non-stressed cells. Evidently the abrupt increase in proteins susceptible to conjugation is capable of temporarily overwhelming the normal homeostatic mechanism regulating the balance between free and conjugated ubiquitin. Levels of the stress-inducible poly-ubiquitin gene also increase in response to the temperature shift (20,22). Subsequent translation of the poly-ubiquitin mRNA during 30-60 minutes of heat stress results in a significant increase in total ubiquitin, a proportional increase in both free and conjugated ubiquitin pools, and a progressive re-establishment of the homeostatic set point for fractional conjugation (20). These latter trends continue for at least an hour following recovery of the cells at 37°C (22). Enhanced rates of degradation are only observed during the recovery period in chicken embryo fibroblasts (20).

Similar patterns of ubiquitin pool changes occur in rat skeletal muscle subjected to various forms of experimentally induced atrophy. The absolute extent and duration of these transient shifts in free and conjugated ubiquitin levels are roughly proportional to the severity of muscle atrophy. During relatively mild suspension-induced disuse atrophy, free and conjugated ubiquitin slowly increase in parallel then decline to control levels over the 14 days required for the affected muscles to adjust to a new steady state level of protein balance (21). These changes in ubiquitin pools are sufficiently gradual to allow maintenance of the homeostatic set point within muscle fibers during the transient enhanced degradation.

Immunohistochemical localization of ubiquitin conjugates reveals that this is a stochastic process in which individual muscle fibers respond independently rather than in a concerted fashion. More acute induction of muscle atrophy in response to either starvation or denervation results in a proportionally more dramatic and shorter time frame for the transient increase in ubiquitin conjugate levels, although the temporal relationship of ubiquitin pools is qualitatively similar to that observed in disuse atrophy (S. Wing, A. Haas, and A. Goldberg, in preparation).

The most dramatic example of enhanced degradation resulting from increases in ubiquitin conjugate pools occurs during the programmed cell death of the intersegmental muscle (ISM) of the tobacco hornworm moth Manduca sexta during metamorphosis (A. Hass, B. Williams, and L. Schwartz, in preparation). Developmental enhanced degradation within the ISM begins at 2 days pre-eclosion (23). Within 28 hours post-eclosion the ISM ceases to exist as a morphologically distinguishable structure within the insect. Schwartz and coworkers have shown that the onset of enhanced degradation within the ISM is characterized by induction of poly-ubiquitin mRNA (23). More detailed biochemical measurements have recently shown that poly-ubiquitin synthesis is accompanied by a significant increase in both the ubiquitin conjugate pool and the fractional level of conjugation within ISM. Enhanced ISM conjugation results from multi-factorial regulation involving polyubiquitin gene induction, increased activity of enzymes responsible for ubiquitin conjugation principally the ubiquitin activating enzyme (E1) and the ubiquitin carrier protein responsible for E3-dependent conjugation ($E2_{14k}$) (15), and induction of a binding protein that counters the effect of a constitutive allosteric inhibitor of conjugation.

Enhanced degradation during erythroid cell maturation is functionally analogous to the developmentally-programmed degradation observed within ISM; however, important foundamental differences exist. Induction of poly-ubiquitin gene expression observed in the above three systems does not occur during erythroid maturation. Pulse labeling of intact cells demonstrates that basal rates of ubiquitin synthesis in anucleate reticulocytes is negligible compared to the substantial rates observed in nucleated cultured cells (24). The difference in ubiquitin synthetic rates between reticulocytes and cultured cells is a direct consequence of the marked stability of ubiquitin within erythroid cells (9). While ubiquitin is a relatively short-lived protein within cultured cells, with a half life ranging from ca 10 hours in HeLa to 28 hours in IMR-90 fibroblasts (11), the ubiquitin contained in reticulocytes is stable for the life of the cell (5,9). This differential stability indicates that the short half life of ubiquitin is not a consequence of its role in protein degradation. However, the short half life is function of ubiquitin-protein conjugation since bovine thymus ^{125}I-ubiquitin (shown to be inactive due to proteolytic cleavage of the carboxyl terminal glycine dipeptide during isolation) (7) microinjected into IMR-90 has a half life of 320 hours (25). The pathway for ubiquitin turnover in cultured cells in unknown but may involve either proline endopeptidase (26) or lysosomal micro-autophagy (27).

It is unlikely that the onset of enhanced degradation via the ATP, ubiquitin-dependent pathway is initiated by induction of ubiquitin synthesis in erythroblasts, as occurs during ISM degradation, since the total ubiquitin content in rabbit reticulocytes and erythrocytes is comparable (40 and 38 pmol/10^6 cells, respectively) (9) and within the range observed for a variety unrelated cell lines (ca 90 pmol/10^6 cells) (5). In addition, the total ubiquitin content in rabbit erythroid cells is similar to that observed in human Friend erythroleukemia cells (47 pmol/10^6 cells) (9), a partially differentiated human erythroid precursor. Although synthesis of ubiquitin is negligible in erythroid cells, they do undergo an increase in total ubiquitin concentration as a consequence of the decrease in absolute cell volume accompanying terminal differentiation (9). The net effect of this reduction in cell size is that reticulocytes possess an intracellular concentration of total ubiquitin that is ca 5-fold higher than the 10-20 μM typically present in other cell lines and intact tissue (5). The intracellular concentration of ubiquitin in erythrocytes is 2-fold higher than reticulocytes because of an additional proportionate reduction in cell volume. At present there is no experimental evidence for a developmentally-linked induction of ubiquitin conjugating enzymes; however, reduction in cell volume will also result in an increase in absolute concentration.

ATP, UBIQUITIN-DEPENDENT DEGRADATION DURING ERYTHROID CELL MATURATION

Most of our knowledge concerning the ATP, ubiquitin-dependent degradative pathway derives from in vitro studies using cell-free rabbit reticulocyte extracts, and to a lesser extent from complimentary observations employing erythrocyte extracts. While this approach has been of considerable value in understanding general features of this degradative pathway, particularly the enzymology of individual steps, there are inherent limitations associated with interpreting data obtained from such samples. Recent evidence suggests that previously unrecognized conjugating enzymes may be lost during the preparation of Fraction II extracts. Since the ubiquitin-dependent degradative system displays a marked preference for the conjugation and degradation of proteins possessing non-native conformations, the effect of unavoidable protein denaturation on functioning of the degradative pathway, is uncertain. Finally, the ATP, ubiquitin-dependent degradative pathway is notably, labile in cell-free extracts. Recent studies have shown that this instability is attributable to a component affecting the degradation of multi-ubiquitin conjugates. Loss of this putative factor results in a progressive decline in the apparent fraction of total ATP-stimulated degradation contributed to by the ubiquitin-dependent pathway (A. Haas and P. Reback, unpublished observation).

Two experimental approaches have been employed for probing the pathway of ATP, ubiquitin-dependent degradation within intact reticulocytes and erythrocytes. Direct, absolute quantitation of free and conjugated ubiquitin pools within intact erythroid cells has provided information on the dynamics of conjugate partitioning in response to various experimental manipulations. Pulse-chase studies have allowed measurements of net degradative flux and correlations with label incorporation into the conjugate pool.

Table 2. Absolute Ubiquitin Pools Within Rabbit Reticulocytes and Erythrocytes[a]

| | Ubiquitin (± S.E.) nmol/ml packed cells | |
	Reticulocytes	Erythrocytes
Free ubiquitin	3.8±0.6	29.3±0.4
Conjugated ubiquitin	19.0±0.7	13.1±0.1
Total ubiquitin	22.8±0.9	42.4±0.4

[a] Used with permission from reference 9.

In nucleated cells a significant fraction of label incorporated into the conjugate pool during even relatively short pulse times is attributable to the ubiquitin moiety because of the rapid rate of ubiquitin synthesis required by its normally short half life (10). Potential ambiguities in interpreting such pulse studies due to contributions from ubiquitin-associated label has been discussed in detail elsewhere (5).

The fraction of total ubiquitin present within the conjugate pool declines during maturation from 83% in reticulocytes to 31% in erythrocytes, Table 2. The difference in absolute concentration of these degradative intermediates, approximated by their content per packed cell volume, is somewhat less bacause of a progressive reduction in cell size during the terminal stages of differentiation. The subcellular distribution of ubiquitin conjugates also changes during erythrocyte maturation (9). The stromal fraction pelleting at 22000 x g accounts for approximately 25% of the total conjugate pool of reticulocytes while less than 10% of this subfraction remains in mature erythrocytes. These quantitative results are consistent with earlier work by Rapoport et al. (28) demonstrating the marked degradation of stromal proteins by the ATP, ubiquitin-dependent pathway of reticulocytes. The decrease in total ubiquitin conjugates between reticulocytes and erythrocytes parallels the characteristic decline in degradative capacity of erythroid cells during terminal differentiation. Absence of stromal conjugates in erythrocytes suggests proteins subject to conjugation have been lost through degradation. The latter observation implies that lower levels of intracellular erythrocyte conjugates result in part from limiting substrates for adduct formation.

On Western blots the distribution of conjugate bands between reticulocytes and erythrocytes in qualitatively similar (Fig. 1, lanes 1 and 3), although the absolute amount of these adducts is substantially greater in reticulocytes as predicted from the quantitative data of Table 2. This indicates that reticulocytes and erythrocytes contain a similar substrate pool. However, the consistently observed skewing of conjugates to higher molecular weights in reticulocytes suggests a greater proportion of these

adducts exists as multiple ubiquitin conjugate. Kinetic studies of
degradative rates in the presence of hemin, a specific inhibitor of
conjugate degradation, show that the high molecular weight conjugates
present in reticulocytes are degradative intermediates while those of lower
molecular weight are relatively stable to proteolysis (14). This difference
in conjugate distribution is not an artifact of ongoing proteolysis within
quenched cell extracts since control studies in which the extract is allowed
to set on ice for extended periods of time yield the same distribution of
adducts, lanes 2 and 4 of Fig. 1. Nor are the conjugates present in
erythrocytes merely residual adducts remaining after loss of one or more of
the ligation enzymes since the conjugates can be quantitatively converted to
free ubiquitin by ATP depletion of intact cells in the presence of
2-deoxyglucose, (7,14). The same pattern of erythrocyte conjugates can be
re-established by repletion of ATP in these cells. Therefore, the ubiquitin
pool dynamics represented by the minimum kinetic model appears to be present
in both reticulocytes and erythrocytes.

By directly monitoring amino acid release, Goldberg and Boches have
shown that energy-dependent degradation within erythrocytes is ca 15-fold
lower than that of reticulocytes (29). In contrast, the erythrocyte
ubiquitin conjugate pool is only 32% lower than that measured within
reticulocytes, Table 2. We have previously noted that this discrepancy
suggests that erythroid cells contain pools of proteolytically labile and
stable conjugates, and that only the latter pool remains within erythrocytes
(5,9). This conclusion agrees with the proposal of Goldberg and Boches that
the lower degradative flux in erythrocytes results from limiting substrate
for degradation. Rates of erythrocyte energy-dependent degradation can be
stimulated to levels approaching those observed in reticulocytes following
in situ denaturation of erythrocyte hemoglobin by incubation with 1 mM
phenylhydrazine (29). Such treatment results in significant induction of a
pattern of ubiquitin conjugates within erythrocytes (9) (Fig. 1, lanes 5 and
6) that are the presumed substrates for the observed stimulation in
degradation. The conjugate bands induced by phenylhydrazine denaturation are
also present but at greatly attenuated levels within untreated erythrocytes
and reticulocytes, Fig. 1. These induced conjugate bands are probably to
hemoglobin since their molecular weight distribution is similar to that
reported by Chin et al. for labeled hemoglobin conjugates in similar
experiments with intact HeLa cells (30). In the latter study, the steady
state level of the induced hemoglobin conjugates was directly proportional
to the rate of energy-dependent degradation exhibited by the cells.

The above results are consistent with a model in which ubiquitin is
continuously ligated to and disassembled from solvent accessible primary
amines present on cytosolic proteins. Steric accessibility of ligation sites
probably contributes to proteolytic specificity through the partitioning
ratio of K_{deg}/K_{dis}. Ubiquitin adducts within erythrocytes must be relatively
stable to degradation, or turnover at very low rates. Structural
perturbation apparently results in enhanced steric accessibility of ligation
sites for the conjugating enzymes as found with phenylhydrazine treatment. A
similar situation occurs with native Dictyostellium calmodulin for which
only the single solvent accessible lysine-115 is conjugated (32). Disruption
of the calmodulin structure results in enhanced conjugation and significant
degradation.

198

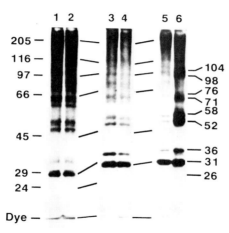

Fig. 1 Immunochemical detection of ubiquitin conjugates. Reticulocyte (lanes 1 and 2) and erythrocyte (lanes 3-6) extracts were resolved by SDS-PAGE and visualized by Western blot immunodetection using ubiquitin-specific antibodies and ^{125}I-protein A. Lanes 1 and 3 represent extracts quenched with SDS sample buffer immediately after lysis. Lanes 2 and 4 are for extracts allowed to stand on ice for 3 hours prior to quench. Lanes 5 and 6 show the change in conjugate pattern present in intact erythrocytes before and after, respectively, in situ hemoglobin denaturation by addition of 1 mM phenylhydrazine. Numbers to the left represent molecular weight markers in kDa. Numbers to the right represent molecular weights for prominent conjugate bands induced by phenylhydrazine. (Data taken by permission form reference 9).

Additional evidence for stable and labile pools of ubiquitin conjugates derives from pulse chase studies with reticulocytes (24). Unambiguous interpretation of pulse studies within reticulocytes is unique to these cells because of the inherently low rate of ubiquitin synthesis. If intact reticulocytes are pulse labeled with $[^3H]$ leucine for 15 min and the endogenous conjugates subsequently isolated with ubiquitin-specific polyclonal antisera, ca 0.6% of the total incorporated radioactivity is present within the immunoreactive fraction (24). When the experiment is repeated in the presence of the lysine analog 4-thialysine to produce globin of abnormal structure, ca 10-fold increase in both degradation and label present within the conjugate pool is observed. If the time course for such an experiment is followed, label incorporation into the conjugate pool parallels and is roughly proportional to incorporation into total protein (principally hemoglobin since this protein represents the bulk of synthesis), Fig. 2. When the data are expressed as the percent of total label present in the conjugate pool (% immunoreactive), it is apparent that label rapidly equilibrates with conjugates, consistent with the steady state partitioning of this pool (Fig. 2, panels A and B).

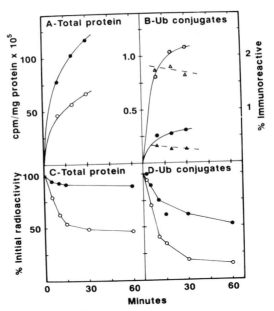

Fig. 2. Dynamics of reticulocyte total protein and conjugate pools during pulse-chase. See text for discussion of the experiment. Closed and open symbols are for control and thialysine cells, respectively. Circles represent [³H] label incorporated into protein. Triangles represent the percent of total incorporated label immunoreactive with ubiquitin-specific antisera. (Adapted with permission from reference 24).

In parallel studies we have quantitated ubiquitin pools and find no significant difference in conjugate levels between control and thialysine-treated reticulocytes even after three hours (14.9±0.5 versus 14.2±0.1 nmol/ml packed cells, respectively). The absence of a significant increase in the conjugate pool suggested that the structural perturbation of globin resulting from incorporation is much more subtle than the in situ denaturation by phenylhydrazine, even though both effect a significant increase in degradation.

The pulse study of Fig. 2 demonstrates that significant conjugation of nascent globin chains occurs even in the absence of an amino acid analog. The fate of these labeled conjugates following cycloheximide block has also been examined by chase studies after 15 min of [³H] leucine incorporation, panels C and D of Fig. 2 (24). Degradation of total protein follows exponential biphasic kinetics having a faster rate and larger absolute extent for analog-containing cells (panel C). Chase of label from the conjugate pool parallels the biphasic response of total protein (Fig. 2). Similar biphasic kinetics for conjugate loss have observed in reticulocyte extracts following ATP depletion. In neither control nor thialysine-containing cells does label quantitatively chase from the conjugate pool (panel D). This indicates that a fraction of the conjugate pool represents proteolytically stable adducts, consistent with conclusions from quantitative pool studies. Approximately half the pool of control conjugates is stable, a fraction similar to that predicted for the reticulocyte pool after subtracting the contribution from adducts to stromal proteins (9). Analog-containing conjugates represent a proportionately larger fraction of labile conjugates (panel D), even though the absolute size of the conjugate pool is identical to that of control cells (above).

In summary the dynamics of ATP, ubiquitin-dependent degradation within these cells are similar to those found in nucleated cells. The steady-state dynamics of erythroid ubiquitin pools suggest that protein substrates compete for ubiquitin ligation. Under conditions leading to the generation of proteins having abnormal structure, these substrates must compete more effectively for ligation relative to proteins of native structure. The proteolytically stable pool of conjugates thus probably represents adducts to proteins of native structure while the proteolytically labile pool likely results from conjugation to proteins of aberrant structure. In addition to mediating the degradative remodeling of these cells during erythrocyte maturation, the system is probably also responsible for maintaining the normal balance between α and ß subunits of globin for correct hemoglobin assembly. This conclusion is supported by studies showing the enhanced conjugation of globin chains in thalassemias (33). The intracellular concentration of the prosthetic heme group appears reciprocally to regulate the balance between rates of subunit synthesis via the heme-sensitive protein kinase and proteolysis via modulation of conjugate degradation (14).

DECAY OF ATP, UBIQUITIN-DEPENDENT DEGRADATION DURING ERYTHROID MATURATION

Reticulocyte maturation is characterized by a progressive loss of degradative capacity to prevent further degradation of cellular constituents required during the life span of the mature cell. Attenuation of the ATP, ubiquitin-dependent pathway appears to be a consequence of two separate mechanisms. Initially, cellular remodeling results in a pool of proteins, also present in reticulocytes, that is subject to limited ubiquitin ligation but whose conjugates are proteolytically stable. Under these substrate-limiting conditions, the observed rates of degradation are significantly lower than those present in reticulocytes. Enhanced rates of degradation can be re-established if the stable pool of proteins is denatured, such as by incubation with phenylhydrazine (Fig. 1) (9,20).

Specific immunohistochemical detection of ubiquitin conjugates present in mature erythrocytes reveals a second mechanism that irreversibly blocks ATP, ubiquitin-dependent degradation at the step of ubiquitin ligation. Using conjugate-specific affinity-purified antibodies, we have shown immunohistochemically that erythrocytes contain lower steady-state levels of conjugates compared to reticulocytes, as expected from pool measurements (Table 2). However, in a mixed population of erythrocytes one observes a considerable range in endogenous conjugate levels (5,34). A small fraction of cells displays signal characteristic of reticulocytes. The majority of cells shows attenuated levels of conjugates expected for mature erythrocytes. A third population of cells shows virtually no endogenous conjugates. The existence of the latter population suggests that erythrocytes progressively lose the ability to conjugate proteins that cannot be attributed to a decline in cellular ubiquitin content, Table 2. That this loss in conjugation results from decay of components required for conjugation is supported by complimentation degradation studies using cell free erythrocyte Fraction II extracts.

Erythrocyte Fraction II shows only modest stimulation in degradation of reductively methylated ^{125}I-BSA in the presence of ATP and ubiquitin. However, degradation is significantly stimulated on addition of a crude pH 9-DTT eluate fraction obtained during the affinity purification of ubiquitin conjugating enzymes on ubiquitin-linked Affigel-10 columns (12,15). The crude pH 9-DTT eluate exhibits no degradation of the model substrate in the absence of erythrocyte Fraction II (not shown). Previous work has shown that this affinity eluate contains ubiquitin activating enzyme (E1), a mixture of ubiquitin carrier proteins (E2), and a trace amount of isopeptide ligase (E3). No stimulation in Fraction II degradation is observed on addition of homogeneous E1 (not shown). That a ubiquitin carrier protein is the missing component in erythrocyte Fraction II was demonstrated by resolving the pH 9-DTT eluate on Sepharose 4B and assaying fractions for stimulating of proteolysis. A peak of activity having a molecular weight of ca 14 KDa, with a minor shoulder at 30 KDa representing a dimeric species, was the only component showing stimulation in assays such as those of Table 3. On thiolester gel analysis this peak corresponded to $E2_{14k}$, the cognate ubiquitin carrier protein for E3-dependent degradation. Similar, more detailed results have been obtained by Pickart and Vella (35). In these studies, E1 and the family of E2 isozymes were found to decay in erythrocytes relative to reticulocytes. The high turnover number for E1 catalysis compared to that of conjugation apparently allows E1 activity to decline considerably without becoming rate limiting with respect to the degradative pathway. Therefore, progressive inactivation of $E2_{14k}$ accounts for the terminal loss in degradative capacity within erythrocytes.

Table 3. Stimulation of ATP, Ubiquitin-Dependent Degradation in Erythrocyte Extracts by Addition of Ubiquitin Carrier Protein[a]

	TCA soluble radioactivity cpm
-ATP	29,600
+ATP	29,600
+ATP,+Ub	36,400
+ATP,+Ub,+E2 fraction	135,000

[a]Assays were conducted with erythrocyte Fraction II as described in Table 1. In the last entry, 5 μl of a crude pH 9-DTT elution containing an unresolved mixture of E2 isozymes was included. Control experiments showed no degradation by the E2 mixture in the absence of added Fraction II.

SUMMARY

Response of the ATP, ubiquitin-dependent system during the enhanced degradation of erythrocyte maturation conforms to the general regulatory features common to several similar but unrelated systems. In erythroid cells enhanced degradation follows three phases: (1) Onset of degradation characterized by an increase in the intracellular concentration of free and conjugated ubiquitin, brought about by reduction in mean cell volume; (2) Active enhanced degradation during cellular remodeling; and (3) Loss of activity as a consequence of spontaneous inactivation of components required for ubiquitin conjugation. The extent of degradative remodeling is probably functionally limited by the loss of these critical ligation enzymes.

ACKNOWLEDGEMENTS

I am grateful to Patricia Bright Reback for her technical contributions and help in the preparation of this manuscript. Supported by United States Public Health Service grant GM34009.

REFERENCES

1. A. Hershko, Ubiquitin-mediated protein degradation, J. Biol. Chem., 263:15237 (1988).
2. B. P. Monia, D. J. Ecker and S. T. Crooke, New prospective on the structure and function of ubiquitin, Biotechnol. 8:209 (1990).
3. A. Ciechanover, H. Heller, S. Elias, A. L. Haas and A. Hershko, ATP-dependent conjugation of reticulocyte proteins with the polypeptide required for protein degradation, Proc. Natl. Acad. Sci. U.S.A. 77:1365 (1980).
4. A. Hershko, A. Ciechanover, H. Heller, A. L. Haas and I. A. Rose, Proposed role of ATP in protein breakdown: Conjugation of proteins with multiple chains of the polypeptide of ATP-dependent proteolysis, Proc. Natl. Acad. Sci. U.S.A. 77:1783 (1980).
5. A. L. Haas, Immunochemical probes of ubiqutin pool dynamics, in: "Ubiquitin," M. Rechsteiner, ed., Plenum Press, New York (1988).
6. C. M. Pickart, Ubiquitin activation and ligation, in: "Ubiquitin," M. Rechsteiner, ed., Plenum Press, New York (1988).
7. A. L. Haas, K. E. Murphy and P. M. Bright, The inactivation of ubiquitin accounts for the inability to demonstrate ATP, ubiquitin-dependent proteolysis in liver extracts, J. Biol. Chem. 260:4694 (1985).
8. J. M. Fagan, L. Waxman and A. L. Goldberg, Skeletal muscle and liver contain a soluble ATP-ubiquitin-dependent proteolytic pathway, Biochem. J. 243:335 (1987).
9. A. L. Haas and P. M. Bright, The immunochemical detection and quantitation of intracellular ubiquitin-protein conjugates, J. Biol. Chem. 260:12464 (1985).
10. A. L. Haas, Role of ubiquitin in protein degradation, in: "Protein Metabolism in Aging," H. L. Segal, M. Rothstein, and E. Bergamini, eds., Wiley-Liss, New York (1990).

11. A. L. Haas and P. M. Bright, The dynamics of ubiquitin pools within cultured human lung fibroblasts, J. Biol. Chem. 262:345 (1987).

12. A. L. Haas and I. A. Rose, The mechanism of ubiquitin activating enzyme, J. Biol. Chem. 257:10329 (1982).

13. V. Chau, J. W. Tobias, A. Bachmair, D. Marriott, D. J. Ecker, D. K. Gonda and A. Varshavsky, A multiubiquitin chain is confined to a specific lysine in a targeted short-lived protein, Science 243:1576 (1989).

14. A. L. Haas and I. A. Rose, Hemin inhibits ATP-dependent ubiquitin-dependent proteolysis: Role of hemin in regulating ubiquitin conjugate degradation, Proc. Natl. Acad. Sci. U.S.A. 78:6845 (1981).

15. A. L. Haas, P. M. Bright and V. Chau, Ubiquitin conjugation by the yeast RAD6 and CDC34 gene products, J. Biol. Chem., in press (1990).

16. A. Hershko, H. Heller, E. Eytan and Y. Reiss, The protein substrate binding site of the ubiquitin-protein ligase system, J. Biol. Chem. 261:11992 (1986).

17. R. L. Dunten and R. E. Cohen, Recognition of modified forms of ribonuclease A by the ubiquitin system, J. Biol. Chem. 264:16739 (1989).

18. A. Bachmair and A. Varshavsky, The degradation signal in a short-lived protein, Cell 56:1019 (1989).

19. A. Haas, R. M. Reback, G. Pratt and M. Rechsteiner, Ubiquitin-mediated degradation of histone H3 does not require the substrate binding protein E3 or attachment of polyubiquitin chains, J. Biol. Chem., in press (1990).

20. U. Bond, N. Agell, A. L. Haas, K. Redman and M. Schlesinger, Ubiquitin in stressed chicken embryo fibroblasts, J. Biol. Chem. 263:2384 (1988).

21. A. L. Haas, The dynamics of ubiquitin pools within skeletal muscle, in: "The Ubiquitin System," M. Schlesinger and A. Hershko, eds., Cold Spring Harbor Laboratory, New York (1988).

22. U. Bond and M. Schlesinger, Ubiquitin is a heat shock protein in chicken embryo fibroblasts, Mol. Cell. Biol. 5:949 (1985).

23. L. M. Schwartz and J. W. Truman, Hormonal control of rates of metamorphic development in the tobacco hornworm Manduca sexta, Devel. Biol. 99:103 (1983).

24. A. Hershko, E. Eytan, A. Ciechanover and A. L. Haas, Immunochemical analysis of the turnover of ubiquitin-protein conjugates in intact cells, J. Biol. Chem. 257:13964 (1982).

25. N. T. Neff, L. Bourret, P. Miao and J. F. Dice, Degradation of proteins microinjected into IMR-90 human diploid fibroblasts, J. Cell. Biol. 91:184 (1981).

26. G. Pratt, R. Hough and M. Rechsteiner, Proteolysis in heat-stressed HeLa cells. Stabilization of ubiquitin correlates with the loss of proline endopeptidase, J. Biol. Chem. 246:12526 (1989).

27. L. Laszlo, F. J. Doherty, N. U. Osborn and R. J. Mayer, Ubiquitinated protein conjugates are specifically enriched in the lysosomal system of fibroblasts, FEBS Lett. 261:365 (1990).

28. S. Rapoport, W. Dubiel and M. Muller, Proteolysis of mitochondria in reticulocytes during maturation is ubiquitin-dependent and is accompanied by a high rate of ATP hydrolysis, FEBS Lett. 180:249 (1985).

29. A. L. Goldberg and F. S. Boches, Oxidized proteins in erythrocytes are rapidly degraded by the adenosine triphosphate-dependent proteolytic system, Science 215:1107 (1982).
30. D. T. Chin, L. Kuehl and M. Rechsteiner, Conjugation of ubiquitin to denatured hemoglobin is proportional to the rate of hemoglobin degradation in HeLa cells, Proc. Natl. Acad. Sci. U.S.A. 79:5857 (1982).
31. L. Gregori, D. Marriott, C. M. West and V. Chau, Specific recognition of calmodulin from Dictylostelium discoideum by the ATP, ubiquitin-dependent degradative pathway, J. Biol. Chem. 260:5232 (1985).
32. L. Gregori, D. Marriott, J. A. Putkey, A. R. Means and V. Chau, Bacterially synthesized vertebrate calmodulin is a specific substrate for ubiquitination, J. Biol. Chem. 262:2562 (1987).
33. J. R. Schaeffer, ATP-dependent proteolysis of hemoglobin α chains in ß-thalassemic hemolysates is ubiquitin-dependent, J. Biol. Chem. 263:13663 (1988).
34. D. A. Riley, J. L. W. Bain, S. Ellis and A. L. Haas, Quantitation and immunohistochemical localization of ubiquitin conjugates within rat red and white skeletal muscles, J. Histochem. Cytochem. 36:621 (1988).
35. C. M. Pickart and A. T. Vella, Levels of active ubiquitin carrier proteins decline during erythroid maturation, J. Biol. Chem. 263:12028 (1988).

ACYLPHOSPHATASE AND CALCIUM TRANSPORT ACROSS ERYTHROCYTE MEMBRANE

Chiara Nediani, Gianfranco Liguri, Niccolò Taddei, Elena Marchetti, Giampietro Ramponi and Paolo Nassi

Dipartimento di Scienze Biochimiche
Università di Firenze, Firenze, Italy

INTRODUCTION

Acylphosphatase is a widespread cytosolic enzyme that catalyzes the hydrolysis of carboxylphosphate bond compounds. Among natural acylphosphates hydrolyzed by the enzyme are 3-phosphoglyceroylphosphate (1) carbamoylphosphate (2), succinoylphosphate (3).

In mammalian tissues acylphosphatase exists in two isoenzymatic forms: one is prevalent in skeletal muscle (4-7), the other in red blood cells (RBCs) (8). The two isoenzymes are very small basic proteins with the same number of aminoacids but differing significantly in primary structure. They exhibit similar substrate specificity and kinetic properties, except that erythrocyte isoenzyme appears to have a higher catalytic power.

The physiological function of acylphosphatase is still debated. It has been postulated that, by hydrolyzing 3-phosphoglyceroylphosphate, acylphosphatase may hasten glycolysis at the expence of ATP formation, when the rate of this pathway is limited by low concentrations of inorganic phosphate and ATP.

In a recent research project we investigated acylphosphatase content and activity during the human erythrocyte lifespan (9). In this study erythrocytes were age-fractioned by isopicnic centrifugation in Percoll density gradients. Acylphosphatase content was determined by a non-competitive enzyme-linked immunoadsorbent assay (ELISA) carried out with policlonal antierythrocyte acylphosphatase antibodies (10). Acylphosphatase activity was measured by a continuous optical test with benzoylphosphate as substrate based on the difference in absorbance at 283 nm between benzoylphosphate and benzoate (11). Acylphosphatase content and activity rose with ageing of red blood cells. Maximum values occurred in mature erythrocytes and these values decreased slightly in the older cells (Fig. 1). This singular behaviour may be attributed to a de-novo synthesis of

Red Blood Cell Aging, Edited by M. Magnani and
A. De Flora, Plenum Press, New York, 1991

the enzyme during reticulocyte stage and its storage, virtually unaffected, in mature erythrocytes. As glycolysis is the only energy source of the red cell, the increase of acylphosphatase during erythrocyte maturation and its persistence at high levels in old cells could contribute to a progressive reduction of the energetic yield of glycolysis due to an <u>uncoupling</u> effect of this enzyme produced by the hydrolysis of the carboxylphosphate bond of 3-phosphoglyceroylphosphate.

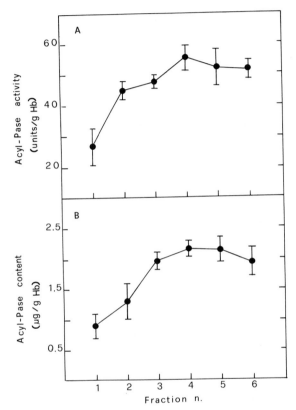

Fig. 1. Acylphosphatase (Acyl-Pase) levels in six classes of increasing age erythrocytes. Each point represents the mean ± S.D. of six determinations. (A) acylphosphatase activity (units/g hemoglobin). (B) acylphosphatase content (μg/g hemoglobin). The changes observed in fractions 2 vs. 1, 3 vs. 2 and 4 vs. 3 were statistically significant ($p < 0.05$).

Acylphosphatase, however, might affect in other was the energy metabolism of erythrocytes. Human red blood cell membrane has a Ca^{2+}-ATPase system that pumps Ca^{2+} ions out of the cell coupling calcium transport to ATP hydrolysis, the latter proceeding through a series of reactions that involve the Ca^{2+} dependent formation of an acylphosphorylated intermediate. It is well known that human RBC Ca^{2+}-ATPase is activated by calmodulin and that this effect is associated with an increased rate in phosphorylated intermediate formation (12). Human RBCs, however, seems to contain other soluble (13) and particulate factors (14) that can modulate Ca^{2+}-ATPase activity.

We supposed that RBC acylphosphatase, on the basis of a potential hydrolytic effect on the acylphosphorylated intermediate, might represent an additional modulator of Ca^{2+}-ATPase that would effect both ATP hydrolysis and calcium transport. In other words, the efficiency of erythrocyte membrane Ca^{2+} pump. The studies that we conducted to examine this hypothesis are here reported.

EXPERIMENTAL

Effect of acylphosphatase on phosphorylated intermediate from RBC membranes

RBC membranes were phosphorylated according to Luthra et al.(15) and the level of phosphoenzyme (EP) was taken as the difference between the amount of ^{32}P incorporated into the membrane protein in a medium with Ca^{2+} and in an identical medium without Ca^{2+}. Labeled membranes were incubated with varying amounts of acylphosphatase from 2 to 10 units per mg membrane protein.

Table 1. Effect of Different Concentration of Acylphosphatase on Phosphorylated Intemediate from RBC Membranes

Sample	Phosphate bound	Phosphate released
	pmol/mg protein	pmol/mg protein/min
Labeled membranes	1.81±0.07	
Control for spontaneous hydrolysis		0.34±0.06
Acylphosphatase-treated labeled membrane		
2U/mg		0.43±0.05
5U/mg		0.61±0.09
10U/mg		0.80±0.12

Labeled membranes (1 mg protein) were incubated in 0.150 M Tris-HCl pH 7.2 at 37°C with varying amounts of acylphosphatase from 2 to 10 units per mg membrane protein (final volume:1 ml). Results are means ± S.E. of five experiments with different membrane preparations. All the changes in ^{32}P release induced by acylphosphatase addition were statistically significant (p < 0.05).

As shown in Table 1, the release of phosphate, net of spontaneous hydrolysis, rose significantly with the increase inacylphosphatase membrane protein ratio. This hydrolytic effect was predictable (given the catalytic properties of our enzyme and the acylphosphate nature of the intermediate) all the more so since the literature has already reported similar effect of acylphosphatase on the acylphosphorylated intermediates of other transport ATPase, such as brain ($Na^+ -K^+$)-ATPase (16) and sarcoplasmic reticulum Ca^{2+}-ATPase (17). However, it is noteworthy that the enzymatic hydrolysis of EP occurred to a considerable extent with acylphosphatase amounts that fall in the physiological range.

In order to kinetically characterize the acylphosphatase effect we incubated a fixed amount of this enzyme (5 units) with variable amounts of ^{32}P-labeled membranes and measured the initial velocity of dephosphorylation as a function of the phosphoenzyme concentration.

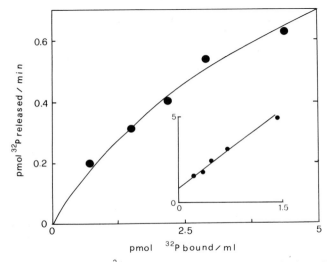

Fig. 2. Initial rate of Ca^{2+}-ATPase intermediate dephosphorylation by acylphosphatase. 5 units of acylphosphatase were incubated in 0.150 M Tris-HCl (pH 7.2 at 37°C) with differing amounts of labeled membranes. EP concentration in the medium was expressed as pmol ^{32}P bound/ml. Phosphate release was measured at two min intervals for 10 min and the initial rate of hydrolysis was estimated by calculating the first derivative value at zero time of the curve that describes the phosphate release, net of spontaneous hydrolysis, as a function of time. Each point represents the mean value of four determinations. The inset shows the double reciprocal plot of the data shown in the Figure.

In these conditions (Fig. 2) we observed that the initial rate of enzymatic dephosphorylation rose with the increase in EP concentration along a hyperbolic curve whose parameters were K_m = 3.41 ± 1.16 (S.E.) nM and v_{max} = 1.19 ± 0.207 (S.E.) pmol ^{32}P released/min. The very low Km value that we found for acylphosphate hydrolis suggests a high affinity in this enzyme for EP, which seems to be consistent with the small number of Ca^{2+}-ATPase units, these latter only accounting in human RBC membranes for about 0.1% of total protein (12).

Effect of acylphosphatase on erythrocyte membrane Ca^{2+}-ATPase

To see if the above action on the acylphosphorylated intermediate resulted in modified functional properties in Ca^{2+}-ATPase, we investigated the effect of acylphosphatase addition on the rate of Ca^{2+} dependent RBC membrane ATP hydrolysis. The mean value of basal Ca^{2+} -ATPase activity in our membrane preparations ranged from 109 to 150 nmol of ATP split/h per mg membrane protein, a value which is close to that obtained by Thakar (18). When added to the assay medium in concentrations that ranged from 0.5 to 10 units per mg membrane protein, acylphosphatase significantly increased the rate of RBC membrane ATP hydrolysis. The increase depended on the amount of acylphosphatase added and maximal stimulation (about two fold over basal value) was observed at 2 units/mg membrane protein (Tab. 2). The effect of acylphosphatase was cumulative with that of calmodulin since Ca^{2+}-ATPase activity observed with optimal calmodulin concentration was further enhanced by the addition of acylphosphatase (Tab. 3). The stimulatory effect of acylphosphatase in the presence of calmodulin was similar to that observed when this enzyme alone was added to the membranes.

Table 2. Effect of Exogenous Acylphospatase on Erythrocyte Membrane Ca^{2+}-ATPase

Acylphosphatase	Ca^{2+}-ATPase activity	
units/mg membrane protein	nmol/h per mg membrane protein	
	A	Δ A
0	109±11	
0.5	134±16	+ 25
1.0	145±15	+ 36
1.5	157±18	+ 48
2.0	213±23	+ 104
5.0	208±22	+ 99
10.0	201±19	+ 92

Results are means ± S.E. of six determinations performed on differing membrane preparations. A represents Ca^{2+}-ATPase with acylphosphatase. Δ A indicates the change in Ca^{2+}-ATPase with acylphosphatase compared to basal activity. Changes in Ca^{2+}-ATPase activity observed with increasing amounts of acylphosphatase were statistically significant ($p < 0.01$ by the one-way analysis of variance).

Table 3. Effect of the Simultaneous Addition of Acylphosphatase and Calmodulin on Ca^{2+}-ATPase

Acylphosphatase	Ca^{2+}-ATPase activity	ΔA
units/mg membrane protein	nmol/h per mg membrane protein	
0	249±31	
1.0	291±35	+ 42
1.5	301±38	+ 52
2.0	348±41	+ 99
2.5	337±39	+ 88

Results are means ± S.D. of six assays performed with different membrane preparations. In all assays calmodulin was added at 40 ng/ml (optimal concentration in our conditions). ΔA indicates the changes in Ca^{2+}-ATPase activity with acylphosphatase compared with the activity without acylphosphatase. The difference observed with increasing concentration of acylphosphatase was statistically significant ($p < 0.01$ by the one-way analysis of variance).

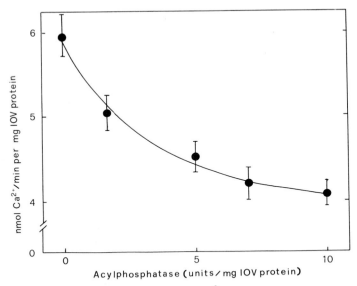

Fig. 3 . Effect of acylphosphatase on Ca^{2+} transport into RBC membrane inside-out vesicles. Ca^{2+} uptake into IOVs was measured at 4 min intervals for 16 min by a rapid filtration technique using 0.6 μm pore Sartorius membrane filter. Each point represents the mean ± S.E. of four determinations. Changes observed with differing amounts of acylphosphatase were statistically significant ($p < 0.01$ by one way analysis of variance).

Effect of acylphosphatase on Ca^{2+} transport into RBC membrane inside-out vesicles (IOVs)

Acylphosphatase also affected Ca^{2+} transport across erythrocyte membranes, but this effect was of opposite type compared with that on the rate of ATP hydrolysis. In fact comparable amounts of acylphosphatase decreased the rate Ca^{2+} transport into inside-out RBC vesicles and more markedly so with higher enzyme amounts. Maximal inhibition of Ca^{2+} transport was obtained with 10 units/mg membrane protein and resulted in a decrease of about 30%, 4.09 ± 0.15 (S.E.) nmol/min per mg IOV protein, with respect to basal value, 5.98 ± 0.27 (S.E.) nmol/min per mg IOV protein (Fig. 3).

CONCLUSION

From the results here reported acylphosphatase appears to have an uncoupling effect on the erythrocyte Ca^{2+} pump. This was confirmed by the diminished Ca^{2+}/ATP ratio that we found in the presence of acylphosphatase when we measured in the same the rate of Ca^{2+} uptake and of ATP hydrolysis by IOVs. This value decreased from 2.01 to 0.91 in the presence of acylphosphatase at a concentration of 10 units/mg IOV protein. As regards the mechanism underlying acylphosphatase action, it is generally accepted that the operation of RBC membrane Ca^{2+} pump proceeds through the sequence of elementary reactions reported in the Scheme (12)

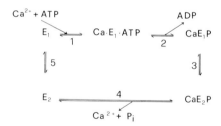

Ca^{2+}-ATPase exists in two conformers, E_1 and E_2. Ca^{2+} and ATP binding to E_1 leads to the Ca^{2+}-E-ATP complex. Phosphorylation by ATP results in the conversion of Ca^{2+}-E$_1$-P to Ca^{2+}-E$_2$-P. After the release of Ca^{2+} and dephosphorylation E_1 is converted into E_2 and the cycle closes.

We propose that acylphosphatase-induced hydrolysis of the phosphoenzyme intermediate occurs before Ca^{2+} transport has taken place, thereby short circuiting the system and giving rise to an accelerated ATP hydrolysis with concomitant inhibition of Ca^{2+} pumping. The effects above described suggest an additional mechanism of energy wasting, which – given the higher acylphosphatase levels observed in mature RBCs with respect to younger erythrocytes – might increase with ageing contributing to the age-dependent decline of the energetic potential of these cells.

REFERENCES

1. G. Ramponi, C. Treves and A. Guerritore, Hydrolytic activity of muscle acylphosphatase on 3-phosphoglyceroylphosphate, *Experientia* 23:1019 (1967).

2. G. Ramponi, F. Melani and A. Guerritore, Azione dell'acilfosfatasi di muscolo su carbamilfosfato, *G. Biochim.* 10:189 (1961).

3. A. Berti, M. Stefani, G. Liguri, G. Camici, G. Manao and G. Ramponi, Acylphosphatase action on dicarboxylic acylphosphates, *Ital. J. Biochem.* 26:377 (1977).

4. G. Cappugi, G. Manao, G. Camici and G. Ramponi, The complete aminoacid sequence of horse muscle acylphosphatase, *J. Biol. Chem.* 255:6868 (1980).

5. G. Camici, G. Manao, G. Cappugi, A. Berti, M. Stefani, G. Liguri and G. Ramponi, The primary structure of turkey muscle acylphosphatase, *Eur. J. Biochem.* 137:269 (1983).

6. G. Manao, G. Camici, G. Cappugi, M. Stefani, G. Liguri, A. Berti and G. Ramponi, Rabbit skeletal muscle acylphosphatase: the aminoacid sequence of form Ra1, *Arch. Biochem. Biophys.* 241:418 (1985).

7. G. Manao, G. Camici, A. Modesti, G. Liguri, A. Berti, M. Stefani, G. Cappugi and G. Ramponi, Human skeletal muscle acylphosphatase. The primary structure, *Mol. Biol. Med.* 2:369 (1984).

8. G. Liguri, G. Camici, G. Manao, G. Cappugi, P. Nassi, A. Modesti and G. Ramponi, A new acylphosphatase isoenzyme from human erythrocytes: purification, characterization and primary structure, *Biochemistry* 25:8089 (1986).

9. G. Liguri, P. Nassi, D. Degl'Innocenti, E. Tremori, C. Nediani, A. Berti and G. Ramponi, Acylphosphatase levels of human erythrocytes during cell aging, *Mech. Ageing Dev.* 39:59 (1987).

10. D. Degl'Innocenti, A. Berti, G. Liguri, M. Stefani and G. Ramponi, Determinazione dei livelli di acilfosfatasi di tipo eritrocitario in tessuti umani con metodo ELISA, 32° Congresso Nazionale della Società Italiana di Biochimica, Messina, 28 Settembre 1986, Riassunti dei Simposi e Poster, p. 400.

11. G. Ramponi, C. Treves and A. Guerritore, Continuous optical assay of acylphosphatase with benzoylphosphate as substrate, *Experientia* 22:705 (1966).

12. N. P. Adamo, A. F. Rega and P. J. Garrahan, Pre-steady-state phosphorylation of the human red cell Ca^{2+}-ATPase, *J. Biol. Chem.* 263:17548 (1988).

13. F. B. Davis, P. J. Davis and S. D. Blas, Role of calmodulin in thyroid hormone stimulation in vitro of human erythrocyte Ca^{2+}-ATPase activity, *J. Clin. Invest.* 71:579 (1983).

14. K. K. W. Wang, A. Villalobo and B. D. Roufogalis, Activation of the Ca^{2+}-ATPase of human erythrocyte membrane by an endogenous Ca^{2+}-dependent neutral protease, *Arch. Biochem. Biophys.* 260:696 (1988).

15. M. G. Luthra, R. P. Watts, K. L. Scherer and H. D. Kim, Calmodulin: an activator of human erythrocyte $(Ca^{2+} + Mg^{2+})$-ATPase phosphorylation, *Biochim. Biophys. Acta* 633:299 (1980).

16. L. E. Hokin, P. S. Sastry, P. R. Galsworthy and A. Yoda, Evidence that a phosphorylated intermediate in a brain transport adenosine triphosphate is an acylphosphate, <u>Proc. Natl. Acad. Sci. U.S.A.</u> 54:177 (1965).

17. M. Stefani, G. Liguri, A. Berti, P. Nassi and G. Ramponi, Hydrolysis by horse muscle acylphosphatase of $(Ca^{2+} + Mg^{2+})$-ATPase phosphorylated intermediate, <u>Arch. Biochem. Biophys.</u> 208:37 (1981).

18. J. H. Thakar, Method for the preparation of human erythrocyte membrane with low basal calcium ATPase, responsive to stimulation, <u>Anal. Biochem.</u> 144:94 (1985).

FREE RADICALS PROMOTE "IN VITRO" A DIFFERENT INTRACELLULAR DECAY OF

RABBIT RETICULOCYTE AND ERYTHROCYTE GLYCOLYTIC ENZYMES

Vilberto Stocchi, Beatrice Biagiarelli, Linda Masat,
Francesco Palma, Fulvio Palma, Giovanni Piccoli,
Luigi Cucchiarini and Mauro Magnani

Istituto di Chimica Biologica "Giorgio Fornaini"
Università degli Studi di Urbino
Urbino, Italy

SUMMARY

Rabbit red blood cells (RBC) were exposed in vitro to an oxygen-radical-generating system represented by iron and acorbic acid. Under these experimental conditions we have investigated the effect of this system on some intracellular rabbit reticulocyte and erythrocyte enzymes. The results obtained have shown a pronounced decay of hexokinase activity both in the erythrocytes and reticulocytes when exposed to these radical species. We have found that the amount of hexokinase inactivated is at least three times higher in a blood sample with a percentage of reticulocytes of 50-60%. This different behaviour of the hexokinase decay in the erythrocytes and reticulocytes could be due to its different intracellular distribution related to the two distinct cells. In addition we have evaluated some important intracellular compounds involved in maintaining the redox and the energetic state of the cell such as the reduced glutathione and the adenine nucleotides and their degradation products, in order to understand if there is any correlation between the hexokinase decay and a change concerning the metabolic conditions of the rabbit reticulocytes and erythrocytes exposed to free radicals.

INTRODUCTION

It is well known that free radicals can damage plasma membrane as well as some intracellular proteins. During recent years various Authors have provided evidence that the autoxidation of hemoglobin may produce oxygen radicals and other activated oxygen species (1-3). The oxygen radicals may promote hemoglobin damage (1-3) lipid peroxidation and stimulate the intracellular proteolysis of proteins (4-12). More recently Davies and

Goldberg (4,6) have shown that the exposure of intact erythrocytes and reticulocytes to various oxygen species induces protein degradation in these red blood cells. For this reason we have investigated in vitro the effect of an oxygen-radical-generating system (ascorbic acid-Fe^{2+}) on some rabbit intracellular reticulocyte and erythrocyte enzymes. Under these experimental conditions we have evaluated the level of enzymes involved in the glycolytic pathway, that in the red blood cell is the only pathway responsible for the energy production. The results obtained have shown that, among the glycolytic enzymes, only the hexokinase activity is susceptible to the action of free radicals, while the activity of the other enzymes remains almost the same.

METHODS

Rabbit red blood cells

Whole blood was collected in heparin as anticoagulant and immediately centrifuged at 3,000 rpm for 10 min at 4°C. After removal of plasma and buffy coat the red blood cells were washed twice with cold isotonic saline solution. The packed erythrocytes were then resuspended at 10% hematocrit in 0.9% (w/v) NaCl solution. This cell suspension was then incubated in presence and not of the oxygen-radical-generating system. Whole blood rich in reticulocytes was obtained from a group of 3 rabbits made anemic by phenylhydrazine administration as previously described (13). In order to exclude some possible interference due to the use of phenylhydrazine reticulocytes were also obtained from rabbits made anemic by repeated bleeding. Blood samples rich in reticulocytes were treated as discussed above for the erythrocytes.

Incubation of rabbit reticulocytes and erythrocytes in presence of ascorbic acid and iron

A 10% (v/v) red blood cell suspension prepared in 0.9% (w/v) NaCl, pH 7.4 was incubated at 37°C for 30 and 60 min, in a shaking water bath. Experiments were carried out adding iron, ascorbic acid and both at various concentrations. At the fixed times an aliquot of RBC suspension was collected and immediately centrifuged at 3,000 rpm for 5 min and the surnatant removed. The packed cells were washed once with cold 0.9% (w/v) NaCl solution and the supernatant discarded. The lysate was prepared adding 3 volumes of cold water to the packed red blood cells.

Evaluation of rabbit red blood cell glycolytic enzymes

The activities of hexokinase (EC 2.7.1.1), glucose phosphate isomerase (EC 5.3.1.9), phosphofructokinase (EC 2.7.1.11), aldolase (EC 4.1.2.1.3), glyceraldehyde phosphate dehydrogenase (EC 1.2.1.1.3), triose phosphate isomerase (EC 5.3.1.1), phosphoglycerate kinase (EC 2.7.2.3), monophosphoglyceromutase (EC 5.4.2.1), enolase (EC 4.2.1.11), pyruvate kinase (EC 2.7.1.4.0), lactate dehydrogenase (EC 1.1.1.2.7) and glucose-6-phosphate dehydrogenase (EC 1.1.1.4.9) were determined in triplicates on red blood cell hemolysates prepared in water. All the enzyme determinations and reduced glutathione (GSH) were performed according to the methods of Beutler (14) with slight modifications.

Hexokinase distribution in rabbit red blood cells

The specific activity of hexokinase in a red blood sample containing 50-60% of reticulocytes is at least 5-6 times higher than that detectable in a normal red blood cell population. In the rabbit reticulcytes the hexokinase activity is present, at least in three distinct forms designated hexokinase Ia, Ia* and Ib separable by ion-exchange hegh performance liquid chromatography on a TSK DEAE 5 PW column (15). Usually, 500 μl of hemolysate were charged onto a (7.5 cm x 7.5 mm I.D.) column, 10 um particles, and the elution of the different forms of hexokinase was obtained using a linear gradient of KCl workings at a flow-rate of 1 ml/min. Fraction of 0.7 ml were collected and assayed for the hexokinase activity.

RESULTS AND DISCUSSION

The exposition of human and rabbit red blood cells to different oxygen-radical-generating systems may promote a plasma membrane lipid peroxidation as well as intracellular degradation of proteins as reflected by liberation of alanine (6). More recently Sals et al. (16) have shown that also some enzymes involved in the detoxification process of the cell, such as the superoxide dismutase, may be inactivated by its product, giving direct evidence that the oxygen radical species may promote an inactivation of some intracellular enzymes (16). At present, no evidence has been reported about the effect of the iron and ascorbate oxygen-radical-generating system on some red blood cell enzymes involved in the glycolytic pathway that in the red blood cell is only pathway responsible for the energy production. We have investigated the effect of an oxygen-radical-generating system represented by ascorbic acid and iron incubating intact rabbit erythrocytes and reticulocytes.

Table 1. Inactivation of Rabbit Hexokinase after Exposure of Intact Erythrocytes to an Oxygen-Radical-Generating System

TIME (min)	A	B
0	0.8	0.8
30	0.8	0.4
60	0.8	0.3

(A), control; (B), 0.1 mM $FeSO_4$ + 10 mM ascorbate.
Data are expressed as U/g Hb determined at 37°C.

219

Table 2. Rabbit Hexokinase Activity after Exposure of Intact Erythrocytes to $FeSO_4$ and Ascorbate

TIME (min)	A	B	C
0	0.8	0.8	0.8
30	0.8	0.8	0.74
60	0.8	0.78	0.65

(A), control; (B), 0.1 mM $FeSO_4$; (C), 10 mM ascorbate.
Data are expressed as U/g Hb determined at 37°C.

The results obtained, reported in Table I, have shown that among the glycolytic enzymes only the hexokinase activity is susceptible to the action of this oxygen-radical-generating system while he activity of the other glycolytic enzymes remains almost the same (data not shown). Table II shows the behaviour of the hexokinase activity incubating the intact red blood cells in presence of iron, ascorbate, and both together. The presence of iron alone, in the external medium, does not influence the decay of the hexokinase activity. The ascorbic acid alone promotes a little decay, but the most pronounced effect, concerning the inactivation of the hexokinase activity, was observed in presence of iron and ascorbate (Table I) suggesting a significant effect of this oxygen-radical-generating system. Further evidence that this system is in some way responsible for the hexokinase decay is shown in Table III where we have tested different concentrations of iron and ascorbic acid. The hexokinase decay increases, increasing the concentration of iron and ascorbic acid in the external medium. A similar situation occurs in a whole blood sample rich in reticulocytes exposed to the oxygen-radical-generating system. Also in this case the decay involves only the hexokinase activity (Table IV). However, the behaviour of this decay is different from that observed in the erythrocytes. This could be due to the fact that the hexokinase level is significantly different in the rabbit reticulocytes and in the erythrocytes being 4-5 times higher in a blood sample with a percentage of reticulocytes of 50-60% and also to the fact that the enzyme shows a different intracellular distribution related to the two distinct cells. In fact, in the rabbit reticulocytes we have at least three different forms of hexokinase that we have designated hexokinase Ia, Ia* and Ib. The hexokinase Ia and Ib are present in soluble form, while the hexokinase Ia* is bound to the mitochondria. During the maturation of the reticulocyte to erythrocyte the chromatographic profile changes significantly, only the hexokinase Ia and a small amount of hexokinase Ib being present in the erythrocyte. The hexokinase Ia* is completely lost during the maturation of the reticulocyte to erythrocyte. This different distribution of the hexokinase activity in the rabbit reticulocytes could also contribute to explain the different decay of the enzyme activity when compared to that observed in a total population of erythrocytes. In fact, the hexokinase Ia* bound to the mitochondria could be less susceptible to the action of free radicals. We have also evaluated the level of compounds in the erythrocytes and in the reticulocytes involved in maintaining the redox and the energetic state of the cell such as the glutathione reduced and the adenine nucleotides and their degradation products. The GSH level in intact erythrocytes (data not shown) undergoes a dramatic fall (corresponding to about 70% of the initial value) incubating the erythrocytes for 60 min in presence of iron and ascorbic acid. Significantly different is the situation in a whole blood sample rich in reticulocytes where the fall of GSH is less pronounced. This is an interesting result because notwithstanding this higher level of GSH the amount of hexokinase inactivated is at least 3 times higher in the reticulocytes. The level of ATP, ADP, and AMP as well as their degradation products ipoxanthine and inosine (data not shown) do not show significant differences in the erythrocytes and reticulocytes incubated with or without ascorbic acid and iron. However, under the experimental conditions used we have to expect a decrease in the level of ATP concomitantly to an increase in the level of ADP, AMP, ipoxanthine and inosine. The results obtained have shown no significant differences between the sample incubated in absence and in presence of iron and ascorbic acid suggesting that the oxygen-radical-generating system is not able to significantly influence the level of these compounds involved in maintaining the energetic state of the cell.

Finally, preliminary experiments, data not given here, have shown that the inactivation process of the hexokinase activity may be reversible under some experimental conditions, suggesting that the inactivation of the hexokinase produces a distinct form of the enzyme without catalitic activity, as could probably happen during the aging of the red blood cell. The inactivated state of the enzyme could also play an important role to explain the highest decay rate observed in the reticulocyte where the ATP-dependent and independent proteolytic systems are active. Experiments to test these hypotheses are in progress in our Lab.

Table 3. Inactivation of Rabbit Hexokinase after Exposure of Intact Erythrocytes to Different Concentrations of Ascorbate and $FeSO_4$

TIME (min)	A	B	C	D	E
0	0.8	0.8	0.8	0.8	0.8
30	0.8	0.6	0.4	0.3	0.03
60	0.8	0.4	0.3	0.05	0.02

(A), control; (B), 0.05 mM $FeSO_4$ + 5 mM ascorbate; (C), 0.1 mM $FeSO_4$ + 10 mM ascorbate; (D), 0.15 mM $FeSO_4$ + 15 mM ascorbate; (E), 0.2 mM $FeSO_4$ + 20 mM ascorbate.
Data are expressed as U/g Hb determined at 37°C

Table 4. Inactivation of Rabbit Hexokinase after Exposure of Intact Reticulocytes to Different Concentrations of Ascorbate and FeSO$_4$

TIME (min)	A	B	C
0	4.2	4.2	4.2
45	4.1	3.1	2.5
90	3.8	2.7	1.9

(A), control; (B), 0.05 mM FeSO$_4$ + 5 mM ascorbate; (C), 0.1 mM FeSO$_4$ + 10 mM ascorbate.
Data are expressed as U/g Hb determined at 37°C.

REFERENCES

1. R. P. Hebbel, J. W. Eaton, J. W. Balasingan and M. H. Steinberg, Spontaneous oxygen radical generation by sickle erythrocytes, J. Clin. Invest. 70:1253 (1982).
2. G. Cohen and P. Hochstein, Generation of hydrogen peroxide in erythrocytes by hemolytic agents, Biochemistry 3:895 (1964).
3. P. Hochstein and S. K. Jain, Association of lipid peroxidation and polymerization of membrane proteins with erythrocyte aging, Fed. Proc. Fed. Am. Soc. Exp. Biol. 40:183 (1981).
4. K. J. A. Davies, Free radicals and protein degradation in human red blood cells, in: "Cellular and Metabolic Aspects of Aging: The Red Cell as a Model" J. W. Eaton, D. K. Konzen and J. G. White, Alan R. Liss, Inc. New York.
5. K. J. A. Davies, The role of intracellular proteolytic systems in antioxydant defenses, in: "Superoxide and Superoxide Dismutase in Chemistry, Biology and Medicine" G. Rotilio, ed) Elsevier North-Holland Biomedical Press, Amsterdam.
6. K. J. A. Davies and A. L. Goldberg, Oxygen radicals stimulate intracellular proteolysis and lipid peroxidation by independent mechanisms in erythrocytes, J. Biol. Chem. 262:8220 (1987).
7. K. J. A. Davies and A. L. Goldberg, Protein damaged by oxigen radicals are rapidly degraded in extracts of red blood cells, J. Biol. Chem. 262:8227 (1987).
8. K. J. A. Davies, Intracellular proteolytic systems may function as secondary antioxidant defenses: an hypothesis, J. Free Radical in Biol. Med. 2:155 (1987).
9. K. J. A. Davies, Protein damage and degradation by oxigen radical. I General aspects, J. Biol. Chem. 262:9895 (1987).
10. K. J. A. Davies, M. E. Delsignore and S. W. Lin, Protein damage and degradation by oxigen radicals. II Modification of amino acids, J. Biol. Chem. 262:9902 (1987).
11. K. J. A. Davies and M. E. Delsignore, Protein damage and degradation by oxygen radicals. III Modification of secondary and tertiary structure, J. Biol. Chem. 262:9908 (1987).

12. K. J. A. Davies, S. W. Lin and R. E. Pacifici, Protein damage and degradation by oxigen radicals. IV Degradation of denatured protein, J. Biol. Chem. 262:9914 (1987).

13. V. Stocchi, M. Magnani, F. Canestrari, M. Dachà and G. Fornaini, Rabbit red blood cell hexokinase evidence for two distinct forms, and their purification and characterization from reticulocytes, J. Biol. Chem. 256:7856 (1981).

14. E. Beutler, Part III. Red cell glycolytic enzymes, in: Red Cell Metabolism. A manual of biochemical methods", Grune & Stratton, New York.

15. V. Stocchi, M. Magnani, G. Piccoli and G. Fornaini, Hexokinase microheterogeneicity in rabbit red blood cells and its behaviour during reticulocytes maturation, Mol. cell. Biochem. 79:133 (1988).

16. D. C. Salo, R. E. Pacifici, S. W. Lin, C. Giulivi and K. J. A. Davies, Superoxide dismutase undergoes proteolysis and fragmentation following oxidative modification and inactivation, J. Biol. Chem. 265:11919 (1990).

ENZYMOLOGY

CLINICAL UTILITY OF FRACTIONATING ERYTHROCYTES INTO

"PERCOLL" DENSITY GRADIENTS

Andrea Mosca[1], Renata Paleari[2], Annalisa Modenese[1],
Silvano Rossini[2], Rosalia Parma[2], Cristina Rocco[2],
Vincenzo Russo[3], Giancarlo Caramenti[4],
Maria Laura Paderi[5] and Renzo Galanello[5]

[1] Dipt. Sc. Tecnol. Biomediche, Univ. Milano
[2] Ist. Sci. H S. Raffaele
[3] Ist. Fisiol. Veter. Biochim., Univ. Milano
[4] I.T.B.A., C.N.R., Milano
[5] Ist. Clin. Biol. Età Evol., Univ. Cagliari

ABSTRACT

Two rapid methods for fractionating the RBC into five or nine layers of
increasing density are reported. These procedures have been used to
monitor the decline of glucose-6-phosphate dehydrogenase (G6PD) and
6-phosphogluconate dehydrogenase (6PGD) activity during the process of red
cell aging in normal subjects and in ß--thal carriers, to study transfused
patients with G6PD and pyruvate kinase (PK) deficiency and to test the
effects of inositol hexaphosphate (IHP) encapsulation on RBC subpopulations.

INTRODUCTION

In the past three decades the human erythrocyte has been object of a
great deal of research directed to understand the mechanisms that cause
cellular aging and death. In most cases the RBC were separated _in vitro_ with
techniques based on density gradient centrifugation. Other, less popular
procedures based upon _in vivo_ age-correlated separation of the red cells,
have been reported (for a review of such techniques see Clark (1) and
Beutler (2).

In this presentation we are not going to concentrate on any aspect of
RBC aging. Instead, we are going to try to demonstrate how one of the
techniques used to fractionate the erythrocytes according to their age can
be used either in the clinic or in the laboratory. In order to do this we
will describe an optimization of a previously reported method (3) and its

Red Blood Cell Aging, Edited by M. Magnani and
A. De Flora, Plenum Press, New York, 1991

application in the analysis of the decay of enzymatic activity in aging erythrocytes. Some aspects of the diagnosis of hemolytic anemia in particular cases of patients suffering from G6PD and PK deficiency will be discussed. Finally, the application of these new techniques to the biochemical characterization of erythrocytes treated for IHP encapsulation will be reported.

MATERIALS AND METHODS

Blood Samples and Clinical Cases

Blood samples from healthy adults were collected in potassium EDTA and processed within three hours from venipuncture. Samples from healthy obligatory ß-thalassemia heterozygotes were collected in ACD at Cagliari, sent in ice boxes by express air mail to Milano, and analyzed within 24h.
The first clinical case regards a 67 year old man admitted into hospital with severe asthenia, scleral icterus and hyperchromic urine. Blood hemoglobin was 8.0 g/dL, total bilirubin 2.7 mg/dL, unconiugated bilirubin 2.4 mg/dL, reticulocytes 87/1000 RBC. In his anamnesis the patient reported several icteric events. Twenty days before admission penicillin had been prescribed to the patient and, approximately at the same time, he had ingested fava beans.
The second case concerns a five year old boy affected by severe chronic anemia and in need of monthly blood transfusions.

Erythrocyte Separation into "Percoll" Density Grandients

The RBC were always isolated from leukocytes and platelets by filtration through α-cellulose and microcrystalline cellulose (4) with the exception of the IHP experiments. Soon after the isolation the red cells were fractionated as described by Salvo et al.(3), except for the following modifications:
1) Morpholinopropanesulfonic acid (MOPS) was used instead of HEPES.
2) Solutions A and B were, respectively, 256 and 310 mOsm/Kg water at 26°C. Both had a pH of 7.40 and contained 4.8% (w/v) BSA.
3) Sol. A and B were filtered through 0.45 μm cellulose acetate filters (Lida Man. Corp., Bensenville, IL, USA) and mixed to form four solutions at final "Percoll" concentrations of 58, 61, 64 and 67% (v/v; d 1.087–1.098 g/mL). For the separation of erythrocytes from ß–thal carriers the "Percoll" concentrations were set at 53, 56, 59 and 62% (d 1.080–1.092 g/mL).
4) A "high resolution" eight–step gradient was obtained by superimposing the following Percoll solutions: 52, 55, 58, 61, 64, 67, 70 and 73%.
5) Centrifugation was carried out in a variable angle rotor centrifuge (Minifuge T, Heraeus Sepatech, Milano) at 1000 g for 10 min at 20°C.

RBC Treatment with IHP

Human red cells from blood donors were treated with IHP by the dialysis membrane method (5) using the hypotonic and hypertonic rejuvenating

solutions described by Zanella (6). At the end of the procedure the cells
were resuspended in isotonic saline solution with 400 mg/dL glucose.

Assays

Reticulocytes were counted on a flow cytometer (Epics 751, Coulter
Scientifics, Milano) after incubation with a thiazole orange reagent
(Becton-Dickinson, Milano) according to Ferguson (7). From a comparison made
between the flow cytometric reticulocyte count (y) and the manual method (x)
performed on blood samples that have reticulocytes in the range 1-35/1000
RBC the following regression equation was elaborated y = 0.862x + 1.3; n =
25; r = 0.841.
Red cell parameters (MCV, MCH and MCHC) were determined with a Cell Counter
(Coulter STKR, Coulter Scientifics).
Glycohemoglobin (Hb A_{1c}) was measured with an HPLC automatic analyser (8).
Enzymatic measurements were obtained in duplicate at 37°C on a differential
pH anlyser (DELPAS CL, Kontron, Milano) by following protocols developed
according.to ICSH recommendations (4,9-11). Typical intra-assay coefficients
of variation relative to G6PD, 6PGD and PK assays were: 3.7% (at G6PD
4.3 U/g Hb); 3.2% (at 6PGD 2.7 U/g Hb); 3.8% (at PK 13.9 U/g Hb).

RESULTS AND DISCUSSION

Characterization of Fractionation Systems

The centrifugation of red cells with the proposed "Percoll" gradients
yields five or nine cell fractions, depending on the type of gradient. The
distribution of the erythrocyte population among these layers is usually
gaussian, approximately centered in the middle of the centrifugation tube
(Fig. 1A). The erythrocytes are separated into classes of increasing
density from the top to the bottom of the tube, as shown by the progressive
decrease in MCV (Fig. 1b) and increment in MCHC (Fig. 1C). The mean
cell hemoglobin content (MCH) remains constant through all the gradient
(content in cell fractions 1 vs. 5, mean±SD: 29.9±1.8 vs. 29.7±2.7 in
normals; 22.2±2.2 vs. 20.9±1.9 in ß-thal carriers). These results agree
with those obtained by everyone who has used the density-based cellular
fractionating systems and prove the reliability of the proposed gradients.

The correlation between density and cell age has been proved in two
ways. The first evidence has been obtained by monitoring the changes in Hb
A_{1c} concentration along the gradient (Fig. 1D). In our experiments a steady
increase in glycohemoglobin concentration from the top fraction to the
bottom has always been evident. Since Hb A_{1c} constantly increases during
the red cell life-span, depending on the subject general glycometabolic
control, a parallel increase of Hb A_{1c} and cell density will be expected if
the density gradient separates the erythrocytes according to their age.
Our data are therefore in agreement with the few observations already
published on this parameter (12,13).
As expected, the glycohemoglobin levels in the ß-thal carrier are similar to
those found in normals (14).

Furthermore, the changes in red cell subpopulations that occur in whole blood upon in vitro storage have also been studied. The experiment was performed by storing a blood unit, collected from a healthy volunteer, in a blood bank refrigerator. The RBC fractionation and further analyses were performed after 0, 15 and 35 days, with the results shown in Fig. 2. It is evident, by examining the distribution patterns in Fig. 2A, the progressive enrichment that occurs in the lowest fractions as the time of storage goes by. Concomitant and consistent changes in MCV and in MCHC

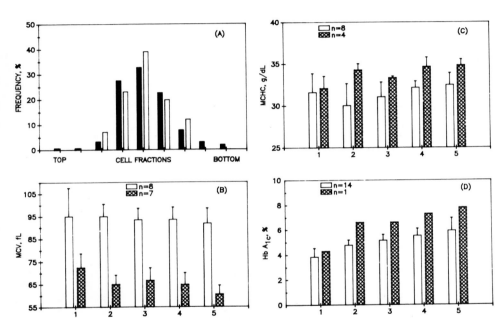

Fig. 1. (A) Percentage of cells applied to the gradient (frequency) vs. cell fractions in red cells from normal subjects (n=8) fractionated by the five (empty bars) and the nine fractions (full) density gradient systems. (B), (C) and (D) Mean cell volume (MCV), mean corpuscolar hemoglobin concentration (MCHC) and glycohemoglobin (Hb A$_{1c}$) in RBC from normals (empty bars) and ß-thal carriers (crosshatch) separated with the five fractions system. One SD error bars are reported. The horizontal axis is the same as in (A)

have also been observed. As far as the enzymatic activities are concerned, the data show that even after one month of storage the G6PD activity is fully preserved. The slight increase in activity measured in the top layer after 15 days of storage (Fig. 2B curve 1) is probably an artefact due to the larger experimental error which is relative to the measurements of this layer. The G6PD activity was also found to be preserved during the whole period of the experiment (not shown).

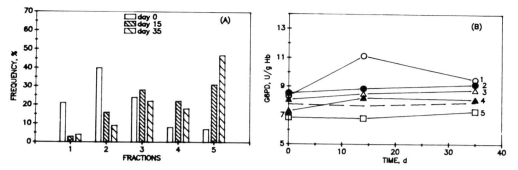

Fig. 2. RBC distribution (A) and G6PD activity (B) in a blood unit stored at
+4°C. The numbers by the curves in (B) refer to the position of the
cell fraction inside the gradient. The dashed line shows the G6PD
activity in unfractionated RBC.

The persistence of unaltered enzymatic activities after one month of
storage of the blood unit is in accordance with the well documented 24-hr
posttransfusion survival values and ATP contents of red cells stored in ACD
or CPD anticoagulants at +4°C for up to 30 days (15).

<u>Decline of G6PD and 6PGD Activities with Erythrocyte Aging</u>

Hexokinase, aldolase and pyruvate kinase are three erythrocyte enzymes
whose activity declines with a half-time significantly shorter than red
cell life-span (16). On the other hand, great majority of erythrocyte enzyme
manifest a very much slower decline in their activity during RBC aging.
These findings are widely accepted among the hematologists, but there is an
open controversy on the exact pattern of the decay in the enzymatic
activity. In fact, while some experimental evidence suggests a single
exponential decay of the activities (16,17), other data indicate a biphasic
tendency (18,19). The latter theory is well supported by clinical data and
has its biological explanation, as reported by Beutler (20). The reasons
of this debate seem to depend on the different methodologies used for the
isolation of the youngest and oldest erythrocytes.

In this work two slightly different methodologies for erythrocyte
fractionation in classes of increasing density have been described. The
first one is a modification of a previously described procedure, the second
one represents a further improvement of this technique. Indeed, the
expansion to the nine layer system permits the harvesting of light fractions
with higher reticulocyte counts. Therefore, in order to compare the two
techniques and to produce new experimental data, we decided to study the
decay of two relevant red cell enzymes, G6PD and 6PGD. Although these
enzymes have activities that are not as age-dependent as those of
hexokinase, aldolase and pyruvate kinase, they differ however from each
other in their half-life (see below).

The study was conducted on two different non G6PD-deficient groups:
normal subjects (four males and four females) and ß-thal carriers (four

females). The data were then elaborated according to Piomelli (17) on the assumption that the mean cell age at the center of the gradient is equal to half of the RBC life-apan (normals: 60 days; ß-thal carriers: 46 days, according to Gallo (21). The results are reported in Fig. 3 and 4. After examining the data of normal subjects analyzed with the five layer gradients (Fig. 3) we found that reticulocytes decrease rapidly from top to bottom. The extrapolation of the regression line (full line in Fig. 3A) on the x-axis in correspondance of 100% reticulocytes gives a point at -5.1 SD. This point corresponds to the position in the gradient where cells of age day 0 are to be theoretically found. The average G6PD activity in these cells can then be extrapolated from the full line in Fig. 3B and from this value a G6PD half-life of 69 days can be calculated. This data is in agreement with the 62 days half-life estimated by Piomelli (17). In another calculation (22) a value of 42 days was reported. The same kind of elaboration performed on G6PD (Fig. 3C) gave us a half-life of 149 days. No other similar calculations have been reported regarding this enzyme. However, a biphasic type of G6PD activity decay has been suggested (22).

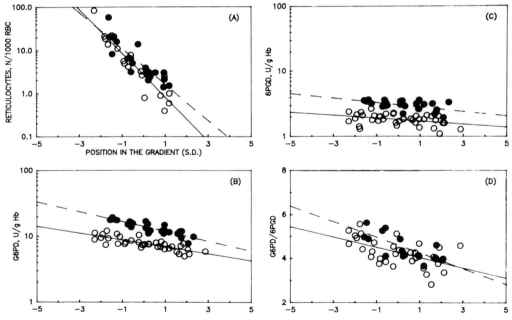

Fig. 3. Rate of decline of reticulocytes (A), G6PD (B), 6PGD (C) and G6PD/6PGD (D) in normals (empty circles, continuous lines) and in ß-thal carriers (full circles, dashed lines). All the analyses were performed by the five layer system.
The x-axis reports the position in the gradient relative to the cumulative distribution function, as proposed by Piomelli (17) and Turner (22). The horizontal scales in (B), (C) and (D) are the same as in (A).

The elaboration of the data collected on subjects heterozygous for ß-thalassemia (dashed lines in Fig. 3) produced the following half-lives: G6PD: 32 days; 6PGD: 71 days. The discrepancies between these values and the ones found for normal subjects are probably due to the calculation procedure based upon the average red cell ages at the center of the gradient.

In Fig. 3D the ratio between the activities of G6PD and 6PGD has been plotted. This ratio, because of the different half-lives of the two enzymes, is not constant in erythrocytes of different age and is subjects to a greater variability in ß-thal carriers than in normals.
This finding should be considered when the G6PD-6PGD ratio is used as a new parameter to individualize heterozigosity in the laboratory for G6PD deficiency among populations with high prevalence of thalassemia (23). Infact the G6PD activity expressed in terms of U/g Hb or of U/RBC has limited clinical use, unless separate reference intervals are established for normals and thalassemics (10).

The results of a study of a limited numer of non-thalassemics (one male and two females) by the "high resolution" method are reported in Fig.4. As illustrated in Fig. 4A a larger amount of reticulocytes in the upper layers can be obtained in comparison with those attainable with the five layer gradient system. Surprisingly, the plot of the G6PD activity decay (Fig. 4B) showed a distinct biphasic trend, in accordance with what has been reported by Jansen (18).

As a last observation we would like to point out that a variable number of elements (between 1 and 3/1000 RBC) positive to the fluorescent dye used in flow cytometry for reticulocyte counting, have always been found in the bottom layer with both gradient systems used in this work. It is unknown if the elements counted by the flow cytometer were reticulocytes or cellular debris. This finding confirms the doubt that Beutler (2) had about the isolation of the aged cells using density-based fractionation systems.

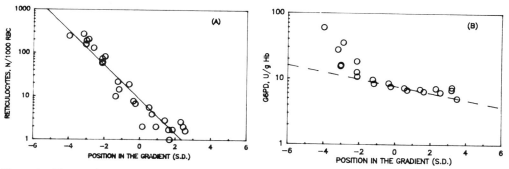

Fig. 4. The rates of decline of reticulocytes (A) and G6PD activity (B) in the nine layer gradients. The dashed line in (B) is the regression line from Fig. 3B, calculated from the data on normal subjects.

Clinical cases

The first clinical case regards a man with hemolytic anemia which was brought to our attention for investigations on G6PD.
The red cell G6PD activity was 1.0 U/g Hb. The interpretation of these laboratory data was hindered by the fact that the enzymatic assay was performed only one week after the transfusion of two blood units. Therefore, in order to overcome this problem, the G6PD activity was also assessed in fractionated erythrocytes, obtainig the results presented in Fig. 5. In this case, in contrast to our routine experience, G6PD activity correlates positively with cell density.

Our explanation is based on the hypothesis that the denser layers contain a vast percentage of erythrocytes from the donors. In fact we have previously illustrated (Fig. 2B) that the enzymatic activity is preserved within the blood unit even one month after blood donation.
If this hypothesis is true the only correct investigation on the G6PD phenotype of our patient would only have to be carried out on the lightest cells. The fact of finding a very low enzymatic activity in the top fraction may indicate the presence of the Gd Mediterranean variant, quite common in Italy and well consistent with the clinical history.

To strengthen our theory we have calculated the red cell G6PD activity that would be expected in a man who has received two blood units collected from healthy non G6PD-deficient donors. This activity should be approximately equal to 0.8 U/g Hb. If we make the hypothesis that the activity is confined to the cells moving in the lower cell fractions, then the average G6PD activity measured in fractions no. 3-5 should be 2.7 U/g (only 30% of the total RBC are distributed within these layers). This calculated number is reasonably close to the average G6PD activity measured in these fractions (2.2 U/g Hb).

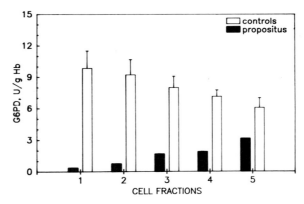

Fig. 5. Distribution of G6PD activity in fractioned erythrocytes from a patient with acute hemolytic anemia after a blood transfusion. The empty bars report the mean (±SD) G6PD activity in fractioned RBC from normals (data from Fig. 3B).

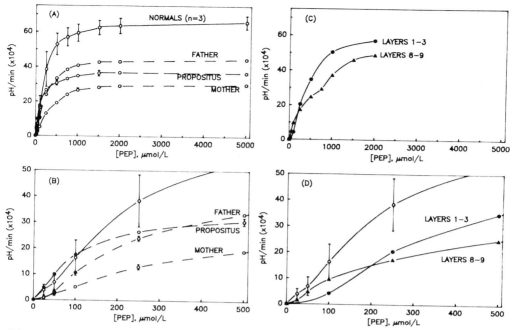

Fig. 6. PK kinetics plots of unfractionated (A and B) and fractionated RBC (C and D) from normal controls and from members of a family with PK deficiency. Curves in (C) and in (D) with full circles show the PK kinetics on pooled lightest (1-3) and heaviest (8-9) cell fractions. Empty symbols in (D) are related to controls.
The lower left corners of plots (A) and (C) are enlarged, respectively, in plots (B) and (D).

The second case concerns a young boy who is probably a compound heterozygote for PK deficiency. Due to his severe hemolytic anemia this patient is chronically transfused. Therefore the study of his red blood cell PK kinetics has to be performed on youngest isolated erythrocytes.

The preliminary results of this investigation are presented in Fig. 6. As illustred (Fig. 6A and B) the kinetics, performed on parents RBC, indicate a reduction in the averange activity and significant changes in the relevant kinetic parameters (not shown here in detail). The same analysis on the propositus gives an intermediate pattern between that of his parents and the normal controls, while the study performed on isolated subpopulations of his erythrocytes seem to indicate kinetic abnormalities, at low PEP concentration, especially in the younger cell fractions (Fig. 6D).

Experience with IHP Encapsulation in Erythrocytes

The entrapment of several substances in erythrocytes of mammalian species can be obtained by using various techniques, of which those based on hypotonic dialysis are probably the most popular (5). The results of the encapsulation process depend on a variety of factors, and therefore adequate quality control procedures are routinely employed in these experiments. The following are the usual parameters: RBC count, MCV, MCH, hematocrit, percentage encapsulation and carrier erythrocyte survival *in vivo*.

In this work a preliminary experience will be reported with the nine layer density gradient fractionation system applied to red cell subpopulations analysis before and after IHP entrapment. The results have been presented in Fig. 7. The outcome of the entrapment was evaluated by testing the shift in the oxygen dissociation curve (ODC), which moved from 25.8 to 32.5 mm Hg (data not shown in detail). The ODC shift was quite modest, in respect to previously published data (6), and therefore the analysis presented here relates more to the effects of the treatment per-se, than to the success of IHP encapsulation.

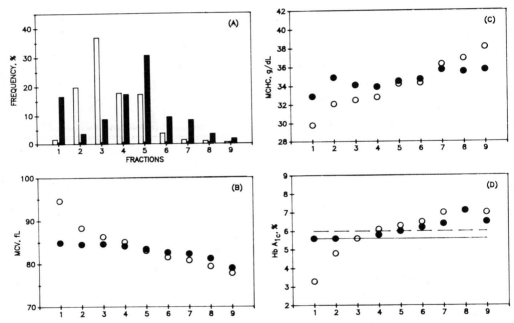

Fig. 7. Red cell distribution (A), MCV (B), MCHC (C) and Hb A_{1c} (D) in fractionated RBC before (empty symbols) and after (full symbols) treatment for IHP encapsulation. The glycohemoglobin concentrations in unfractionated red cells before (continuous line) and after (dashed) IHP treatent is reported in (D). The horizontal scale in (B), (C) and (D) is the same as in (A).

As shown in Fig. 7A the IHP treatment strongly modifies the RBC distribution, by increasing the proportion of the heaviest and top fractions. The MCV and MCHC patterns (Fig. 7B and C) indicate a reduction in MCV in the upper layers, with a correlated raise of their MCHC, together with an overall flattening of these parameters. Finally, from the analysis of Hb A_{1c} data a remarkable reduction of the fractions with the lowest glycohemoglobin concentration (layers 1 and 2) is apparent.

In conclusion, it seems that the red cell handling for IHP entrapment produces a modification in the subpopulation pattern, characterized by a significant reduction in the younger cell fractions.

CONCLUSIONS

The use of a resolutive density gradient system for the fractionation of erythrocytes, such as the nine layer system here described, instead of a traditional system, such as the five layer gradient system, allows for the deduction of different conclusions. That is the case in the G6PD activity decay pattern which occurs during red cell aging.

These kinds of fractionation techniques may be clinically useful in cases where, for various reasons, the laboratory tests can only be performed on the younger cell fractions. Examples of such cases may be those of transfused patients and of subjects with very unstable hemoglobin variants. Finally, we have been able to confirm that Hb A is a very important parameter to test the correlation between cell density and age.

ACKNOWLEDGMENT

Work partially supported by the C.N.R. Target Project on Biotechnology and Bioinstrumentation (grant no. 89.00189.70.115.14555) and by Regione Autonoma della Sardegna (Legge Regionale No. 11, 30-4-1990). The skillful assistance of Enrico Rosti (C.N.R., Milano) is gratefully acknowledged.

REFERENCES

1. M. R. Clark, Senescence of red blood cells: progress and problems, Physiol. Rev., 68:503 (1988).
2. E. Beutler, Isolation of the aged, Blood Cells, 14:1 (1988).
3. G. Salvo, P. Caprari, P. Samoggia, G. Mariani and A. M. Salvati, Human erythrocyte separation according to age on a discontinuous "Percoll" density gradient, Clin. Chim. Acta 122:293 (1982).
4. E. Beutler, K. G. Blume, J. C. Kaplan, J.W. Lhor, B. Ramot and W. N. Valentine, International Committee for Standardization in Haematology: recommended methods for red-cell enzyme analysis, Br. J. Haematol. 35:331 (1977).
5. G. L. Dale, D. G. Villacorte and E. Beutler, High-yield entrapment of proteins into erythrocytes, Biochem. Med. 18:220 (1977).
6. A. Zanella, L. Sabbioneda, F. Rossi, M. B. Colombo, V. Russo, G. Aguggini, G. Fiorelli, S. Villa, A. Brovelli, M. A. Castellana, C. Seppi, C. Barlassina and G. Sirchia, IHP-loaded red cells: in vitro and in vivo properties, Advances in the Biosciences (in press).
7. D. J. Ferguson, S. F. Lee and P. A. Gordon, Evaluation of reticulocyte counts by flow cytometry in a routine laboratory, Am. J. Haematol 33:13 (1990).

8. A. Carpinelli, A. Mosca and P. A. Bonini, Evaluation of a new semi-automated high performance liquid chromatography method for glycosylated haemoglobins, J. Autom. Chem. 8:192 (1986).

9. R. Paleari, F. Ceriotti, P. A. Bonini and A. Mosca, Standardization problems relevant to quantitative laboratory methods for glucose-6-phosphate dehydrogenase deficiency detection, Giorn. It. Chim. Clin. 15:191 (1990).

10. A. Mosca, M. Paderi, A. Sanna, R. Paleari, A. Cao and R. Galanello, Preliminary experience with the differential pH technique for glucose-6-phosphate dehydrogenase (G6PD) measurement in whole blood: application to an area with high prevalence of thalassaemia and G6PD deficiency, Haematologica 75:397 (1990).

11. M. Lakomek, H. Winkler, S. Linne and W. Schroter, Erythrocyte pyruvate kinase deficiency: a kinetic method for differentiation between heterozygosity and compound-heterozygosity, Am. J. Hematol. 31:225 (1989).

12. J. F. Fitzgibbons, R. D. Koler and R. T. Jones, Red cell age-related changes of hemoglobins Ala+b and Alc in normal and diabetic sunjects, J. Clin. Invest. 58:820 (1976).

13. K. Nakashima, O. Nishzaki, Y. Andoh, H. Takei, A. Iati and Y. Yoshida, Glycated hemoglobin in fractionated erythrocytes, Clin. Chem. 35:958 (1989).

14. S. Panzer, G. Kronik, K. Lechner, P. Bettelheim, E. Neumann and R. Dudczak, Glycosylated hemoglobins (GHb): an index of red cell survival, Blood 59:1348 (1982).

15. C. R. Valeri, "Blood banking and the Use of Frozen Blood Products", CRC Press, Cleveland (1976).

16. C. Seaman, S. Wyss and S. Piomelli, The decline in energetic metabolism with aging of the erythrocyte and its relationship to cell death, Am. J. Hematol. 8:31 (1980).

17. S. Piomelli, L. M. Corash, D. D. Davenport, J. Miraglia and E. L. Amorosi, In vivo lability of glucose-6-phosphate dehydrogenase in Gd A- and Gd Mediterranean deficiency, J. Clin. Invest. 47:940 (1968).

18. G. Jansen, L. Koenderman, G. Rijksen, B.P. Cats and G. E. J. Staal, Characteristics of hexokinase, pyruvate kinase, and glucose-6-phosphate dehydrogenase during adult and neonatal reticulocyte maturation, Am. J. Haematol. 20:203 (1985).

19. M. Magnani, V. Stocchi, L. Chiarantini, G. Serafini, M. Dachà and G. Fornaini, Rabbit red blood cell hexokinase decay mechanism during reticulocyte maturation, J. Biol. Chem. 261:8327 (1986).

20. E. Beutler, The relationship of red cell enzymes to red cell life-span, Blood Cells 14:69 (1988).

21. E. Gallo, P.G. Pich, G. Ricco, G. Scaglio, C. Camaschella and U. Mazza, The relationship between anemia, fecal stercobilinogen, erythrocyte survival and globin synthesis in heterozygotes for ß-thalassemia, Blood 46:693 (1975).

22. B. M. Turner, R. A. Fisher and H. Harris, The age related loss of activity of four enzymes in the human erythrocyte, Clin. Chim. Acta 50:85 (1974).

23. P. Arese, D. Ghigo and S. Traves, Le enzimopatie ereditarie del globulo rosso umano, in: "Aggiornamenti di Patologia Clinica", UTET, Torino (1987).

THE ROLE OF RED CELL AGING IN THE DIAGNOSIS OF GLYCOLYTIC

ENZYME DEFECTS

Gerard E. J. Staal and Gert Rijksen

University Hospital Utrecht
Department of Haematology Lab. Med. Enzymology
Utrecht, The Netherlands

INTRODUCTION

During the development from stem cell to mature erythrocyte, the human red cell loses several of its metabolic and synthetic capacities. The mature red cell is devoid of DNA and RNA and therefore lacks the possibility of protein synthesis. With respect to metabolism, the disappearance of mitochondria and ribosomes deprives the erythroid cell from obtaining energy through oxydative phosphorylation and from renewed enzyme required in metabolic processes. As a consequence the mature red cell is almost completely dependent upon anaerobic glycolysis as a source of energy supply during its 120 day mean life span. Despite the limited metabolic capacity of the glycolytic pathway in the mature erythrocyte sufficient ATP is available for processes involved in its optimal functioning i.e. keeping the haem iron in the reduced state, synthesis of glutathione, maintaining the electrochemical gradients over the plasma membrane, salvage of purine nucleotides and protection of hemoglobin against methemoglobin formation and oxidative denaturation. In reticulocytes the energy consumption pattern is completely different. First of all the production of ATP is about 100 times higher than in mature erythrocytes, partly by increased glycolytic enzyme activities and further by a still intact oxidative phosphorylation. The increased activities of glycolytic enzymes in young red cell fractions predominantly concern those of hexokinase, pyruvate kinase and glucose-6-phosphate dehydrogenase. The increased enzyme activities are linearly correlated with the reticulocyte counts in these fractions. This finding suggests that the presumed age dependency of these enzymes is mainly related to reticulocyte maturation. In patients with hemolytic anemia the diagnosis of a deficiency of one of these enzymes may be masked by an increased reticulocyte number. On the other hand in patients with severe transfusion-dependent hemolytic anemia, the diagnosis of an erythrocyte enzymopathy is often troubled by the presence of relatively large amounts of donor erythrocytes with normal enzyme activities. In this paper we present a few

Red Blood Cell Aging, Edited by M. Magnani and
A. De Flora, Plenum Press, New York, 1991

examples, mainly concerning abnormal pyruvate kinase, illustrating the above mentioned problems, in the diagnosis of an enzymopathy.

RESULTS

Hexokinase Deficiency

Hexokinase (ADP:D:hexose-6-phosphotransferase EC 2.7.1.1, HK) is one of the rate limiting enzymes in erythrocyte glycolysis. A few years ago we reported on a patient with hexokinase deficiency (2). The proband was a 19-yr-old woman, anemic since birth, with marked reticulocytosis. At the age of 2 yr, splenectomy was performed. The hematologic data of the patient are shown in Table I. The patient was slightly anemic with a pronounced reticulocytosis. Haptoglobin was low. Blood platelets and leukocytes were increased. The father and mother of the patient were first cousins.

Table 1. Routine Hematologic Data (from ref. 2)

	Patients	Normals
Erythrocytes ($\times 10^{12}$/liter)	2.33	3.7- 5.0
Hemoglobin (g/dl)	9.82	11.9- 15.5
Haptoglobin (mg/dl)	10	30 - 240
Fetal hemoglobin (%)	2.4	0.1- 1.5
Packed cell volume (ml/100 ml)	30	36 -
Mean corpuscular volume (fl)	134	84 - 104
Mean corpuscular hemoglobin (pg)	42.0	28.2- 35.7
Mean corpuscular hemoglobin concentration (g/100 ml erythrocytes	31.1	32.2- 37
Reticulocytes (%)	50	0.4- 1.8
Leukocytes ($\times 10^{9}$/liter)	17.8	4 - 10
Blood platelets ($\times 10^{9}$/liter)	475	150 - 300

When hexokinase activity is plotted against the number of reticulocytes (obtained from normal healthy individuals by discontinuous Percoll density gradient centrifugation) a linear relationship is found (fig. 1).

The higher the reticulocyte number, the higher the enzyme activity. The same is true for the enzyme pyruvate kinase (fig. 1) In the erythrocytes of the patient, the absolute activity of HK was normal or near the lower limit of normal, when compared to normal control. However when HK activity was plotted against the number of reticulocytes (obtained at different times), together with the pyruvate kinase activity in the same cell fraction and control reticulocytes, a real HK deficiency is detected as shown in fig. 1. Although the absolute HK activity is only about 25% of normal, when the degree of reticulocytosis and the activity of pyruvate kinase are taken into account. The parents and three of the sibs were found to be heterozygous for the defect. They showed decreased HK activities ranging from 50% to 67% of normal, whereas pyruvate kinase and G6PD were in the normal range. As HK deficiency is very rare and as the parents consanguineous, it is very probable that the patient is a true homozygote, rather than compound heterozygous for two different mutations. One has to realize that the HK activity in the homozygous patient is even higher than in the apparently heterozygous parents and sibs. This parents example clearly demonstrates that the HK deficiency may be masked by an increased reticulocyte number.

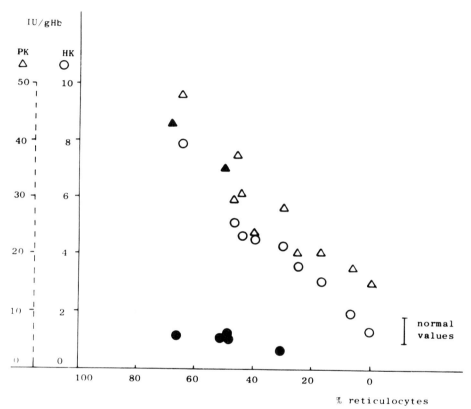

Fig. 1. Red cell activities of hexokinase (o,●) and pyruvate kinase (△,▲) of total cell fraction and density separated fractions as a function of the reticulocyte count. The open symbols represent normal controls, the closed symbols represent data from a hexokinase deficient patient.

Pyruvate kinase deficiency

Pyruvate kinase (ATP:pyruvate 2-0-phosphotransferase, EC 2.7.1.40, PK) is a tetramer consisting of identical or nearly identical subunits of about 60 kDa of which four forms exist in mammals named the M (or M_1)-, K (or M_2)-, L- and R-types (3). The L-type is mainly found in liver and to a lesser extent in kidney and intestine. R-type is present in erythrocytes.

There are important similarities between the L- and R-type. First, an antiserum raised in rabbits against pure human L-type PK does not discriminate between the liver and the erythrocyte enzyme (4).This indicates a large structural similarity between the enzymes from the two sources. Secondly, a deficiency of the human erythrocyte PK is also expressed in the liver (5). It has been demonstrated that two different mRNAs encode for the L- and R-type PK (6). The two types of mRNAs are transcribed from a single gene, presumably involving gene rearrangement or different processing of a common nuclear RNA precursor. In more recent studies it was shown that the L-type gene possesses two tissue specific promoters generating heterogenous mRNAs,(7-8). The same phenomenon is valid for the M and K type PK i.e. one gene but two mRNAs are involved in the synthesis of M- and K-type isozymes. This was proved by the determination of the complete nucleotide sequences for both M- and K-type pyruvate kinase from rat by sequencing the cDNAs (9). The derived amino acid sequences turned out to be identical except for one region of 45 residues (9). The K-type may be regarded as the prototype of PK and can be detected in early stages of development: i.e. during erythroid differentiation the K-type is replaced by the L-type pyruvate kinase.

L-type pyruvate kinase deficiency is a rather common erythroenzymopathy associated with chronic non-spherocytic hemolytic anemia. In patients with severe, transfusion-dependent, hemolytic anemia, the diagnosis of an erythrocytic enzymopathy is often troubled by the presence of relatively large amounts of donor erythrocytes with normal enzyme activities. Recently we could detect a pyruvate kinase deficiency in a 2-yr old child with hemolytic anemia, who has to be transfused every 4-6 weeks (10). As a result of these multiple transfusions her red cells consist for the major part of donor erythrocytes. Examination of red cell enzymes showed pyruvate kinase activities just above or at the lower limit of the normal range both in the patient and her parents, whereas hexokinase and glucose-6-phosphate dehydrogenase activities were quite normal. Therefore we tried to separate patient's own red cells from the donor cells in order to establish a possible diagnosis of pyruvate kinase deficiency. Separation of red cells into fractions of different density was achieved by using a discontinuous Percoll density gradient, prepared as described by Rennie et al. (11). The cellular distribution over the density layers is shown in table 2.

Table 2. Distribution of Red Cells After Density Fraction
(from ref. 10)

Percoll fraction	Percentage of cells		
(%)	Controls (n=19)		Patients
40		−	1.5
57	0.5	− 2	4.6
65	20	− 25	2.3
72	45	− 70	82
80	5	− 10	9

In controls hardly any cells were recovered from the 40% Percoll fraction, whereas a minor, but substantial amount of the patient's cells ended up in this fraction. Reticulocyte counts in the 57% Percoll fraction ranged up to 50% in normal controls and even higher in reticulocytosis controls. The activities of hexokinase, pyruvate kinase and glucose-6-phosphate dehydrogenase of fractionated control samples showed a steep decline with increasing density (Figs. 2,3 and Table 3).

The enzyme pattern in cell fractions prepared from the patient sample was clearly different. Hexokinase and glucose-6-phosphate dehydrogenase activities in the lowest density fractions were extremely high. On the contrary pyruvate kinase in the lowest density fractions were severely deficient (fig. 3). This conclusion was further substantiated by comparison of the ratio of pyruvate kinase over hexokinase activity (Fig. 4).

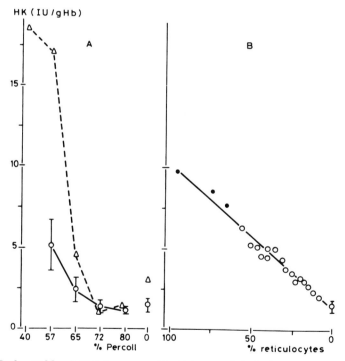

Fig. 2. A. Red cell hexokinase activities of total populations (0%) and density separated fractions of controls (o) and patient (△). The control values are expressed as mean ± SD (n=19).
B. Individual activities from the low density fractions are replotted against the reticulocyte counts in this fraction (r = 0.98); (o) represent data from normal controls, (●) from reticulocytosis controls. The brackets at 0% reticulocytes indicate the normal range of unfractionated controls (Mean ± SD).

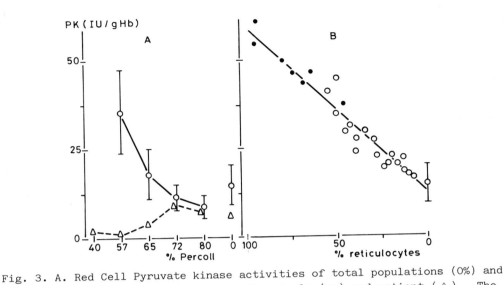

Fig. 3. A. Red Cell Pyruvate kinase activities of total populations (0%) and density separated fractions of controls (o) and patient (△). The control values are expressed as mean ± SD (n=25).
B. Individual activities from the low density fractions are replotted against the reticulocyte counts in this fraction (r=0.98); (o) represent data from normal controls, (•) from reticulocytosis controls.
The brackets at 0% reticulocytes indicate the normal range of unfractionated controls (Mean ± SD).

Table 3. Glycolytic Enzyme Activities of Whole Red Cell Populations and Density Separated Fractions (from ref. 10)

	Pyruvate Kinase[a]	Hexokinase[a]	Glucose-6-PD[a]
Patient			
Total cells	8.5	3.1	15.0
40% Percoll	1.9	18.6	50.0
58%	1.2	17.1	45.3
65%	4.4	4.6	33.9
72%	9.5	1.3	10.1
80%	7.6	1.4	5.3
Father	8.3	1.3	8.3
Mother	8.1	1.5	14.3
Normal controls[b]	11.4±3.0	1.32±0.40	8.9±2.5

[a]Enzyme activities are expressed in Units/gr Hb
[b]Normal values are mean ± SD (n = 30)

244

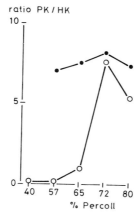

Fig. 4. Ratio of pyruvate kinase over hexokinase activity in density fractionated samples of controls (●) and patient (○). The values are derived from Figs 2 and 3.

In the lowest density fractions this ratio is approx. 70 times lower than might be expected suggesting that the residual pyruvate kinase activity in these cells is<2% of normal.

Detection of K-type pyruvate kinase

The cell fractions containing only enzyme deficient cells were too small to allow a complete enzymological and immunological characterization. Therefore, we confined ourselves to the investigation of the isozyme distribution by immunofluorescence of intact cells and by cellulose acetate electrophoresis of hemolysates. Fig. 5 shows the results of immunofluorescence experiments using monoclonal antibodies specific for K-type pyruvate kinase (ref. 12) and polyclonal antibodies against L-type. The enzyme deficient cells (40% Percoll fraction) showed only a faint immunoreaction with anti L-type antibodies, which was,however, significantly above background levels indicative for the presence of a small amount of enzyme protein. A much stronger immunoreactivity was apparent with anti K-type antibodies, whereas a completely negative reaction was obtained in fraction of comparable density from control patients with hemolytic anemia and reticulocytosis. These results strongly suggest that the residual pyruvate kinase activity in the patient's cells mainly originates from the presence of K-type pyruvate kinase. Cellulose acetate electrophoresis showed only the presence of K-type pyruvate kinase activity in the patient's sample, confirming and extending the above suggestion. Thus, there is a relatively small amount of immunoreactive L-type protein present in the patient's cells, which cannot however be detected by activity staining after electrophoresis, suggesting that it is enzymatically inactive.

So, in conclusion, using density fractionation erythroenzymopathies can be diagnosed in spite of the presence of massive numbers of donor cells.

Table 4. Erythrocyte Activities of Hexokinase, Glucose-6-P Dehydrogenase, and Pyruvate Kinase in Four Age-separated Fractions for Controls and Propositus (from Ref. 13)

	Percoll				
	0	58%	65%	72%	80%
Age-dependent separation					
Controls (n=6)					
Hexokinase	1.17 ± 0.19	2.61 ± 1.6	1.55 ± 0.30	1.16 ± 0.16	0.94 ± 0.20
Glucose-6-P dehydrogenase	13.7 ± 1.0	19.5 ± 8.0	15.7 ± 1.5	13.2 ± 1.5	9.8 ± 0.9
Pyruvate kinase	12.6 ± 1.6	22.8 ± 3.7	15.6 ± 2.4	11.5 ± 1.7	9.5 ± 1.9
Propositus					
Hexokinase	1.33	2.91	1.46	1.16	0.79
Glucose-6-P dehydrogenase	13.3	19.2	14.4	12.6	10.1
Pyruvate kinase	20.1	27.8	22.5	17.4	16.3
Cellular distribution (%)					
Controls (n=12)		1.8 ± 0.6	33.4 ± 15.0	58.5 ± 13.0	6.4 ± 6.2
Propositus		1	38	55	6

The enzyme activities (IU/g HB) of the unseparated samples (0) and the 58, 65, 72, and 80% Percoll fractions, respectively, are expressed as mean ± SD.

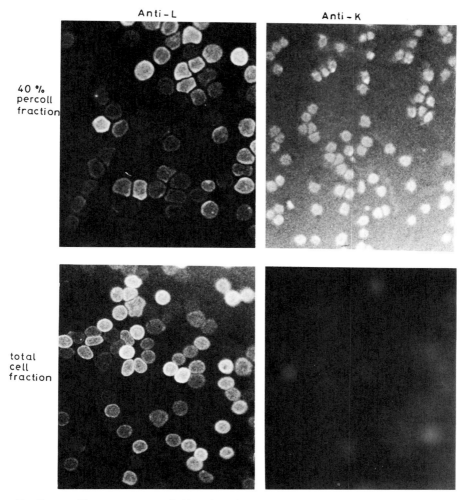

Anti-L   Anti-K

40 %
percoll
fraction

total
cell
fraction

Fig. 5. Innumofluorescence of the total cell population and the 40% Percoll
    fraction of the pyruvate kinase deficient patient using antibodies
    to L-type and K_2-type pyruvate kinase. Amplification was 125-fold,
    except for the upper right panel (40-fold).

Pyruvate kinase and the High ATP syndrome

    A few years ago (13) we described a family with an abnormality
characterized by high erythrocyte ATP (180-200% of the normal value),
increased pyruvate kinase activity (about twice the normal level), normal
lactate production, and decreased 2,3-DPG content. The mode of inheritance
was autosomal dominant; no hemolysis was observed. The high pyruvate kinase
activity was not related to an elevated reticulocyte count. This is clearly
illustrated in table IV; after separation according to age by discontinuous
Percoll density gradient centrifugation, pyruvate kinase was specifically
increased in the different fraction. In contrast, normal values were found
for the enzymes hexokinase and glucose-6-phosphate dehydrogenase. Also, the
cellular distribution of the patient's erythrocytes over the gradient was
normal (table IV).

It is well known that R-type pyruvate kinase from human erythrocytes shows allosteric properties (14). The kinetic behavior of pyruvate kinase can be explained by the R $\rightleftharpoons$ T model of Monod. In this model, the substrate phospho-enol-pyruvate, as well as Fru-1,6-P_2, cooperate in shifting the equilibrium towards the R-form, whereas the T-form is favored by ATP. The R-form is catalytically more active than the T-form. The patient's enzyme was characterized by an increase in the affinity for the substrate phospho-enol-pyruvate (the $K_{0.5}$ for this substrate was found to be decreased more that sixfold compared with controls); whereas the enzyme was not inhibited by the negative effector ATP. This means a shift in the Monod model. The decreased stimulation by Fru-1.6-P_2, is in agreement with this assumption. So, in conclusion, the increased pyruvate kinase detected in this particular case is not due to an increased reticulocyte count, but is probably caused by a mutation in the gene resulting in a structural change in the enzyme leading to a shift to the R-form.

REFERENCES

1. H. Chr. Benöhr, and H. D. Waller, Metabolism in hemolytic states, Clin. Haematol., 4:45 (1975).
2. G. Rijksen, J.-W. N. Akkerman, A. W. L. van den Wall, D. Pott hofstede, and G. E. J. Staal, Generalized Hexokinase deficiency in the blood cells of a patient with non-spherocytic hemolytic anemia, Blood, 61:12 (1983).
3. G. E. J. Staal and G. Rijksen, Regulation of pyruvate kinase in normal and pathological conditions, in "Regulation of Carbohydrate metabolism", R. Beitner, ed., vol. I:143, CRC Press, Boca Raton (1985).
4. K. Imamura, K. Taniuchi and T. Tanaka. Multimolecular forms of pyruvate kinase, J. Biochem. 71:1001 (1972).
5. G. E. J. Staal, G. Rijksen, A. M. C. Vlug, B. Vromen-van den Bos, J.-W. N. Akkerman, G. Gorter, J. Dierick and M. Petermans. Extreme deficiency of L-type pyruvate kinase with moderate clinical expression, Clin. Chim. Acta 118:241 (1982).
6. J. Marie, M. P. Simon, J. C. Dreyfus and A. Kahn, One gene, but two messenger RNAs encode liver L and red Cell L' pyruvate kinase subunits, Nature 292:70 (1981).
7. M. Cognet, Y. C. Lone, S. Vaulont, A. Kahn and J. Marie, Structure of the rat L-type pyruvate kinase gene, J. Mol. Biol. 196:11 (1987).
8 G. L. Tremp, J. Boquet, A. Kahn and D. Doegelen, Expression of the rat L-type pyruvate kinase gene from its dual erythroid and liver specific promoter in transgene mice, J. Biol. Chem. 264:19904 (1989).
9. T. Noguchi, H. Inone and T. Tanaka, The M - and M -types isozymes of rat pyruvate kinase are produced from the same gene by alternative RNA splicing, J. Biol. Chem. 261:13807 (1986).
10. G. Rijksen, A. J. P. Veerman, G.P.M. Schipper-kester and G. E. J. Staal, Diagnosis of pyruvate kinase deficiency in a transfusion dependent patient with severe hemolytic anemia, American J. of Hematology (1990) in press.
11. C. M. Rennie, S. Thompson, A. C. Parker, A. Maddy, Human erythrocyte

fraction in "Percoll" density gradients, Clin. Chim. Acta 98:119 (1979).

12. H. E. van Erp, P. J. M. Roholl, E.D. Sprengers, C. W. M. van Veelen, G .E. J. Staal, Production and characteriztion of monoclonal antibodies against human L-type pyruvate kinase, Eur. J. Cell Biol. 47:388 (1988).

13. G. E. J. Staal, G. Jansen and D. Roos, Pyruvate kinase and the "High ATP syndrome", J. Clin.Invest. 74: 231 (1984).

14. G. E. J. Staal, J. F. Koster, H. Hamp, L. Van Milligen-Boersma and C. Veeger, Human erythrocyte pyruvate kinase. Its purification and some properties. Biochim. Biophys. Acta 227:86 (1971).

MOLECULAR CHARACTERIZATION OF PLASMA MEMBRANE CALCIUM PUMP ISOFORMS

Emanuel E. Strehler, Roger Heim and Ernesto Carafoli

Laboratory for Biochemistry
Swiss Federal Institute of Technology (ETH)
Zurich, Switzerland

INTRODUCTION

The ability to precisely control the free intracellular Ca^{2+} concentration is of utmost importance for eukaryotic cell function. Membrane-intrinsic transport systems play an essential role in this intracellular Ca^{2+} regulation (1). Among the Ca^{2+} transport system of the plasma membrane, ATP-driven Ca^{2+} pumps have attracted increasing attention over the past few years. Because of their very low abundance ($\leq 0.1\%$ of the total membrane protein) these enzymes have traditionally been difficult to study on a biochemical level. In this situation, however, human red blood cells have proved to be an invaluable source for the analysis and the purification of these molecules. This is mainly due to the absence in mature human erythrocytes of an extensive intracellular membrane network as well as to the easy availability of large quantities of these cells. The rapid accumulation of molecular details concerning the regulation, structure and function of plasma membrane Ca^{2+} pumps has thus been largely due to the development of an efficient purification procedure for this enzyme from human erythrocytes (2), and to the recent success in the application of refined methods of protein biochemistry and recombinant DNA technology (3). In this report we shall begin with a brief summary of the general characteristics of typical plasma membrane Ca^{2+} pumps and will then describe the recent advances made in the molecular characterization of these enzymes. Particular attention will be paid to the molecular genetic basis of isoform diversity and to comparative aspects of their structural, functional and regulatory properties.

GENERAL PROPERTIES AND UBIQUITY OF PLASMA MEMBRANE CALCIUM PUMPS

Plasma membrane Ca^{2+} pumps (PMCAs) belong to the class of P-type ion transport ATPases (4) that form a phosphorylated intermediate during the reaction cycle and are inhibited by low concentrations of vanadate (5,6).

Among the P-type ATPases (other members of this class include SR/ER Ca^{2+} pumps, Na^+/K^+ pumps, H^+ pumps) (4) PMCAs are distinguished by their large molecular weight ($Mr \approx 140,000$, compared with 90,000 to 110,000 for the other P-type pumps) and by their stimulation by direct interaction with Ca^{2+}-calmodulin (5,6). This latter property has allowed the development of an efficient purification procedure for the PMCA, using calmodulin affinity chromatography as final step to achieve essentially homogeneous protein preparations (2). Purified PMCA from erythrocytes has been shown to be fully functional, as demonstrated by its high affinity for Ca^{2+} in the presence of calmodulin and its proper function as an ATP-dependent Ca^{2+} transport system in reconstituted liposomes (7). The stoichiometry of Ca^{2+} transported per ATP split has been shown to be 1:1, and experimental evidence suggests that protons are exchanged for Ca^{2+} during the transport cycle although the question of whether the stoichiometry for this exchange is 1:1 or 2:1 has not yet been conclusively answered (8,9). Besides its stimulation by calmodulin, the PMCA (notably that from cardiac sarcolemma (10) has been shown to be activated by phosphorylation via the cAMP-dependent protein kinase, and protein kinase C phosphorylation of the pump has also been reported (11). In addition, acid phospholipids as well as limited proteolysis are known to stimulate the activity of PMCAs (12) of particular interest is the activation of the pump by the Ca^{2+}-dependent protease calpain (13). Considering the importance for eukaryotic cells to be able to reach and maintain low intracellular free Ca^{2+} levels (typically around 200 nM in the resting state), and the role played by PMCAs in achieving this goal, it can be assumed that PMCAs are ubiquitous in eukaryotic cells. Plant (14), the protozoan "Leishmania donovani" (15) and even yeast (16) have been shown to contain high affinity Ca^{2+} ATPases.

PEPTIDES SEQUENCING AND MOLECULAR CLONING REVEAL THE PRESENCE OF A MULTIGENE FAMILY FOR MAMMALIAN PLASMA MEMBRANE CALCIUM PUMPS

Within less than three years since the establishment of the first amino acid sequence data (17,18) complete primary structures have become available for a number of different rat and human PMCAs, and it has become clear that PMCAs are encoded by multigene families in these organisms (19-22). Extensive peptide sequencing efforts and screenings of various rat, bovine and human cDNA libraries have led to the identification, on the nucleotide level, of a minimum of four but probably even five different mammalian PMCA isoforms encoded by as many different genes (reviewed in ref 23). "Corresponding" isoforms in rats and humans are $\geq 99\%$ identical, whereas the different PMCA gene products within a species show only 75% to 85% identity and 85% to 90% similarity (23). At least two different PMCAs (PMCA1 and PMCA4) are present in human erythrocytes (22), with isoform 4 being the major species (>70%) and isoform 1 -which appears to be a ubiquitous, "housekeeping" PMCA (21) - accounting for most of the remaining erythrocyte PMCA. PMCA isoform diversity is further increased via alternative splicing of the primary RNA transcripts (21,23,24). cDNA cloning and sequencing suggest the occurrence of alternative RNA splicing at up to four different locations within the primary transcript (23). An overview of currently available cDNA-deduced data on PMCA isoforms is presented in Table 1. These data show that PMCAs may vary significantly in size, with calculated Mr values ranging from 127,300 to 138,800. Most of the alternative splicing variants have not yet been shown to exist on the protein level; however, the location of the splicing pathway-dependent divergences within the encoded protein sequence indicate a functional significance for such isoform subtypes because these divergences coincide with important domains of the pump molecule (see below).

Table 1. cDNA Deduced Data on mammalian Plasma Membrane Ca^{2+} Pump isoform

Isoform[a]	Source	Splicing variants[b]	Amino acid residues[c]	Calculated M_r
PMCA 1	rat, human	1a	1176	129,500
		1b	1220	134,700
		1c	1249	137,800
		1d	1258	138,800
		1e[d]	1209	133,500
PMCA 2	rat, human[e]	2(b)	1198	132,600
		2(f)[e,f]	1099[e]	121,300[e]
PMCA 3	rat	3a	1159	127,300
PMCA 4	human	4a	1170	129,400
		4b	1205	133,900
		4g[g]	1169	130,300
(PMCA 5)	Bovine[h]	5 (?)	(71)[h]	(7,700)[h]

[a] See refs 21,22 for the rationale behind the general PMCA nomenclature.
[b] The nomenclature for alternative splicing variants uses lower case letters (following their "historical" pattern of detection) added to the corresponding isoform gene number.
[c] Full-length cDNA sequences have only been published for rat PMCA1a, human PMCA1b, rat PMCA2b, rat PMCA3a and human PMCA4b. All other isoform sizes are extrapolations assuming identity in the remaining regions with the corresponding known full-length sequence.
[d] Hypothetical isoform generated from an mRNA lacking putative alternatively spliced sequences at sites "C" and "D" (see below, Figure 2).
[e] Human sequence incomplete.
[f] Lacking putative alternatively spliced sequence at site "A" (Figure 2).
[g] Lacking putative alternatively spliced sequence at site "B" (Figure 2).
[h] Ony C-terminal 71 residues published (30).

IDENTIFICATION OF STRUCTURAL, FUNCTIONAL AND REGULATORY DOMAINS

Protein labeling techniques as well as limited proteolysis, and reconstitution experiments have been used in combination with biochemical, biophysical and immunological methods to learn more about the molecular proteins and the domain organization of PMCAs. These studies allowed a crude mapping of the location of several important domains in the protein (25). Direct peptide sequencing has allowed the elucidation of the primary structures of the region encompassing the aspartate residue that forms the phosphorylated intermediate (18), the FITC-sensitive site throught to be part of the ATP binding region (17), the high affinity calmodulin binding site (26), the site of phosphorylation by the cAMP dependent protein kinase (27) and the candidate region of interaction of the pump with acidic phospholipids (28).

Comparisons among differnt P-type pumps as well as computer-assisted secondary structure predictions reveal remarkable similarities in the overall arrangement of several domains in these proteins (20). Although the degree of overall sequence identity between different P-type pump types averages 25% or less there are several highly conserved regions in all of these proteins that correspond to structurally and functionally important domains (29). The available information has been used to build a planar model for the possible structural arrangement of ion-motive P-type ATPases and an example of such a model, as adapted for PMCA 1b, is shown in Figure 1. In this model the PMCA polypeptide may transverse the lipid bilayer up to ten times; four of the transmembrane regions are clustered towards the N-terminus whereas the remaining ones are located close to the C-terminus. Only small parts of the pump are exposed on the extracellular side of the membrane and both the N -and the C-terminus are located intracellularly. The major hydrophilic regions of the pump are located on the cytoplasmic side of the membrane and can be separated into three major sequence blocks. The first is located between the putative transmembrane regions 2 and 3 and contains the "transduction" (or phospatase) domain (29) as well as the putative phospholipid-sensitive region of the pump (see Figure 1).

The second and largest cytoplasmic domain contains the active site (encompassing the aspartate residue that is phosphorylated during the reaction cycle), the nucleotide binding domain and the "hinge" region possibly involved in mediating the effects of conformational changes of the former domains relative to each other. In contrast to the above domains which are highly conserved among all P-type cation transporters known to date (31) the third major block of cytoplasmic sequence (corresponding to the ≈150 C-terminal residues of PMCAs) is found only in the plasma membrane calcium pumps. This "regulatory" region accounts for the bulk of additional molecular mass of PMCAs when compared to other P-type pumps and contains the calmodulin binding domain and the site of cAMP dependent phosphorylation (Figure 1). Since both the catalytic domain and the "regulatory" C-terminal domain must be assumed to be located on the cytoplasmic face of the plasma membrane, an even number of bilayer-spanning segments is expected to be present between these two regions. The precise transmembrane topology has not yet been fully elucidated for any of the P-type ATPases but very recent studies using antibodies directed againyst synthetic peptides (F. Hofmann, personal communication) as well as the possible alternative splicing product encoding isoform hPMCA4g (see Table 1 and Figure 2) indicate that only four transmembrane segments may in fact be present in the C-terminal part of PMCAs.

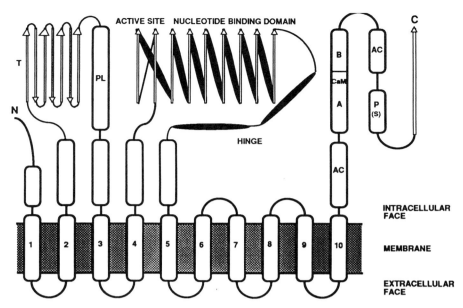

Fig. 1 Model for the topology of PMCAs. In this model a total of ten
transmembrane segments (TM 1 to 10) are shown (but see discussion of
this issue in the text), and the location of important domains has
been indicated. Open rods and filled bars represent putative alpha-
helices, and arrows denote beta-sheet structures. AC, acidic regions
flaking the calmodulin binding domain; C, C-terminus; CaM, calmodulin
binding domain separated into subdomains A and B; N, N-terminus; T,
trasduction domain; P(S), region containing the serine residue
susceptible to phosphorylation by the cAMP dependent protein kinase;
PL, phospholipid-sensitive region.

ELUCIDATION OF STRUCTURE-FUNCTION RELATIONSHIPS

    The importance for ion transport and ATPase activity of specific amino
acids has begun to be tested in the SR Ca^{2+} pump by site-directed
mutagenesis of the corresponding cDNA followed by expression and analysis
of the altered protein. These studies have, for example, clearly
established the importance of the active site aspartate residue and of the
FITC-sensitive lysine residue for proper ATPase activity (32). An essential
question concerns the identification of the amino acid residues involved in
high affinity Ca^{2+} binding. Site-directed mutagenesis studies on the SR Ca^{2+}
pump (33,34) demonstrate the importance of charged and of several proline
residues in the putative transmembrane segments M4, M5, M6 and M8. Although
comparable studies have not yet been performed on any PMCA many of the
findings reported for the SR Ca^{2+} pump will probably also be true for the
PMCA.

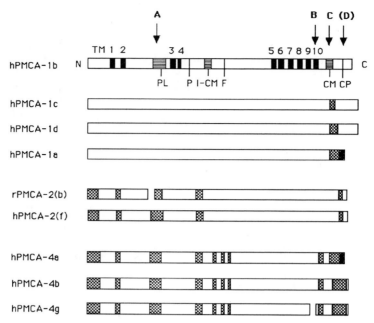

Fig. 2 Schematic comparison of PMCA isoforms. Arrows labeled A, B, C and D
denote regions of potential isoform variability due to alternative
RNA splicing. The possibility of alternative splicing at site D has
been suggested on the basis of a single rat cDNA clone (19) but
appears rather unlikely according to recent data on the human gene
structure (M.A. Strehler-Page, personal communication). Important
domains are indicated in the top scheme for isoform hPMCA1b: N,
N-terminus; TM 1 to 10, putative membrane spanning segments; PL,
phospholipid-sensitive region; P, location of the active site
aspartate residue; I-CM, region that interact with the CaM binding
domain in the absence of CaM (leading to the "inhibited" state); F,
FITC binding site; CM, calmodulin binding domain; CP, location of the
serine residue sensitive to phosphorylation by the cAMP dependent
protein kinase. The shaded areas in all other isoform schemes
indicate areas of particularly high amino acid sequence divergence
from hPMCA1b. Note that for hPMCA1a, 1c, 1d, hPMCA2(b), hPMCA4a and
hPMCA4g no complete cDNA sequences are available and that the
corresponding data represent extrapolations based on comparisons with
the known full-length sequences.

The structural and functional properties of the regulatory domain of
PMCAs are of particular interest. Studies of the calmodulin binding domain,
using synthetic peptides corresponding to the "wild-type" or to specifically
altered sequences show that a 20-residue peptide (the N-terminal subdomain
A in Fig. 1) still displays high affinity CaM binding and that a conserved
aromatic (tryptophan) residue in the N-terminal half of the CaM binding
region may play an important role in the binding process (35,36). Various
experimental data indicate that the CaM binding region serves as an
"inhibitory" domain that prevents high affinity Ca^{2+} binding of the pump in

the absence of Ca^{2+}/CaM (25). Accordingly, very recent studies suggest that in the absence of Ca^{2+}/CaM the CaM binding domain interacts with a specific region located between the phosphorylation and the FITC-binding site in the enzyme itself (ref. 35 and R. Falchetto, T. Vortherr, J. Brunner and E. C., manuscript submitted; see also Figure 2).

ANALYSIS OF PMCA ISOFORMS MAY YIELD INSIGHTS INTO FUNCTIONAL SPECIALIZATIONS

No rigorous study of the pattern of expression of PMCA isoforms in different mammalian tissues has been carried out yet, but Northern analysis of rat RNA indicates that PMCA1 is probably a ubiquitous isoform, whereas other PMCA gene products may be expressed in a more tissue-restricted manner (21). Of the differentially spliced human PMCA1 mRNAs hPMCA1c appears to be particularly abundant in adult skeletal muscle (24) whereas hPMCA4 is the major isoform of human erythrocytes (22).

Inspection of the areas of high sequence divergence among PMCAs (schematically summarized in Figure 2) shows that the N-termini are poorly conserved, but nothing is known yet about their function. A highly charged stretch in the transduction domain (around residue 300) is neither conserved in primary sequence nor in absolute length (23). This region appears to be involved in the phospholipid sensitivity of PMCAs (see figures 1 and 2) and isoform variability in this region may thus well be of functional significance. Alternative splicing at site A (Figure 2) would lead to additional variations in this region, further underlining the potential importance of the corresponding sequences for isoform-specific specializations. Interestingly, the region that may be involved in the intramolecular "inhibitory" interaction with the calmodulin binding domain is also not very well conserved between isoforms (Figure 2).

If they truly exist in vivo, isoforms generated from mRNAs with a spliced-out exon at site B (Figure 2) would be highly informative. Since such pumps would lack one of the predicted transmembrane segments in their C-terminal part (see hPMCA4g in Figure 2), the model shown in Figure 1 would necessarily have to be revised (see also above, discussion of that model).

Further striking PMCA isoform variability is apparent in the C-teminal region starting immediately after the highly conserved calmodulin binding subdomain A (Figure 2) (24). Although work with synthetic peptides has shown that a molecule of only 20 residue length (essentially corresponding to the CaM binding subdomain A) display CaM binding with a remarkably high affinity, synthetic 28mers corresponding to the entire CaM binding sequence (subdomains A and B) but containing either the hPMCA1b or the hPMCA1a/c/d subdomain B appear to differ significantly in their affinity towards CaM (J.T. Penniston, personal communication). In addition, the site of cAMP-dependent phosphorylation is differently spaced in the PMCA 1b vs. 1c and 1d subtypes, and it is absent altogether from the 1a isoform (24). Judged from amino acid sequence inspection and from in vitro phosphorylation studies a corresponding cAMP dependent phosphorylation site may also be lacking in other isoforms, most notably in the major enzyme (hPMCA4) of human erythrocytes (27). Alternative splicing at site D in Figure 2 might provide an alternative way to generate PMCA isoforms lacking the consensus cAMP dependent phosphorylation sequence. However, the evidence for alternative splicing at site D is weak and very recent data on the structural organization of the human PMCA1 gene even seem to preclude the possibility of splicing at that site (M.A. Strehler-Page, personal communication). A possible role of the phosphorylation by cAMP dependent protein kinase may consist in protecting the calmodulin binding "regulatory" (inhibitory) domain of the corresponding PMCA isoform from being attacked by the protease calpain. Calpain has been shown to constitutively activate the PMCA by specifically removing the "inhibitory" C-terminal region (13,37). While constitutive activation of the Ca^{2+} pump may not be desirable in many circumstances, conditions may also exist in which a maximally active Ca^{2+} extrusion system provides an advantage to the cell, e.g. to prevent premature ageing due to damaging reactions occurring at persistently elevated intracellular Ca^{2+} levels.

To obtain more detailed information on the functional and structural properties of the various C-terminal regulatory regions present in different PMCA isoform, studies have been initiated to overexpress the truncated C-termini of all products of alternative splicing at site C (Figure 2) of the human PMCA1. The corresponding cDNA fragments have been inserted in a suitable prokaryotic expression vector (pJLA 502 (38)) and the protein products synthesized in "E. coli" have recently been purified to homogeneity (F. Kessler, personal communication). Analyses of the precise CaM binding properties and the extent of phosphorylation by the cAMP dependent protein kinase, as well as the determination of the three-dimensional structure of the C-termini of each isoform subtype are now realistic goals for the near future. Such studies are expected to shed further light on the significance underlying the intriguing diversity of mammalian PMCA isoforms.

ACKNOWLEDGEMENTS

Supported by the Swiss National Science Foundation (grants 31-25285.88 and 31-28772.90). E.E.S. is the recipient of a START research career development award (31-27103.89) from the Swiss National Science Foundation.

REFERENCES

1. E. Carafoli, Intracellular calcium homeostasis, <u>Annu. Rev.Biochem.</u> 56:395 (1987).
2. V. Niggli, J. T. Penniston and E. Carafoli, Purification of the (Ca^{2+} Mg^{2+}) ATPase from human erythrocytes using a calmodulin affinity column, <u>J. Biol. Chem.</u> 254:9955 (1979).
3. E. Carafoli, A. K. Verma, P. James, E. Strehler and J. T. Penniston, The calcium pump of the plasma membrane: structure-function relationships, <u>in</u>: "Calcium Protein Signaling", H. Hidaka ed., Plenum Press, New York (1989).
4. P. L. Pedersen and E. Carafoli, Ion-motive ATPases, <u>Trends in Biochem. Sci.</u> 12:146 (1987).
5. H. J. Schatzmann, The calcium pump of erythrocytes and other animal

cells, in: "Membrane Transport of Calcium", E. Carafoli ed., Academic Press, London (1982).

6. J. T. Penniston, Plasma membrane Ca^{2+} pumps, in: "Calcium and Cell Function", N. Y. Cheung ed., Academic Press, New York (1983).

7. E. Carafoli and M. Zurini, The Ca^{2+} -pumping ATPase of plasma membranes. Purification, reconstitution and properties, Biochim. Biophys. Acta Rev. Bioenerg. 683:279 (1982).

8. V. Niggli, E. Sigel and E. Carafoli, The purified Ca^{2+} pump of human erythrocyte membranes catalyses an electroneutral Ca^{2+}:H^+ exchange in recostituted liposomal systems, J. Biol. Chem. 257:2350 (1982).

9. B. Gassner, S. Luterbacher, H. J. Schatzmann and A. Wüthrich, Dependence of the red blood cell calcium pump on the membrane potential, Cell Calcium 9:95 (1988).

10. L. Neyses, L. Reinlib and E. Carafoli, Phosphorylation of the Ca^{2+} -pumping ATPase of heart sarcolemma and erythrocyte plasma membrane by the cAMP-dependent protein kinase, J. Biol. Chem. 260:10283 (1985).

11. J. I. Smallwood, B. Gügi and H. Rasmussen, Regulation of erythrocyte Ca^{2+} pump activity by protein kinase C, J. Biol. Chem. 263:2195 (1988).

12. V. Niggli, E. S. Adunyah and E. Carafoli, Acidic phospholipids, unsaturated fatty acids and proteolysis mimic the effect of calmodulin on the purified erythrocyte Ca^{2+} -ATPase, J. Biol. Chem. 256:8588 (1981).

13. K. K. Wang, B. D. Roufogalis and A. Villalobo, Calpain I activates Ca^{2+} transport by the reconstituted erythrocyte Ca^{2+} pump J. Membr. Biol. 112:233 (1989).

14. C. Robinson, C. Larsson and T. J. Buckhout, Identification of a calmodulin-stimulated (Ca^{2+}-Mg^{2+})-ATPase in a plasma membrane fractions isolated from maize (Zea mays L.) leaves, Physiol. Plant 72:177 (1988).

15. J. Ghosh, M. Ray, S. Sarkar and A. Bhaduri, A high affinity Ca^{2+} -ATPase on the surface membrane of "Leishmania donovani promastigote, J. Biol. Chem. 265:11345 (1990).

16. H. K. Rudolph, A. Antebi, G. R. Fink, C. M. Buckley, T. E. Dorman, J. LeVitre, L. S. Davidow, J.-I. Mao and D. T. Moir, The yeast secretary pathway is perturbed by mutations in PMR1, a member of a Ca ATPase family, Cell 58:133 (1989).

17. A. G. Filoteo, J. P. Gorski and J. T. Penniston, The ATP-binding site of the erythrocyte membrane Ca^{2+} pump. Amino acid sequence of the fluorescein isothiocynate-reactive region J. Biol. Chem. 262:6526 (1987).

18. P. James, E. I. Zvaritch, M. I. Shakhparonov, J. T. Penniston and E. Carafoli, The amino acid sequence of the phosphorylation domain of the erythrocyte Ca^{2+} ATPase, Biochem. Biophys. Res. Commun. 149:7 (1987).

19. G. E. Shull and J. Greeb, Molecular cloning of two isoforms of the plasma membrane Ca^{2+}-transporting ATPase from rat brain. Structural and functional domains exhibit similarity to Na^+, K^+- and other cation transport ATPases, J. Biol. Chem. 263:8646 (1988).

20. A. K. Verma, A. G. Filoteo, D. R. Stanford, E. D. Wieben, J. T. Penniston, E. E. Strehler, R. Fischer, R. Heim, G. Vogel, S. Mathews, M.-A. Strehler-Page, P. James, T. Vorherr, J. Krebs and E. Carafoli,

Complete primary structure of a human plasma membrane Ca^{2+} pump, <u>J. Biol. Chem.</u> 263:14152 (1988).

21. J. Greeb and G. E. Shull, Molecular cloning of a third isoform of the calmodulin-sensitive plasma membrane Ca^{2+}-transporting ATPase that is expressed predominantly in brain and skeletal muscle, <u>J. Biol. Chem.</u> 264:18569 (1989).

22. E. E. Strehler, P. James, R. Fischer, R. Heim, T. Vorherr, A. G. Filoteo, J. T. Penniston and E. Carafoli, Peptide sequence analysis and molecular cloning reveal two calcium pump isoform in the human erythrocyte membrane, <u>J. Biol. Chem.</u> 265:2835 (1990).

23. E. E. Strehler, Recent advances in the molecular characterization of plasma membrane Ca^{2+} pumps, <u>J. Membr. Biol.</u>: in press.

24. E. E. Strehler, M.-A. Strehler-Page, G. Vogel and E. Carafoli, mRNAs for plasma membrane calcium pump isoforms differing in their regulatory domain are generated by alternative splicing that involves two internal donor sites in a single exon, <u>Proc. Natl. Acad. Sci. USA</u> 86:6908 (1989).

25. E. Carafoli, R. Fischer, P. James, J. Krebs, M. Maeda, A. Enyedi, A. Morelli and A. De Flora, The calcium pump of plasma membrane: recent studies on the purified enzyme and on its proteolytic fragments, with particular attention to the calmodulin binding domain, <u>in</u>: "Calcium Binding Proteins in Health and Disease", A. W. Norman, T. C. Vanaman and A. R. Means, eds., Academic Press, San Diego (1987).

26. P. James, M. Maedda, R. Fischer, A. K. Verma, J. Krebs, J. T. Penniston and E. Carafoli, Identification and primary structure of a calmodulin binding domain of the Ca^{2+} pump of human erythrocytes, <u>J. Biol. Chem.</u> 263:2905 (1988).

27. P. H. James, M. Pruschy, T. E. Vortherr, J. T. Penniston and E. Carafoli, Primary structure of the cAMP-dependent phosphorylation site of the plasma membrane calcium pump, <u>Biochemistry</u> 28:4253 (1989).

28. E. Zvaritch, P. James, T. Vorherr, R. Falchetto, N. Modyanov and E. Carafoli, Mapping of functional domains in the plasma membrane Ca^{2+} pump using trypsin proteolysis, Biochemistry 29: in press (1990)

29. N. M. Green and D. H. MacLennan, ATP driven ion pumps: an evolutionary mosaic, <u>Biochem. Soc. Trans.</u> 17:819 (1989).

30. P. Brandt, M. Zurini, R. L. Neve, R. E. Rhoads and T. C. Vanaman, A C-terminal, calmodulin-like regulatory domain from the plasma membrane Ca^{2+}-pumping ATPase, <u>Proc. Natl. Acad. Sci. USA</u> 85:2914 (1988).

31. N. M. Green, ATP-driven cation pumps: aligment of sequences, <u>Biochem. Soc. Trans.</u> 17: 970 (1989).

32. K. Maruyama and D. H. MacLennan, Mutation of aspartic acid-351, lysine-352, and lysine-515 alters the Ca^{2+} transport activity of the Ca^{2+}-ATPase expressed in CAS-1 cells, <u>Proc. Natl. Acad. Sci. USA</u> 85:3314 (1988).

33. D. M. Clarke, T. W. Loo, G. Inesi and D. H. MacLennan, Location of high affinity Ca^{2+}- binding sites within the predicted transmembrane domain of the sarcoplasmic reticulum Ca^{2+}-ATPase, <u>Nature</u> 339:476 (1989).

34. B. Vilsen, J. P. Andersen, D. M. Clarke and D. H. MacLennan, Functional consequences of proline mutations in the cytoplasmic and

transmembrane sectors of the Ca^{2+} - ATPase of sarcoplasmic reticulum, J. Biol. Chem. 264:21024 (1989).

35. A. Enyedi, T. Vorherr, P. James, D. J. McCormick, A. G. Filoteo, J. T. Penniston and E. Carafoli, The calmodulin binding domain of the plasma membrane Ca^{2+} pump interacts with calmodulin and with another part of the pump J. Biol. Chem. 264:12313 (1989).

36. T. Vorherr, P. James, J. Krebs, A. Enyedi, D. J. Mc Cormick, J. T. Penniston and E. Carafoli, Interaction of calmodulin with the calmodulin binding domain of the plasma membrane Ca^{2+} pump, Biochemistry 29:355 (1990).

37. P. James, T. Vorherr, J. Krebs, A. Morelli, G. Castello, D. J. McCormick, J. T. Penniston, A. DeFlora and E. Carafoli, Modulation of erythrocyte Ca^{2+} ATPase by selective calpain cleavage of the calmodulin-binding domain, J. Biol. Chem. 264:8289 (1989).

38. B. Schauder, H. Blocker, R. Frank and J. E. G. McCarthy, Inducible expression vectors incorporating the "Escherichia coli atpE" translational initiation region, Gene 52:279 (1987).

HUMAN ERYTHROCYTE D-ASPARTYL/L-ISOASPARTYL METHYLTRANSFERASES:

ENZYMES THAT RECOGNIZE AGE-DAMAGED PROTEINS

Diego Ingrosso[1,2] and Steven Clarke[2]

[1] Istituto di Biochimica delle Macromolecole
I Facoltà di Medicina e Chirurgia
Università di Napoli
Napoli, Italy
[2] Department of Chemistry and Biochemistry
and the Molecular Biology Institute
University of California Los Angeles
Los Angeles, California, U.S.A.

Among the large number of protein post-biosynthetic modifications described so far are a group of non-enzymatic reactions that reflect the spontaneous, intrinsic, decomposition of these macromolecules as they age in cells. These alterations include oxidation (1), formation of advanced glycosylation end products (2), and linked deamination/isomerization/racemization reactions (3). Our interest has been focused on the latter reactions that lead to the loss of L-aspartyl and L-asparaginyl residues in proteins and the recognition of the damaged proteins by enzymes that can lead to their cellular removal by repair or degradation reactions (4-9).

The chemical pathways of the spontaneous degradation of normal aspartyl and asparaginyl residues, leading to the succinimide-mediated formation of altered isoaspartyl or D-aspartyl residues, are shown in Fig. 1. Deamination of asparaginyl residues has been long recognized as a factor in protein aging (10), although it has only recently been appreciated that intermediate succinimide formation leads to isomerized, as well as deamidated products (3,11,12). It has also recently become clear that at least a subset of isoaspartyl residues in proteins arises from normal aspartyl residues rather than deamidation of asparaginyl residues (13,14). Racemization has also been regarded as an important component in the aging process (15), generating D-aspartyl residues from Asn/Asp, particularly in long-lived proteins of certain organs, tissues and cells, such as the lens, and erythrocytes (16). The mechanism for racemization appears to be common in part to the deamidation/isomerization, since it also occurs through a succinimide intermediate (Fig. 1).

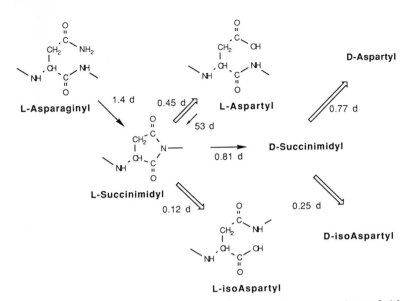

Fig. 1. Major chemical pathways for the spontaneous degradation of L-aspartyl and L-asparaginyl residues in proteins. An intramolecular reaction results in the five-membered succinimide ring, which can spontaneously open to give either aspartyl or isoaspartyl derivatives. The succinimide form is also racemization prone, leading to the formation of products in the D-configuration. For each reaction, a half-time (in days) is given for a model peptide (Val-Tyr-Pro-Asx-Gly-Ala) at 37°C, pH 7.4 (3).

Not all asparaginyl/aspartyl residues are equally susceptible to succinimide formation. The factors which promote these reactions include the presence of a glycyl or seryl residue as the following residue (3,12) and a conformation that allows the intramolecular reaction (11). The deamidation isomerization processes of several natural and recombinant proteins have been described, including epidermal growth factor (17,18), triose phosphate isomerase (19), seminal ribonuclease (20), calmodulin (14,21), glucagon (13), growth hormone (22), and serine hydroxymethyltransferase (23). Once formed, the isospartyl and D-aspartyl products of succinimide breakdown may induce significant changes in protein structure with the possibility of consequent functional impairment (3,11). For example brain calmodulin treated under conditions mimicking the _in vivo_ aging process and resulting in the appearance of isospartyl residues has been found to be less effective as an activator of the Ca^{2+}/calmodulin-dependent protein kinase (24).

The growing interest in these deamidation and isomerization reactions has been fueled by the discovery of an S-adenosylmethionine dependent D-aspartyl/L-isoaspartyl methyltransferase activity that selectively methyl esterifies protein substrates at these abnormal aspartyl residues (25-27).

The specificity of this methyltransferase has been studied in detail, using both synthetic and natural peptides. L-isoaspartyl residues have been

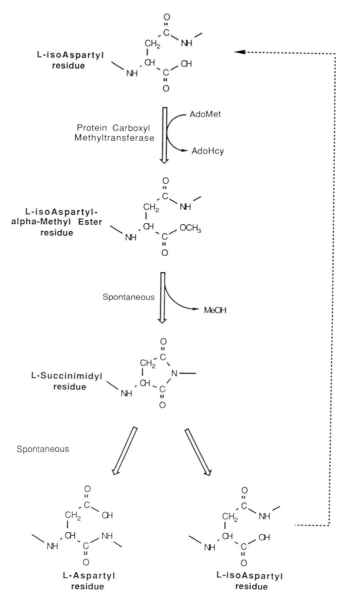

Fig. 2. Scheme for the repair of aspartyl isomerization with the D-aspartyl/L-isoaspartyl methyltransferase. The enzymatic methylation of L-isoaspartyl residues results in enhanced non-enzymatic succinimide formation. The hydrolysis of the succinimide results in normal aspartyl formation about 20% of the time in peptide model systems. The isoaspartyl residue that is formed can then be recycled through the system. In practice, racemization of the succinimide would also result in the formation of D-aspartyl and D-isoaspartyl residues.

found methylated *in vitro* in peptides and proteins (13,14,17,18,20,21,22, 26,27). The formation of D-aspartyl ß-methylesters has also been detected but to date only in red blood cell membrane and cytosolic proteins (25,28-31).

It has been proposed that this methyltransferase catalyzes the first step of a "repair" pathway, which is only partially enzymatic and would be effective in the conversion of isopeptide bonds into their normal counterpart (25). An example of such a repair pathway is given in Fig. 2. Although some L-isoaspartyl peptide is regenerated in this methylation/ demethylation "repair" pathway, repeated cycles of this sequence of reactions can lead to the complete removal of the isopeptide bond. No evidence, however, exists for enzymes that might regenerate the asparaginyl residue. On the other hand the complete "repair" of isopeptide bonds derived from aspartyl isomerization is possible through this mechanism alone.

The feasibility of this repair has been demonstrated *in vitro*, using both synthetic (20,32) and natural peptides (24,33). However some limitations do exist regarding the general applicability of this mechanism to all isoaspartyl-containing peptides and proteins. It has been shown that the three-dimensional structure of the protein substrate plays a major role in its methyl-accepting activity. For example isoaspartyl-containing deaminated forms of mouse epidermal growth factor (17) and bovine seminal ribonuclease (20) are not good substrates for this methyltransferase unless unfolded. Recent studies also indicate the existence of some sequence determinants for the methyl-acceptor activity (34).

We review here recent studies on this enzyme with special emphasis on the human erythrocyte model system and also discuss some original results from our laboratory. For further discussion of these topics we also refer the reader to the article by Galletti et al. in this volume.

S-ADENOSYLMETHIONINE SYNTHESIS AND PROTEIN METHYLATION IN ERYTHROCYTES

A remarkable amount of information on the D-aspartyl/L-isoaspartyl protein methyl transfer reactions has been obtained by using the red blood cell model (for reviews see refs. 4-9). These cells are no longer capable of replacing aged proteins by *de novo* synthesized molecules, but are still metabolically active in both S-adenosylmethionine synthesis (35,36) and protein carboxyl methylation (37). Human red blood cells actively synthesize S-adenosylmethionine from methionine and ATP through the action of an enzyme S-adenosylmethionine synthetase (35). Protein methyl esterification, catalyzed by the D-aspartyl/L isoaspartyl protein methyltransferase appears to be the most important adenosylmethionine-consuming reaction in this cell type (36,38). Endogenous membrane substrates include the cytoskeletal components of bands 2.1 and 4.1, as well as the integral anion transporter band 3 protein (31,37). Methyl esterification of cytosolic proteins including hemoglobin and carbonic anhydrase has also been found (29).

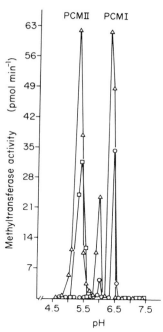

Fig. 3. Isoelectric focusing analysis of the isoenzymes of the D-aspartyl/
L-isoaspartyl methyltransferase at different purification steps. The
isoelectric focusing analysis was performed as described (48).
Enzyme preparations from both Sephadex G75 and MonoQ chromatography
(see Table I) were analyzed. Sephadex G75 gel filtration
chromatography was performed as described by Gilbert et al. (45).
MonoQ anion-exchange chromatography was performed according to
Ingrosso et al. (47). Open triangles: analysis of a G75 gel
filtration chromatography purified fraction. Open squares: analysis
of the isoenzyme II (PCM II) obtained from a MonoQ anion-exchange
chromatography. Open circles: analysis of both isoenzyme I (PCM I)
and the intermediate peak from MonoQ anion-exchange chromatography.

For sometime the understanding of the biological role of this
protein methylation was hampered by the scarce information on the site
specificity of this methyltransferase. The initial hypothesis of a
regulatory role for this type of enzyme did not fit evidence including the
fact that no methylation of "normal" L-aspartyl or L-glutamyl residues had
ever been shown (5). Moreover in intact erythrocytes, no regulation by
methylation either of band 3 anion transporter (39) or the response to
cytoskeletal stress induced by the ionophore A23187 (40) could be
demonstrated. The A23187-induced decrease in membrane protein methylation
results from the depletion of the intracellular pool of adenosylmethionine
due to a drop of ATP concentration (40,41).

An initial insight into the physiological role of the D-aspartyl/
L-isoaspartyl methyltransferase was provided by the demonstration that
the methyl esterification levels of the major methyl accepting membrane
proteins are significantly increased in density separated red cells of

increasing age (42,43). Just previously, D-aspartyl ß-methyl esters were isolated from hydrolysates of human erythrocyte membrane proteins methylated both in vitro and in intact cells, while no L-aspartyl ß-methyl ester could be detected (25,28). A small but significant increase in the amount of membrane protein D-aspartyl residues could also be detected in intact erythrocytes incubated for 24 h in the presence of methylation inhibitor precursors, thus supporting a repair role for the methyltransferase (44).

It still remains unexplained, however, why D-aspartyl residues in peptides do not appear to be good methyl-accepting sites for this enzyme. It is possible that the D-aspartyl/L-isoaspartyl methyltransferase has a relatively wide specificity toward isoaspartyl peptides and a much narrower specificity toward D-aspartyl-containing peptides.

CHARACTERIZATION OF THE RED CELL D-ASPARTYL/L-ISOASPARTYL METHYLTRANSFERASE

Erythrocytes (45) and brain (46) are major sources suitable for the purification of D-aspartyl/L-isoaspartyl methyltransferase and each tissue contains two major isoenzymes. These isoenzymes, purified to homogeneity from human red cells (Table I), show molecular weights of approximately 25,000 (48). These two species have isoelectric points that differ by about 1 pH unit, which corresponds to about 3-4 charge differences (48). At least one additional minor form is detectable by isoelectric focusing, with a pI value intermediate between the two major species (Fig. 3). This minor form can be effectively separated from the major isoenzyme I by a Mono Q-Fast Protein Liquid Chromatography step under the conditions described in Fig. 4.

In 1989 we determined the complete sequence of human erythrocyte D-aspartyl/L-isoaspartyl methyltransferase isoenzyme I (47,49). This structure includes 226 residues and an N-acetyl N-terminus (Fig. 5). Two alternative residues were detected at two positions, Ile/Leu[22] and Ile/Val[119] respectively, in peptides obtained from the enzyme prepared from different blood donors (47). These variations have been interpreted with the existence of two or more related genes and/or allelic variants for this methyltransferase, in analogy to other erythrocyte enzymes (47).

The sequence Asn[19]-Gly[20] was found susceptible to hydroxylamine cleavage and could be therefore in vivo prone to deamidation of isoenzyme I resulting in isoaspartyl formation through succinimide intermediate (47). It is interesting to note that this deamidation could explain the appearance of minor forms like peak III in Fig. 4, particularly during cell aging. It will be therefore interesting to test this enzyme form with respect to its catalytic activity and structural features, particularly in age-separated red blood cells.

It is noteworthy that the sequence of D-aspartyl/L-isoaspartyl methyltransferase isoenzymes I from human erythrocytes is very similar to the homologous sequences for the bovine brain (50) and rat brain (51) enzymes. This degree of similarity, which is comparable to that of highly conserved proteins like calmodulin (45), suggests that these enzymes play important roles in cellular metabolism.

Table I. Purification of Human Erythrocyte D-Aspartyl/L-Isoaspartyl Methyltransferase

Step	Volume (ml)	Units/ml[a]	total units	Yield[b] (%)	protein[c] (mg/ml)	specific activity[d]	x-fold purification
Cytosol	214	176	37664	100	38	4.6	1
55% amm. sulfate	9	1157	10413	27.6	13.1	88.3	19.2
Sephadex G-75	2	856	1712	4.5	0.6	1426.6	310.1
MonoQ Isoenzyme I	4	169	676	1.8	0.048	3520.8	765.4
MonoQ Isoenzyme II	4	106	424	1.1	0.093	1139.8	243.5
MonoQ peak III	2	70	140	0.4	0.050	1400.0	304.3
Vydac HPLC Isoenz. I	2	11[f]	22[f]	0.06	0.0012	9166.6[e]	1992.7[f]
Vydac HPLC Isoenz. II	2	18[f]	36[f]	0.09	0.0012	15000.0[e]	3260.8[f]

The purification was performed according to the general procedure reported by
Ingrosso et al. (47). In particular in the MonoQ anion exchange chromatography the linear elution gradient was
replaced by the non-linear gradient as in Fig. 4. [a] Units are picomoles of methyl esters formed on ovalbumin
per minute at 37°C, according to the enzyme assay described by Gilbert et al. (45), except that S-adenosyl
[methyl-^{14}C]-L-methionine (Amersham, specific activity 100 cpm/pmol) was used instead of the tritiated
compound. [b] Not corrected for the small amount of sample reserved for the assays and not applied at the next
step. [c] Protein concentration was measured by the BioRad Coomassie dye binding assay in all steps but in the
HPLC on Vydac column, where protein was determined by integrating the area of the peaks. [d] Picomoles per minute
per milligram of methyltransferase protein. [e] Active fraction eluted from the Sephadex G75 gel filtration column
were pooled and concentrated by ultrafiltration. [f] These values are probably underestimated, because of the
possible partial inactivation of the enzyme under the conditions of the HPLC.

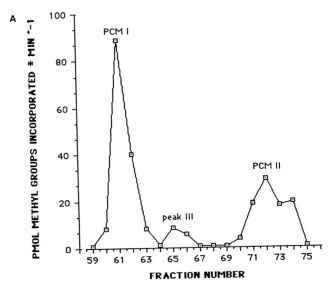

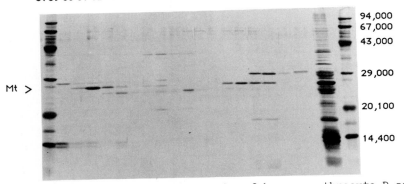

Fig. 4. MonoQ anion-exchange chromatography of human erythrocyte D-aspartyl/
L-isoaspartyl methyltransferase. Enzyme eluted from a Sephadex G75
column (45) was further purified by a MonoQ FPLC chromatography
(Pharmacia). In order to improve the resolution between PCM I and
the intermediate activity peak, the conditions for the MonoQ
chromatography, described in ref. (47), were slightly modified by
replacing the linear elution gradient with a non-linear gradient
formed by a Waters Model 680 Automated Gradient Controller (gradient
curve no. 8), A: activity elution profile showing isoenzymes I (PCM
I), isoenzyme II (PCM II), and the intermediate activity peak (peak
III) between the two major isoenzymes. B: SDS-PAGE analysis of the
individual peak fractions. Polypeptide bands were visualized by
silver stain. Mt indicates the position of the D-aspartyl/
L-isoaspartyl methyltransferase. St: molecular weight markers
including phosphorylase b (94,000), bovine serum albumin (67,000),
ovalbumin (43,000), carbonic anhydrase (29,000), soybean trypsin
inhibitor (20,100), and α-lactalbumin (14,400).

```
                10                  20                  30
   Ac A W K S G G A S H S E L I H N L R K N G I L K T D K V F E V M L A T D R

      A W K
              S G G A S H S E L I H N L R · · · · · · ·
                                              D K V F E V M L A T D R

       40                  50                  60                  70
   S H Y A K C N P Y M D S P Q S I G F Q A T I S A P H M H A Y A L E L L F

   S H Y A K
              C N P Y M D S P Q S I G F Q A T I S A P H M · · Y
                                                        A L E L L F >

               80
   D Q L H E G A K A L D V G S G S G I L T A C F A R M V G Q T G K V I G I

   > D Q L H E G A K
                      A L D V G S G S G I L T A C F A R
                                              M V G · T G K
                                                        V I G I >

       110                 120                 130                 140
   D H I K E L V D D S I N N V R K D D P T L L S S G R V Q L V V G D G R M

   > D H I ·
               E L V D D S I N N V R
                                   K D D P T L L S S G R
                                                  V Q L V V G D G R
                                                              M >

           150                 160                 170                 180
   G Y A E E A P Y D A I H V G A A A P V V P Q A L I D Q L K P G G R L I L

   > G Y A E E A P Y D A I H V G A A A P V V P Q A · · · · · ·
                                                    · · · ·
                                                          L I L >

           190                 200                 210
   P V G P A G G N Q M L E Q Y D K L Q D G S I K M K P L M G V I Y V P L T

   > P V G P A G G N Q M L E Q Y D K L Q D G S I K
                                      M K P L M G V I Y V P L T >

       220
   D K E K Q W S R W K

   > D K
          · · · · · · · ·
```

Fig. 5. Sequence comparison of human erythrocyte D–aspartyl/L–isoaspartyl methyltransferase isoenzyme I and II. The complete sequence of isoenzyme I is underlined. The sequences of peptides from the proteolytic cleavage of isoenzyme II are shown below the corresponding regions of isoenzyme I. The sequence of PCM I was determined as reported by Ingrosso et al. (47). Isoenzyme II peptides were obtained from the cleavage of isoenzyme II by either trypsin or "S. aureus" V8 protease. Dots indicate parts of the sequence of isoenzyme II which have not yet been determined.

We have also determined the sequence of proteolytic fragments accounting for approximately 90% of the sequence of the more basic isoenzyme II from human erythrocytes. We detected no sequence differences in these regions (Fig. 5). However, we have recently obtained evidence that there are differences at the C-termini that can account for the difference in the pI between these two major isoenzymes. Preliminary results from Southern blotting experiments (not shown) support a differential splicing mechanism for isoenzyme generation rather than the presence of multiple genes.

D-ASPARTYL/L-ISOASPARTYL METHYLATION IN HUMAN ERYTHROCYTES: A REPAIR MECHANISM OR A SIGNAL FOR THE DISPOSAL OF DAMAGED PROTEINS?

As detailed above, it has been hypothesized that enzymatic protein methylation at the level of abnormal aspartyl residues may be part of a mechanism for the repair of these altered residues.
Several studies provided the evidence that this repair is possible in vitro (20,24,32,33). However, it still needs to be demonstrated that this mechanism is also operative in vivo. In vivo studies have been made possible recently in Xenopus oocytes microinjected with isoaspartyl-containing peptides (52). In this system it has been shown that these isopeptides are methyl esterified and hydrolyzed. Only more indirect evidence is available in the erythrocyte model. However computer models based on the kinetic constants actually measured in vitro for both the deamidation/isomerization and the repair reactions would indicate that a repair pathway can be effective in preventing the accumulation of membrane protein damage in the red cell due to loss of asparaginyl and normal aspartyl residues as a result of deamidation and isomerization (7). This scheme is, however, still incomplete because of the lack of information about the further metabolism of D-aspartyl ß-methyl esters.

Another hypothesis about the physiological role of the D-aspartyl/ L-isoaspartyl methylation is that this post-biosynthetic protein modification could function as a signal for the selective disposal of proteins containing altered aspartyl residues. In this direction can be interpreted, for example, the recent discovery of a post-proline endopeptidase from human erythrocytes, which can also recognize and cleave succinimide-containing peptides (53). In our opinion, the recent improvements in the understanding of the structures of both the methyl accepting substrates and the methylating enzymes open new perspectives for testing these hypotheses in not only the erythrocyte model but in other cell systems as well.

REFERENCES

1. K. J. A. Davies, A. G. Wiese, A. Sevanian and E.H. Kim, Repair systems in oxidative stress, in "Molecular biology of aging", (C.E. Finch and T.E. Johnson, eds.), Wiley-Liss, New York, 1990, pp. 123-141.
2. H. Vlassara, Advanced non-enzymatic tissue glycosylation: mechanism implicated in the complications associated with aging, in "Molecular Biology of Aging", (C.E. Finch and T.E. Johnson, eds.), Wiley-Liss, New York, 1990, pp. 171-185.

3. T. Geiger and S. Clarke, Deamidation, isomerization and racemization at asparaginyl and aspartyl residues in peptides: succinimide-linked reactions that contribute to protein degradation, J. Biol. Chem. 262:785 (1987).

4. S. Clarke, The role of aspartic acid and asparagine residues in the aging of erythrocyte proteins: cellular metabolism of racemized and isomerized forms by methylation reactions, in "Cellular and Molecular Aspects of Aging: The Red Cell as a Model", (J.W. Eaton, D.K. Konzen and J.G. Write eds.) Alan R. Liss Inc., New York, 1985, pp. 91-103.

5. S. Clarke, Protein carboxyl methyltransferases: two distinct classes of enzymes, Ann. Rev. Biochem. 54:479 (1985).

6. J. Lowenson and S. Clarke, Does the chemical instability of aspartyl and asparaginyl residues in proteins contribute to erythrocyte aging? The role of protein carboxyl methylation reactions, Blood Cells 14:103 (1988).

7. J. D.Lowenson and S. Clarke, Spontaneous degradation and enzymatic repair of aspartyl and asparaginyl residues in aging red cell proteins analyzed by computer simulation, Gerontology, 1990 in press.

8. P Galletti, D. Ingrosso, C. Manna, P. Iardino and V. Zappia, in "Protein Metabolism in Aging", (H.L. Segal, M. Rothstein and E. Bergamini eds.) Wiley-Liss, New York, (1990), pp. 15-32.

9. I. M. Ota and S. Clarke, The function and enzymology of protein D-Aspartyl/L-Isoaspartyl methyltransferases in eukaryotic and prokaryotic cells, in "Protein Methylation", (W. K. Paik and S. Kim., eds.), CRC Press, Boca Raton, (1990) p. 179-194.

10. A. B. Robinson and C. J. Rudd, Deamidation of glutaminyl and asparaginyl residues in peptides and proteins, Curr. Top. Cell. Regul. 8:24 (1974).

11. S. Clarke, propensity for spontaneous succinimide formation from aspartyl and asparaginyl residues in cellular proteins, Int. J. Peptide Protein Res. 30:808 (1987).

12. R. C. Stephenson and S. Clarke, Succinimide formation from aspartyl and asparaginyl peptides as a model for the spontaneous degradation of proteins, J. Biol. Chem. 264: 6164 (1989).

13. I. M. Ota, L. Ding and S. Clarke, Methylation at specific altered aspartyl and asparaginyl residues in glucagon by the erythrocyte protein carboxyl methyltransferase, J. Biol. Chem. 262:8522 (1987).

14. I. M. Ota and S. Clarke, Enzymatic methylation of L-isoaspartyl residues derived from aspartyl residues in affinity-purified calmodulin: the role of conformational flexibility in spontaneous isoaspartyl formation, J. Biol. Chem. 264:54 (1989).

15. J. L. Bada, In vivo racemization in mammalian proteins, Methods Enzymol., 106:98 (1984).

16. L. S. Brunauer and S. Clarke, Age-dependent accumulation of protein residues which can be hydrolyzed to D-aspartic acid in human erythrocytes, J. Biol. Chem. 261:12538 (1986).

17. P. Galletti, P. Iardino, D. Ingrosso, C. Manna and V. Zappia, Enzymatic methyl esterification of a deamidated form of mouse epidermal growth factor, Int. J. Peptide Protein Res. 33: 397 (1989).

18. C. George-Nascimento, J. Lowenson, M. Borissenko, M. Calderon, A. Medina-Selby, J. Kuo, S. Clarke and A. Randolph, Replacement of a

labile aspartyl residue increases the stability of human epidermal growth factor, Biochemistry 29:9584 (1990).

19. K. U. Yuksel and R. W. Gracy, In vitro deamidation of human triosephosphate isomerase, Arch. Biochem. Biophys. 248: 452 (1986).

20. P. Galletti, A. Ciardiello, D. Ingrosso, A. Di Donato and G. D'Alessio, Repair of isopeptide bonds by protein carboxyl O-Methyltransferase: seminal ribonuclease as a model system, Biochemistry 27:1752 (1988).

21. I. M. Ota and S. Clarke, Multiple sites of methyl esterification of calmodulin in intact human erythrocytes, Arch. Biochem. Biophys. 279: 320 (1990).

22. B. A. Johnson, J. M. Shirokawa, W. S. Hancock, M. W. Spellman, L. J. Basa and D. W. Aswad, Formation of isoaspartate at two distinct sites during in vitro aging of human growth Hormone, J. Biol. Chem. 264:14262 (1989).

23. A. Artigues, A. Birkett and V. Schirch, Evidence for the in vivo deamidation and isomerization of an asparaginyl residue in cytosolic serine hydroxymethyltransferase, J. Biol. Chem. 265:4853, 1990.

24. B. A. Johnson, E.L. Langmach and D. W. Aswad, Partial repair of deamidation-damaged calmodulin by protein carboxyl methyltransferase, J. Biol. Chem. 262: 12283 (1987).

25. P. N. McFadden and S. Clarke, Methylation at D-aspartyl residues in red cells: a possible step in the repair of aged membrane proteins, Proc. Natl. Acad. Sci. U. S. A. 79:2460 (1982).

26. E. D. Murray Jr. and S.Clarke, Synthetic peptide substrates for the erythrocyte protein carboxyl methyltransferase: detection of a new site of methylation at isomerized L-aspartyl residues, J. Biol. Chem. 259:10722 (1984).

27. D. W. Aswad, Stoichiometric methylation of porcine adrenocorticotropin by protein carboxyl methyltransferase requires deamidation of asparagine 25: evidence for methylation at the alpha-carboxyl group of atypical L-isoaspartyl residues, J. Biol. Chem. 259: 10714 (1984).

28. C. M. O'Connor and S. Clarke, Methylation of erythrocyte membrane proteins at extracellular and intracellular D-aspartyl sites in vitro J. Biol. Chem. 258:8485 (1983).

29. C. M. O'Connor and S. Clarke, Carboxyl methylation of cytosolic proteins in intact human erythrocytes: Identification of numerous methyl accepting proteins including hemoglobin and carbonic anhydrase, J. Biol. Chem. 259: 2570 (1984).

30. C. M. O'Connor, D. W. Aswad and S. Clarke, Mammalian brain and erythrocyte carboxyl methyltransferases are similar enzymes that recognize both D-aspartyl and L-isoaspartyl residues in structurally altered protein substrates, Proc. Natl. Acad. Sci. U. S. A. 81:7757 (1984).

31. L. L. Lou and S. Clarke, Enzymatic methylation of band 3 anion transporter in intact human erythrocytes, Biochemistry, 26:52 (1987).

32. P. N. McFadden and S. Clarke, Conversion of isoaspartyl peptides to normal peptides: implications for the cellular repair of damaged proteins, Proc. Natl. Acad. Sci. U.S.A., 84:2595 (1987).

33. B. A. Johnson, E. D. Murray Jr., S. Clarke, D. B. Glass and D. W. Aswad, Protein carboxyl methyltransferase facilitates conversion of atypical L-isoaspartyl peptides to normal L-aspartyl peptides, J. Biol. Chem. 262: 5622 (1987).

274

34. J. D. Lowenson and S. Clarke, Identification of isoaspartyl-containing sequences in peptides and proteins that are unusually poor substrates for the class II protein carboxyl methyltransferase, J. Biol. Chem. 265:3106 (1990).
35. K. L. Oden and S. Clarke, S-adenosyl-L-methionine synthetase from human erythrocytes: role in the regulation of cellular S-adenosylmethionine levels, Biochemistry 22:2978 (1983).
36. J. R. Barber, B. H. Morimoto, L. S. Brunauer and S. Clarke, Metabolism of S-adenosyl-L-methionine in intact human erythrocytes, Biochim. Biophys. Acta 886:361 (1986).
37. C. Freitag and S. Clarke, Reversible methylation of cytoskeletal and membrane proteins in human erythrocytes, J. Biol. Chem. 256:6102 (1981).
38. C. A. Ladino and C. M. O'Connor, Protein carboxyl methylation and methyl ester turnover in density-fractionated human erythrocytes, Mech. Ageing Develop. 55:123 (1990).
39. L. L. Lou and S. Clarke, Carboxyl methylation of human erythrocyte band 3 in intact cells: Relation to anion transport activity, Biochem. J. 235:183 (1986).
40. J. R. Barber and S. Clarke, Membrane protein carboxyl methylation does not appear to be involved in the response of erythrocytes to cytoskeletal stress, Biochem. Biophys. Res. Commun. 123:133 (1984).
41. P. Galletti, D. Ingrosso, C. Manna, G. Pontoni and V. Zappia, Enzymatic basis for the calcium-induced decrease of membrane protein methyl esterification in intact erythrocytes: evidence for an impairment of S-adenosylmethionine synthesis, Eur. J. Biochem. 154:489 (1986).
42. J. R. Barber and S. Clarke, Membrane protein carboxyl methylation increases with human erythrocyte age: evidence for an increase in the number of methylatable sites, J. Biol. Chem. 258:1189 (1983).
43. P. Galletti, D. Ingrosso, A. Nappi, V. Gragnaniello, A. Iolascon and L. Pinto, Increased methyl esterification of membrane proteins in aged red blood cells: preferential esterification of ankyrin and band 4.1 cytoskeletal proteins, Eur. J. Biochem. 135:25 (1983).
44. J. R. Barber and S. Clarke, Inhibition of protein carboxyl methylation by S-adenosyl-L-homocysteine in intact erythrocytes: physiological consequences, J. Biol. Chem. 259:7115 (1984).
45. J. M. Gilbert, A. Fowler, J. Bleibaum and S. Clarke, Purification of homologous protein carboxyl methyltransferase isozymes from human and bovine erythrocytes, Biochemistry 27:5227 (1988).
46. D. W. Aswad and E. A. Deight, Purification and characterization of two distinct isozymes of protein carboxymethylase from bovine brain, J. Neurochem. 40:1718 (1983).
47. D. Ingrosso, A. W. Fowler, J. Bleibaum and S. Clarke, Sequence of the D-aspartyl/L-isoaspartyl protein methyltransferase from human erythrocytes: evidence for protein, DNA, RNA and small molecule S-adenosylmethionine-dependent methyltransferases, J. Biol. Chem. 264:20131 (1989).
48. I. M. Ota, J. M. Gilbert and S. Clarke, Two major isozymes of the protein D-aspartyl/L-isoaspartyl methyltransferase from human erythrocytes, Biochem. Biophys. Res. Commun. 151:1136 (1988).
49. D. Ingrosso, A. W. Fowler, K. Bleibaum and S. Clarke, Specificity of endoproteinase Asp-N (Pseudomonas fragi): Cleavage at glutamyl

residues in two proteins, <u>Biochem. Biophys. Res. Commun.</u> 162:1528 (1989).

50. W. J. Henzel, J. T. Stults, C. A. Hsu and D. W. Aswad, The primary structure of a protein carboxyl methyltransferase from bovine brain that selectively methylates L–isoaspartyl sites, <u>J. Biol. Chem.</u> 264: 15905 (1989).

51. M. Sato, T. Yoshida and S. Tuboi, Primary structure of rat brain protein carboxyl methyltransferase deduced from cDNA sequence, <u>Biochim. Biophys. Res. Commun.</u> 161:342 (1989).

52. E. A. Romanik and C. M. O'Connor, Methylation of microinjected isoaspartyl peptides in xenopus oocytes: competition with protein carboxyl methylation reactions, <u>J. Biol. Chem.</u> 264:14050 (1989).

53. J. Momand and S. Clarke, Rapid degradation of D– and L–succinimide –containing peptides by a post–proline endopeptidase from human erythrocytes, <u>Biochemistry</u> 26:7798 (1987).

THE ISOENZYMES OF MAMMALIAN HEXOKINASE: TISSUE SPECIFICITY AND
IN VIVO DECLINE

Koko Murakami and Sergio Piomelli

Division of Pediatric Hematology/Oncology
Columbia University
College of Physicians and Surgeons
New York, U.S.A.

INTRODUCTION

The reticulocyte loses most of its intracellular organelles during
maturation into the red blood cell (RBC). The RBC remains dependent on the
preformed enzymatic machinery throughout its 120 day life-span, as it is
incapable of protein synthesis. This provides an excellent model of
molecular aging.

Previous studies from our laboratory have suggested that RBC senescence
reflects a summation of metabolic failures, that may play an important role
in the removal of the senescent RBC, rather than molecular aging of a single
enzyme (1).

HK (EC 2.7.1.1.) appears to contribute most to the metabolic RBC
senescence: it is a rate-limiting step with marked regulatory properties,
and sharply declines during the aging processes (2-5).

Heterogeneous forms of HKI in the RBC were described originally by
electrophoresis (6,7). With column chromatography, a subtype was well
separated from HKI in rabbit RBC (8). In humans, however, the RBC isoenzymes
have been reported by some as heterogeneous forms of HKI (9-10), while
others have reported two isoenzymes (11).

HK$_R$: an RBC-specific HK isoenzyme

We have employed a fast protein liquid chromatography (FPLC) system
using a MonoQ column (Pharmacia) to separate the HK isoenzymes of human
tissues. This method proved extremely efficient to characterize the human
RBC HK isoenzymes because of its high resolution and rapid separation. Our
results indicate that HK$_R$, an isoenzyme present in the RBC along with HK$_I$,
is specific to the RBC and rapidly decays during cell aging.

The HK from RBC hemolysates was first adsorbed onto the MonoQ column in a buffer with no salt, then eluted by a linear gradient of KCl. The HK of human RBC was well separated into two peaks, one eluting at 120 mM KCl (HK_I) and another at 160 mM KCl (HK_R).

To verify the specificity of the RBC enzyme, the HK of human platelets and WBC and of liver extracts was similarly analyzed by MonoQ chromatography. The platelet HK consisted only of HK_I, while tha WBC contained also KH_{II}. Therefore, HK_R, that eluted at a lower salt concentration than HK_{II}, cannot represent either WBC or platelet contamination.

The HK from liver was separated by MonoQ column into two major forms: HK_I and HK_{III}: this eluted later than HK_{II} and was inhibited by glucose at a high concentration. Thus, HK_R, eluting between HK_I and HK_{II}, appeared different from any HK isoenzyme in other blood cell types or other tissues tested. It most likely represents a unique isoenzyme of the RBC. It behaved kinetically as HK_I.

When the hemolysate of human RBC was first directly applied on DEAE-cellulose to wash out hemoglobin and later chromatographed on a BioGel column, the HK eluted in a sharp peak. The fractions from this were individually analyzed on MonoQ columns. The first of these fractions contained most of the activity with a small percentage of the activity due to HK ; HK dominated in the last Biogel fraction. Thus, HK_R that clearly migrated faster than HK_I on BioGel, appears to be larger than HK_I by several KDa.

The particulate-bound HK in a mitochondria-enriched fraction from human reticulocytes was solubilized with G6P. The bound form eluted specifically with G6P. The G6P-solubilized HK was analyzed on a MonoQ column; it eluted in a closely overlapping peak with HK_I, and no HK_R was found: thus, also in the human RBC, HK_R is not a mitochondria-bound isoenzyme.

The HK isoenzyme MonoQ profile of human and rabbit RBC were compared: the two isoenzymes of human RBC eluted at 120 and 160 mM KCl, whereas those of rabbit RBC eluted at 120 and 180 mM KCl respectively. Thus the rabbit RBC isoenzymes were even further apart from each other than those of human RBC.

Our study demonstrates that HK_R, the second form of human RBC HK, is eluted in a position different from any other HK isoenzyme present in other tissues. The HK_R isoenzyme appears therefore to be unique to the RBC.

The age dependency of the RBC HK isoenzymes

The HK isoenzyme patterns of normal and reticulocyte-rich RBC suspensions were compared. RBC from normal blood contained approximately equal activities of HK_I and HK_R. On the contrary, in reticulocyte-rich RBC suspensions, HK_R was predominant, representing 75-80% of the total HK activity.

RBC were then ultracentrifuged by buoyant density on a discontinuous gradient of arabinogalactane, yielding fractions of increasing mean cell age. In the lightest cell fraction (mean age=22 days), HK_R predominated constituting 82% of total activity. HK_R then decreased dramatically in the successive layers, in inverse proportion to the age of the cells. In the heaviest cells (mean age=79 days) HK_R was negligible. It was thus clear that HK_I was relatively stable and that HK_R rapidly decayed during aging of the cells.

In normal RBC, the activities of HK_I and HK_R were approximately equal. In reticulocyte-rich RBC suspensions, HK_R was predominant. When RBC were separated on buoyant density gradients into age-dependent layers (12), HK_R appeared to be predominat in the youngest cell population, and negligible in the oldest RBC. Thus, HK_3, the RBC-specific isoenzyme, appears to be present at a high level in the reticulocyte and to degrade rapidly.

From the relative specific acticity in the layers, the t1/2 of both isoenzymes could be estimated (13). While HK_R decreased very rapidly, with a t1/2 of ~10 days, HK_I was much more stable with a t1/2 of ~ 66 days. It appears therefore that the rapid <u>in vivo</u> decay of HK reflects the rapid decay of HK . Our findings for human RBC HK closely parallel the findings by Magnani et al in the rabbit. In that species as well, in fact, the second RBC isoenzyme decreases during cell aging.

Therefore, it appears likely that the presence of an isoenzyme of HK in RBC precursors, that rapidly decays upon entry into the circulation, is a general phenomenon.

The genetic control of the RBC HK isoenzymes

Both isoenzymes of human RBC HK appear to behave kinetically as HK_I, as do also the two isoenzymes of rabbit RBC. Magnani et al. suggested that the subtypes of RBC HK result from post-translational modifications of HKI (14).

Our studies indicated that HK is an isozyme specific to the RBC, not found in any other tissues and cells. HK_I, but not HK_R, binds to the mitochondrial fraction. HK_R is clearly larger by several KDa than HK_I; moreover, the two forms independently decay at different rates. Thus, our data suggest that HK_I and HK_R are separate gene products independently regulated, and that the HK_R gene is expressed only in the RBC. Although HK_I is smaller than HK_R, it does not appear to be a degradative intemediate of HK_R. In fact, as HK_R decreases, HK_I does not increase and the two forms decay at independent rates.

The observations in patients with non-spherocytic hemolytic anemia associated with HK-deficiency also support the hypothesis of two independent genes.

In the RBC of a case reported by Rijksen et al the cathodal subtype (HK_I) was missing, but the anodal subtype (HK_R) was present (15).

Similarly, in the RBC of the case reported by the group of Magnani and Stocchi as HK Melzo, HK_R was by far the most preponderant subtype (16). In the platelets, in both cases, HK activity appeared also reduced. In cultured fibroblasts from HK Melzo abnormal kinetics were observed (17).

In another patient reported by the group of Magnani and Stocchi as HK Napoli, abnormal kinetics were reported in both RBC and fibroblasts (18). (It must be noted that this individual is most likely just a carrier of a variant HKI and certainly does not suffer from chronic hemolytic anemia).

Thus, as expected, in all these cases, the abnormality of HKI was expressed in all tissues that contain it.
The opposite situation was observed, however, in an individual without hemolysis reported by Altay et al. (19).

In her RBC HK_R was missing, but HK_I was present; there was no abnormality in other tissues, and the defect appeared limited to the RBC, as it is to be expected if HK_R is RBC-specific.

The observations in all these patients appear therefore precisely consistent with our hypothesis that the HK_R gene is independently regulated from the gene for HK_I and is specific to the RBC.
There appear to be two distinct congenital defects for HK_I and HK_R respectively, as expected from mutations at two independent loci:

when HKI is affected, the defect can be expressed in the RBC as well as in other tissues that usually contain HK_I :

instead when HK_R is affected, the defect can be expressed only in the RBC.

The hypothesis of post-translational modification is thus not supported by the current evidence.

The hexokinase degrading activity

An activity which degrades HK (HKDA) was demonstrated in crude tissue extracts. HKDA resulted in the decay of both HK function, measured by enzymatic assay, and HK protein, silver-stained SDS-PAGE digestion, regardleess of the source of isozyme.

Reticulocyte HK_R and HK_I, brain HK_I, and placenta HK_I all showed similar HKDA. The specific activities of HKDA (units HK activity degraded per units HK activity present initially) were also comparable among these preparations.

The inactivation of the HK enzymatic activity induced by HKDA was irreversible and the decrease of the HK protein, measured by SDS-PAGE analysis, parallelled it. The decay of HK activity seems due to proteolysis, and not to inactivation, and HKDA appears to be a proteolytic activity.
During an extensive process of purification, HKDA was always associated exclusively with the HK-containing fractions. HKDA, thus, appeared to induce proteolysis either as a function of HK itself or of a protease very tightly associated with it.

The role of ubiquitin in the HKDA

The degradation of HK protein has been observed in vitro in the presence of MgATP (20). Among various mechanisms hypothesized to explain the MgATP requirement, one proposes that ubiquitin is conjugated to the protein, resulting in a substrate which is then recognized by protease(s) (21). This pathway has been reported to decline during the aging of the red blood cell (22,23).

Magnani et al. using an antibody have shown that HK is degraded in a reticulocyte-rich lysate. Using ^{125}I-labeled HK, they suggested that HK degradation takes place through a ubiquitin-dependent mechanism. However, as their study shows the native HK and ^{125}I-labeled HK (inactive) are degraded by different mechanisms, the significance of the degradation of the inactive ^{125}I-labeled HK remains unclear (24).

Other proteases which require only ATP have also been reported in the reticulocyte; thus a ubiquitin-dependent pathway is not the only proteolytic system present that requires ATP (25,26).

The supernatant from a reticulocyte-rich red blood cell lysate (containing predominantly HK_R) was adsorbed on DEAE52, and the hemoglobin together with the ubiquitin was washed out. The residual proteins were then eluted with 400 mM KCl and precipitated with ammonium sulfate (27). Dialyzed aliquots of these preparations were incubated in the presence of glucose, with or without 5 mM ATP. The HK activity decayed only in the presence of ATP, but was stable without ATP. When ubiquitin was included with ATP, the HK activity did not decay any faster.

In these rather crude preparations, some residual ubiquitin conjugated to endogenous proteins may be already present. It is therefore possible that it be released and then activate ubiquitin-dependent proteolysis.

These observations indicate that additional exogenous ubiquitin does not affect HK decay. These observations do not absolutely rule out a possible role for endogenous ubiquitin, but make it appear highly unlikely.

In the reticulocytes, other ATP-requiring proteolytic systems have been shown to be responsible for the degradation of ornithine decarboxylase and other proteins (25,26). Thus, the degradation pathway for cellular proteins may be more complex than is presently thought.

HKDA is associated with the HK molecule itself

All preparations of HK tested, even when purified to near homogeneity, showed HKDA. During purification, HKDA was always isolated in the same fractions as the HK enzymatic activity itself. The other fractions devoid of enzymatic activity did not degrade HK, whether ATP was present or absent.

Since these preparations were extensively purified, these experiments therefore suggest that either HK itself contains its own hydrolytic activity or that HKDA results from a protease very closely associated with HK.

Ovalbumin, included as an internal standard, was neither degraded along with HK itself, nor did it inhibit HK decay. Also other proteins, such as SCMBSA and pyruvate kinase, were not degraded along with HK preparations.

Thus, the proteolytic activity of HKDA that remained associated with all purified HK preparations appeared to be specific for the HK molecule itself.

Since the HKDA was exclusively associated with the HK enzymatic activity it could be due to a protease tightly bound to the HK molecule or to a property of the HK molecule itself. It appears possible that HKDA is a function of HK itself, since homogeneous preparations of HK all contained HKDA, but showed no proteolytic activities against other proteins.

The different rates of decline in vivo of HKR and HKI

We have shown that in the RBC in vivo, the rate of degradation of HK_R is much faster than that of HK_I.

This difference cannot be explained from the present observations, since the rate of decay for all HK isozymes in vitro was the same, regardless of the tissue of origin. Thus the difference cannot be due to a greater intrinsic instability of HK_R or HK_I.

Magnani et al. reported that HK_I and HK_R degraded at similar rates in reticulocyte extracts (28). They suggested that HK_I may be stabilized in vivo by binding to mitochondria, while HK_R, which does not bind to subcellular structures may degrade more rapidly.

This appears, at present, the most plausible hypothesis for the different rates of decline in vivo of the two HK isozymes.

CONCLUSIONS

Two distinct HK isozymes are present in the RBC: HK_R and HK_I.

HK_R is RBC-specific, is predominant in the reticulocytes and declines in vivo at a very rapid rate ($t\frac{1}{2}$ = ~10 days).

HK_I is common to many other body tissues, and declines in vivo in the RBC at a much slower rate ($t\frac{1}{2}$ = ~66 days).
HK_I and HK_R are controlled by two separate genes: a defect of HK_I manifests itself in many tissues, a defect of HK_R manifests itself only in the RBC.

HKDA, a specific HK degrading activity present in tissue extracts, appears to be a proteolytic activity.

HKDA activity appears not to be ubiquitin dependent. It is most likely a property of the HK molecule itself.

HK_R is not bound to mitochondria, while HK_I is. This difference probably explains their different rates of in vivo decay.

ACKNOWLEDGEMENTS

These studies were supported by Grant# DK26793-11 from the National Institute for Diabetes, Digestive and Kidney Diseases, and by grant HL-28381, from the Institute of Heart, Lung and Blood Diseases, National Institutes of Health, Bethesda, MD

The authors are indebted to Liana Apelis for help in the preparation of this manuscript.

REFERENCES

1. C. Seaman, S. Wyss and S. Piomelli, The decline in energetic metabolism with aging of the erythrocyte and its relationship to cell death, Am. J. Hematol. 8:31 (1980).
2. R. G. Chapman, M. A. Hennessey, A. M. Watersdorph, F. M. Huennekens and B. W. Gabrio, Erythrocyte metabolism: level of glycolytic enzymes and regulatin of glycolysis, J. Clin. Invest. 41:1249 (1962).
3. F. Brok, B. Ramot, E. Zwag and D. Danon, Enzyme activities in human blood cells of different age groups, Israel J. Med. 2:291 (1966).
4. S. Piomelli and S.R. Wyss, Metabolic death of the red blood cell, Blood 38:833a (1971).
5. T. A. Rapoport, R. Heinrich, G. Jacobash and S. Rapoport, A linear steady-state treatment of enymatic chains: a mathematical model of glycolysis of human erythrocyte, Eur. J. Biochem. 42:107 (1974).
6. E. W. Jr Holmes, J. I. Malone, A. I. Winegrad and F. A. Oski, Hexokinase isoenzymes in human erythrocytes: association of type II with fetal hemoglobin, Science 156: 646 (1967).
7. J. C. Kaplan and E. Beutler, Hexokinase isoenzymes in human erythrocytes, Science 159:215 (1968).
8. V. Stocchi, A. Stulzini and M. Magnani, Chromatographic fractionation of multiple forms of red blood cell hexokinases, J. Chromatogr. 237:330 (1982).
9. M. Magnani, G. Serafini and V. Stocchi, Hexokinase type I multiplicity in human erythrocytes, Biochem. J. 254:617 (1988).
10. G. Rijksen, G. Vansen, R. J. Kraaijenhagen, M. J. M. van der Vlist, A. M. C. Vlug and G. E. J. Staal, Separation and characterization of hexokinase I subtypes from human erythrocytes, Biochim. Biophys. Acta 659:292 (1981).
11. M. Gahr, Different biochemical properties of foetal and adult red cell hexokinase isoenzymes, Hoppe – Seyler's Z. Physiol. Chem. 361:829 (1980).
12. L. M. Corash, S. Piomelli, M. C. Chen, C. Seaman and E. Gross, Separation of erythrocytes according to age on a simplified density gradient, J. Lab. Clin. Med. 84:147 (1974.)
13. S. Piomelli, L. M. Corash, D. D. Davenport, J. Miraglia and E. L. Amorosi, In vivo lability of glucose-6-phosphate dehydrogenase in Gd and Gd Mediterranean deficiency, J. Clin. Invest. 47: 940, 1968.
14. I. A. Rose, J. V. B. Warms: Mitochondrial hexokinase. Release, rebinding, and location. J. Biol. Chem. 242:1635, 1967.
15. G. Rijksen, J. W. N. Akkerman, A. W. L. van den Wall Bake , D. Pott

Hofstede, G. E. J. Staal: Generalized hexokinase deficiency in the blood cells of a patient with nonspherocytic hemolytic anemia. Blood 61:12, 1983.

16. M. Magnani, V. Stocchi, L. Cucchiarini, G. Novelli, S. Lodi, L. Isa and G. Fornaini, Hereditary nonspherocytic hemolytic anemia due to a new hexokinase variant with reduced stability, Blood 66:690 (1985).

17. M. Magnani, L. Chiarantini, V. Stocchi, M. Dachà and G. Fornaini, Glucose metabolism in fibroblasts from patients with erythrocyte hexokinase deficiency, J. Inher. Metab. Dis. 9:129 (1986).

18. M. Magnani, V. Stocchi, F. Canestrari, M. Dachà, P. Balestri, M. A. Farnetani, D. Giorgi, A. Fois and G. Fornaini, Human erythrocyte hexokinase deficiency: a new variant with abnormal kinetic properties, Brit. J. Haematol. 61:41 (1985).

19. C. Altay, C. A. Alper and D. G. Nathan, Normal and variant isoenzymes of human blood cell hexokinase and the isoenzyme patterns in hemolytic anemia, Blood 36:219 (1970).

20. A. Hershko and A. Ciehanover, A heath stable polypeptide component of an ATP-dependent proteolytic system from reticulocytes, Biochem. Biophys. Res. Commun. 81:1100 (1978).

21. A. L. Goldberg and A. C. John, Intracellular protein degradation in mammalian and bacterial cells, Ann. Rev. Biochem. 45, 747 (1976).

22. S. Speicer and J. D. Etlinger, Loss of ATP-dependent proteolysis with maturation of reticulocytes and erythrocytes, J. Biol. Chem. 257, 14122 (1982).

23. O. Raviv, H. Heller and A. Hershko, Alterations in components of ubiquitin-protein ligase system, following maturation of reticulocyte to erythrocyte, Biochem. Biophys. Res. Commun. 145:658 (1987).

24. M. Magnani, V. Stocchi, L. Chiarantini, G. Serafini, M. Dachà and G. Fornaini, Rabbit red cell hexokinase. Decay mechanisms during reticulocyte maturation, J. Biol. Chem. 261:8327 (1986).

25. Z. Bercovich, Y. Rosenberg-Hasson, A. Ciehanover and C. Kahana, Degradation of ornithine decarboxylase in reticulocyte lysates is ATP-dependent and ubiquitin-independent, J. Biol. Chem. 264:15949 (1989).

26. R. Hough, G. Pratt and M. Rechsteiner, Purification of two high molecular weight proteases from rabbit reticulocytes lysate, J. Biol. Chem. 261:2400 (1986).

27. M. Magnani, V. Stocchi, M. Dachà and G. Fornaini, Rabbit red cell hexokinase. Intracellular distribution during reticulocyte maturation, Mol. Cell. Biochem. 61:83 (1984).

PHAGOCYTOSIS OF PHENYLHYDRAZINE OXIDIZED AND G-6-PD DEFICIENT RED

BLOOD CELLS: THE ROLE OF SUGARS AND CELL-BOUND IMMUNOGLOBULINS

Sara Horn, Nava Bashan, Shimon Moses, Jacob Gopas

Dept. of Biochemistry, Microbiology and Immunology
Faculty of Health Science, Ben-Gurion University
Beer-Sheva, Israel

INTRODUCTION

Senescent or damaged red blood cells (RBCs) are selectively removed
from the blood by macrophages in the spleen and liver (1). Changes observed
in these cells involve various membranal modifications such as desialyzation
(2-4), surface galactosyl exposure (5-7), degradation or configural
aggregation of band 3 (8-10) and changes in membrane phospholipid asymmetry
(11,12). These modifications have been reported to be recognized by
macrophages either directly (4-7,11,12), or indirectly, by binding of
autoantibodies and complement components to the cells (2,8-10,13). Similar
recognition mechanisms have been reported in some damaged RBCs such as in
thalassemia and sickle cell anemia (10,11,14,15).

The molecular basis of premature removal of oxidatively damaged RBCs,
as found in various hemolytic anemias, is less clear. Models for studying
the effect of oxidation on RBCs include *in vitro* treatment of the cells with
various oxidative agents (9,16,17). Phenylhydrazine (Phz) has been widely
used for this purpose since this drug is known to be an effective inducer of
hemolysis (18). Reactions between Phz and hemoglobin are known to generate
both superoxide and hydrogen peroxide (9) which cause various kinds of
cellular damage, such as: decrease in intracellular reduced glutathione
(GSH), methemoglobin production, and disulfide bond formation between GSH or
hemoglobin and membranal proteins (16,20). As a result of the cellular
changes induced by Phz, the RBC membrane is affected: membranal aggregates
between proteins are formed, including polymerization of spectrin and
aggregation of the anion channel Band 3 (21,22); dysfunction of the
membranal Ca^{++} ATPase and elevated cellular Ca^{++} levels are observed (23);
all leading to decreased deformability of the membrane and increased uptake
of the Phz-oxidized cells in the spleen and the liver (24,25).

Red Blood Cell Aging, Edited by M. Magnani and
A. De Flora, Plenum Press, New York, 1991

Similar membrane damage as in Phz-oxidized RBCs was observed also in RBCs from some hemolytic anemias and unstable hemoglobins (26,27). Among them, glucose-6-phosphate dehydrogenase (G6PD) deficiency is highly prevalent in certain communities and is characterized by a variety of mutants (28,29). Since in most mutants the activity of the enzyme is very low, GSH and NADPH levels in these RBCs become limiting factors, leading to chronic or acute hemolytic anemia under physiological or oxidative conditions. In the chronic mutants membranal aggregates of high molecular weight, elevation of Ca^{++} intracellular levels and decreased deformability were also reported (30-32). In the Mediterranean G6PD type, these aggregates were found only during the acute hemolytic phase obtained when patients RBCs were exposed to oxidative stress (infections, hemolytic agents such as naphtaline or metabolites of fava beans) (16,30,32).

In order to correlate these changes with ultimate macrophage-related RBC destruction in vivo, it has to be assumed that the membranal modifications on Phz treated RBCs or G6PD deficient RBCs can be recognized by macrophages, leading to increased phagocytosis. The erythrophagocytosis assay has been useful in studying the interaction between human RBCs and autologous or xenogeneic macrophages in vitro in cases of normal aging RBCs (6,7,33,34) and in hemolytic anemias such as thalassemia or sickle cell anemia (11,14).

The present studies were undertaken in order to analyse the mechanism(s) of recognition and removal of human normal Phz-oxidized and G6PD deficient RBCs (Mediterranean type) by murine thyoglycollate elicited macrophages.

We show that G6PD deficient and Phz-oxidized RBCs are efficently phagocytosed by macrophages. The phagocytosis involves recognition of galactosyl/mannosyl residues on the damaged red cell membrane by macrophage lectin-like receptors.

The results also show that phagocytosis of Phz-oxidized and G6PD deficient RBCs occurs in the absence of serum but involves macrophage Fc receptors. Since a basal level of autologous antibodies was found to be bound on control and damaged red cell membranes, it is suggested that during oxidation, RBC-bound antibodies undergo configurational or topographical changes, which enable recognition and phagocytosis of the oxidized cells through Fc receptors.

We conclude that more than one mechanism is involved in the recognition and phagocytosis of Phz-oxidized and G6PD deficient RBCs: therefore removal of oxidatively damaged RBCs seems to be a multifactor process.

MATERIALS AND METHODS

Red blood cell treatments

Preparation: red blood cell preparations were performed as described before (35). Briefly, freshly heparinized human blood was depleted of leukocytes (36), washed thoroughly with cold phosphate buffered saline (PBS) and resuspended to 10% hematocrit (% Hct) in PBS containing 6 mM glucose. The RBC suspensions were then incubated with 3 mM (unless otherwise specified) Phz (Sigma, St. Louis) for 1 hr at 37°C. After incubation the cells were washed and suspended in cold RPMI (10% Hct).

Incubation with serum: One volume of packed RBCs (PC) (Phz-treated, G6PD or control) was incubated with one volume of fresh autologous serum, for 30 min at 37°C. The suspension was then diluted 1:5 (10% Hct) in RPMI 1640 (Beth Haemeck, Israel) before its addition to macrophage cultures or washed and resuspended in RPMI for other assays.

Opsonization: Control RBC suspensions (10% Hct), were incubated with anti-D antibodies (Gamma Biological, Incs.) at a final dilution of 1:2 for 30 min at 37°C. The cells were then washed and resuspended in cold RPMI (10% Hct).

Enzymatic treatment: Suspensions of Phz-treated or G6PD RBCs (10% Hct) were incubated with saturating concentrations of the enzyme ß-galactosidase (from E. Coli; Sigma, St. Louis) (20U/ml PC) for 60 min. at 37°C. After incubation the cells were washed and resuspended in cold RPMI (10% Hct).

Treatment with F(ab')$_2$ fragments of anti human Ig: Blocking of the Fc portion of antibodies bound to RBCs was obtained by incubating Phz-treated or G6PD RBCs (10% Hct) with saturating concentrations of rabbit anti human Ig antibodies-F(ab')$_2$ fragments (Sigma, St. Louis) (1 µg/100 µl suspension) for 45 min at room temperature. The RBCs were then washed four times with PBS and resuspended in RPMI.

^{125}I-Protein A binding assay

The number of red cell-bound IgG molecules was estimated by protein A binding assay as described by Yam and co-workers (37). Briefly, following the various treatments, RBCs were washed with PBS containing 0.2% BSA, transferred to BSA-coated tubes (10^8 cells/100 µl) and incubated with 3 ng of affinity purified ^{125}I-protein A (Amersham, UK, specific activity 30mCi/mg protein) for 30 min at 37°C. The cells were then washed four times, transferred to new tubes and counted in a Beckman Gamma 5500 counter.

Phagocytosis assays

Macrophages were recovered from mouse peritoneal cavities (Balb/c mice, Annilab Co.). (4-5 days after i.p. injection of 3 ml of 3% sterile thioglycollate, obtained from Bacto, Difco Labs., Detroit MI, USA.) by washing cavities twice with RPMI containing 10U heparin/ml. The cells were then adhered to glass slides, as previously described (35). 0.1 ml of RBC suspension (10% Hct) was added to the cultured macrophages and incubated for 1 hr at 37°C. The medium was subsequently removed, the cell layer was washed twice with PBS and the non-internalized RBCs were removed by 20 sec lysis with double distilled water. The macrophages were then stained with Giemsa and examined by light microscopy. The number of RBCs ingested in 600 randomly-observed macrophages was counted.

Inhibition of lectin-like receptor mediated phagocytosis: Sugars at 1 mM final concentration were incubated with macrophage cultures for 30 min prior to RBC addition and during phagocytosis assay (35).

Inhibition of Fc receptor mediated phagocytosis: (a) Cultured adherent macrophages were incubated for 30 min at 37°C with saturating concentrations of aggregated Ig-antibodies (aIg) (4g/100 µl) (38) or anti-Fc receptor monoclonal antibodies (gift from Prof. R. Apte, Dept. of Microbiology and Immunology, Ben-Gurion Univ.) (1µg/100µl), solved in RPMI. The cells were then washed gently with fresh RPMI and RBCs were added for phagocytosis. As controls, macrophages were incubated with similar concentrations of specific and non-specific class II MHC antibodies (anti I-A^d and anti I-A^K, respectively) (Becton-Dickinson, USA). (b) Peritoneal macrophages were adhered on slides coated with fixed complexes of BSA-anti BSA antibodies (Sigma, St. Louis) (2µg/100 µl), as described by Kaplan et al. (39). Control coverslips were treated with BSA or PBS only. The percentage of phagocytosis inhibition was determined as follows: [(test cells/untreated cells)-1] x100.

Data analysis

All assays were performed in triplicate and all experiments were repeated at least three times. The results are expressed as mean ± standard deviation (SD). Statistical differences between control and test values were analysed by unpaired Student's t test.

RESULTS

Phagocytosis of normal Phz-oxidized and G6PD deficient RBCs

Phagocytosis of untreated RBCs by murine macrophages was found to be less than 2 RBCs/100 macrophages (% phagocytosis). Pretreatment of normal RBCs with 3 mM Phz for 1 hour caused a significant increase in phagocytosis up to values of 53.5 RBCs/100mac, which was comparable to the values obtained with anti-D opsonized RBCs. (Fig. 1) G6PD deficient RBCs were also found to be efficiently phagocytosed (average values of 19.4 RBCs/100mac.) without any addition of oxidants and achieved values of 45 RBCs/100mac. in some anemic patients (see in Table 1).
This phagocytosis was obtained in serum-free medium and was not affected by incubation of the cells with fresh autologous serum.

We have previously shown that the phagocytosis observed was not a result of non specific activation of the macrophages by oxygen derived products in the assay (35). The specific mechanism(s) of recognition and phagocytosis of the damaged RBCs were therefore studied by: (a) examining the macrophage receptor(s) involved in this process. (b) examining the component(s) on the RBC oxidized membrane that are responsible for this recognition.

In all subsequent experiments, normal RBCs oxidized by 3mM and unoxidized G6PD deficient RBCs were used (unless otherwise specified), since at these conditions significant phagocytosis without any apparent hemolysis is observed (the viability of the cells was 90-95%).

The involvement of sugars in phagocytosis of Phz-treated and G6PD deficient RBCs

Macrophages have the ability to recognize sugar derivatives such as D-galactose or D-mannose on various cells (6,7,34,40-42). In order to determine whether carbohydrate recognition is involved in the interaction between macrophages and the oxidatively damaged RBCs, phagocytosis inhibition studies using various sugars were performed. In our previous report (35) incubation of macrophages with very low concentration (0.1 mM) of sugars such as D-galactose or D-mannose specifically inhibited the phagocytosis of Phz-treated RBCs in a concentration-dependent manner (up to a maximal inhibition of 63% at 1mM sugar addition).

As can be seen in Fig. 2, the phagocytosis of G6PD deficient RBCs was similarly inhibited by the addition of 1mM galactose or mannose (but not

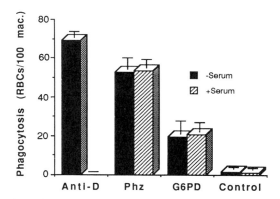

Fig. 1. Phagocytosis of Phz-oxidized and G6PD deficient RBCs. Normal human RBCs exposed to 3mM Phz and RBCs from G6PD patients were incubated in the presence (+serum) or absence (-serum) of 50% autologous serum, and phagocytosis was determined. As controls, phagocytosis of untreated normal (control) and opsonized (Anti-D) RBCs was examined.

glucose) to the macrophages, whereas phagocytosis of Anti-D opsonized RBCs, which is mediated by Fc receptors, was not affected by the addition of the same sugars. These results indicate that lectin-like macrophage receptor(s) are involved in the recognition and phagocytosis of these damaged RBCs.

Treatment of the Phz-oxidized and G6PD deficient RBCs with the enzyme ß-galactosidase, which removes galactosyl residues on the membrane, also reduced the phagocytosis to a degree similar to that obtained with D-galactose and D-mannose (about 50% inhibition), suggesting that galactosyl residues on the modified cell membrane are responsible for the phagocytosis obtained. We thus conclude that the mechanism by which Phz-oxidized and G6PD deficient RBCs are phagocytosed involves recognition of exposed galactosyl/ mannosyl membranal residues by the lectin-like receptors.

The involvement of macrophage Fc receptor in the phagocytosis of Phz-oxidized and G6PD deficient RBCs

Since phagocytosis of the oxidized and G6PD deficient RBCs was only partially inhibited by sugars, additional mechanisms could be responsible for the phagocytosis observed. Opsonization of RBCs with serum immunoglobulins has been largely reported in mediating phagocytosis of senescent RBCs through recognition of the cell by macrophage Fc receptors. The involvement of macrophage Fc receptor(s) in the recognition of Phz-

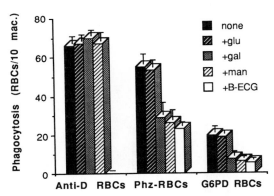

Fig. 2. Inhibition of erythrophagocytosis by sugars and ß-galactosidase: RBCs (Anti-D opsonized, Phz-oxidized and G6PD deficient) were added to macrophages previously incubated with 1mM galactose/mannose (gal/man) as well as glucose (glu). The effect of ß-galactosidase (ß-ECG) was examined by incubation of the RBCs with the enzyme (20U/ml PC).

oxidized and G6PD deficient RBCs was therefore investigated. As shown in Fig. 3, phagocytosis was specifically but partially inhibited by incubating macrophages with Fc receptor blockers such as BSA-anti BSA complexes and aggregated immunoglobulins (aIg) (38,39). As controls, phagocytosis of Anti-D opsonized RBCs were almost totally inhibited by the same blockers whereas phagocytosis of latex particles, which is not immunoglobulin-mediated, was not affected (data not shown).

Similar inhibition values of phagocytosis were obtained when macrophages were incubated with monoclonal antibodies against the Fc receptor (Fig. 3). This antibody specifically inhibited Fc receptor mediated phagocytosis, since the addition of non-relevant monoclonal antibodies (anti IA^K) or antibodies directed to another macrophage membrane protein (anti IA^d) did not significantly affect the phagocytosis of Phz or G6PD RBCs. These results indicate that macrophage Fc receptors are involved in the recognition and phagocytosis of these damaged RBCs.

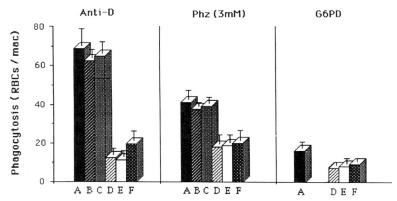

Fig. 3. The effect of macrophage Fc receptor blockers on phagocytosis of Phz
and G6PD deficient RBCs: Anti-D opsonized, Phz-oxidized and G6PD
deficient RBCs were added to macrophages previously incubated with
the following Fc-receptor blockers: aggregated Ig (D), BSA-anti BSA
complexes (E), monoclonal anti Fc receptors (F). As controls were
used: untreated macrophages (A), monoclonal anti IA^k (B), monoclonal
anti IA^d (C).

The additive effect of macrophage lectin-like and Fc receptors in the
phagocytosis of Phz-RBCs

In order to determine the relative role of lectin-like and Fc receptors
in the recognition of Phz-oxidized RBCs, aIg and galactose or mannose were
simultaneously added to macrophage cultures and phagocytosis was measured.
As can be seen in Fig. 4, an additive inhibitory effect up to 90% was
obtained in the presence of both blockers, thus excluding the possibility of
phagocytosis inhibition due to steric hindrance and suggesting the specific
involvement of both lectin-like and Fc receptors in the recognition of
Phz-RBCs.

Determination of bound immunoglobulins on RBCs

Since serum immunoglobulins seemed not to be relevant for the
phagocytosis observed the amount of immunoglobulins bound to the cells was
examined by ^{125}I-Protein A binding assay (Fig. 5). Marked binding of
^{125}I-protein A to control "stored" cells was observed, which is in agreement
with other reports (43,44). The amount of ^{125}I-protein A molecules bound to
3mM Phz-oxidized and G6PD deficient RBCs was low (2967±218 and 3418±726
cpm/10^8 cells respectively) and similar to the amount found on untreated
cells (2984±256 cpm/10^8 cells). This amount of bound-protein A was not
increased after incubation with autologous serum (Fig. 5). Flow cytometric
analysis of Phz-oxidized RBCs with FITC conjugated anti-human IgG antibodies
confirmed these observations and showed that enhanced binding of serum
immunoglobulins to RBCs occurs only after severe oxidation of the cells with
high concentrations of Phz (>9mM) (45).

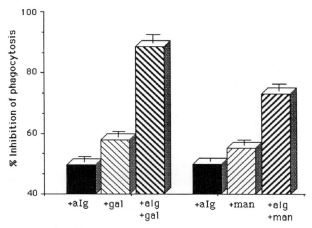

Fig. 4. The additive inhibitory effect of sugar and aIg on Phz-oxidized
RBCs. Phz-oxidized RBCs were added to macrophages previously treated
with 1mM galactose/mannose (gal/man), with aggregated Ig (aIg) or
additively aIg with each sugar (aIg+gal/man). Phagocytosis of RBCs
was then determined.

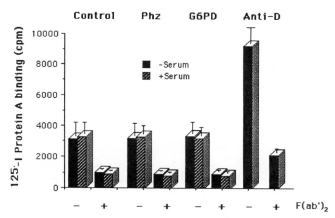

When RBCs were incubated with anti-human Ig F(ab') fragments, which
bind to immunoglobulins and prevent their binding to protein A, the binding
of protein A both to normal and opsonized cells was 70% inhibited,
indicating a specific blocking effect of these values, a low number of
28 ± 2.5, 27 ± 2.1 and 33 ± 6.5 IgG molecules/cell was estimated to be bound on
control, oxidized and G6PD deficient cells respectively.

Fig. 5. Binding of ^{125}I-protein A to Phz-oxidized and G6PD deficient RBCs.
Phz-oxidized and G6PD deficient RBC were incubated with/without
serum (± serum). The washed cells were then incubated with anti
human IgG-F(ab')$_2$ fragments and the binding of ^{125}I-protein A was
assessed (cpm).

Table 1. The Effect of F(ab')$_2$ Anti-Human Ig Fragments on Phagocytosis of Phz-Treated RBCs.

Treatment of Macrophages	Treatment of RBCs	Phagocytosis Phz-RBCs(3mM)	(RBCs/100 mac) anti-D RBCs
ut[a]	ut[a]	41.1±4.5	68.7±8.6
	+aIg	19.4±3.5	12.4±3.1
	+gal	23.1±4.8	71±10
+F(ab')$_2$	ut[a]	17.3±3.0	7.2±2.1
	+aIg	22.1±1.2	8.6±1.3
	+gal	6.1±0.6	9.2±1.6

[a]ut-untreated.

[b]Phz-oxidized RBCs were incubated with anti human IgG-F(ab')$_2$ fragments, which lack the Fc portion. The cells were then washed and added to macrophages previously treated with aIg (0.026 mg/ml) or with galactose (gal) (1mM).

Basal amounts of immunoglobulins bound to Phz-RBCs are recognized by macrophages

Incubation of RBCs with anti-human Ig F(ab')$_2$ fragments prevented the binding of protein A to the cells (Fig. 5). In accordance, addition IgG F(ab')$_2$ fragments resulted in 90% inhibition of phagocytosis of anti-D opsonized RBCs. In comparison, the same treatment caused only 56% decrease in phagocytosis of Phz-RBCs (Tab.1). Interestingly, an additive inhibitory effect was obtained when Phz-RBCs, previously exposed to F(ab')$_2$ fragments, were added to macrophages that were incubated with galactose but not with aIg. These results suggest that in conjugation with exposed galactosyl/ mannosyl residues, the existing amount of membrane bound molecules IgG is also involved in the recognition and phagocytosis of Phz-oxidized RBCs.

Erythrophagocytosis and ^{125}I-protein A binding to RBCs from various G6PD patients the effect of anemic conditions

The effect of anemic conditions (as expressed by hemoglobin values) on erythrocytosis values and Protein A binding to RBCs was assessed by analysing various G6PD patients.

As shown in Tab. 2, the mean G6PD erythrophagocytosis values was enhanced as compared to the control, whereas hemoglobin (Hb) and protein A binding values were not significantly different from the controls. When each G6PD patient was analysed, the degree of anemia of the various patients was found to be in direct correlation with erythrophagocytosis values and with binding of protein A to the cells. In this way, patients with low Hb values (BS or NU in Tab. 2) showed a relative high erythrophagocytosis values and increased binding of protein A. Thus we suggest that enhanced binding of immunoglobulins may play a role in recognition of severely damaged RBCs, as observed in G6PD anemic patients or highly oxidized RBCs.

DISCUSSION

In this paper the mechanism of recognition and phagocytosis of oxidative damaged RBCs was studied in order to further our understanding of the basis of premature removal of RBCs in G6PD deficient hemolytic anemia. For this purpose, a heterologous system was used involving mouse thioglycollate-elicited peritoneal macrophages and G6PD deficient RBCs from the mediterranean type that are characterized by a nearly normal G6PD activity in the young cells, which declines drastically with aging, leading to in vivo premature removal of the cells or acute hemolysis after oxidative stress (28,29).

G6PD deficient RBCs, as well as RBCs oxidized by Phz, were efficiently phagocytosed by macrophages (Fig. 1) (24,25,35,46). However, the mechanism(s) by which these cells are recognized is still unclear.

Table 2. The Effect of Anemic Conditions on Erythrophagocytosis and [125]I Protein A Binding in G6PD Deficient Patients.

RBCs	Hb[a]	Phag.[b]	[125]I Protein A[c]
Mean Controls	15.5±2.7	1.5±0.9	2984±256
Mean G6PD	13.7±3.5	19.4±9.0	3418±72
Patients:			
LS	18.1	6.1±0.1	3000±157
DY	18.0	7.2±2.6	3023±174
DI	15.2	19.9±2.7	3020±120
GM	13.6	18.3±3.1	3200±323
BR	12.0	19.5±2.7	n.d.
BS	8.7	45.3±3.6	4868±410
NU	10.7	38.4±2.8	3400±320

[a]Hemoglobin (Hb) content of RBCs from different G6PD patients is expressed in gr/100 ml blood.
[b]Phagocytosis (RBCs/100 mac.) of G6PD deficient RBCs is determined as previously described.
[c][125]I-protein A binding (cpm) to RBCs from G6PD deficient patients is determined by incubating 10^8 cells with the protein and counting cpm after extensive washings.

One of the known mechanisms of interaction between cells involves recognition of membrane sugars by lectin-like receptors. This mechanism has also been reported in the recognition of senescent RBCs, bacteria or yeast by macrophages (6,34,40,41). It was therefore reasonable to test whether macrophage lectin-like receptors are also involved in the recognition of oxidatively damaged RBCs. Phagocytosis of G6PD and Phz-treated, but not opsonized RBCs, was found to be inhibited by several sugars, most efficiently by D-galactose and D-mannose (Fig. 2). These results imply the involvement of galactosyl/mannosyl lectin-like receptor(s) in the recognition of the oxidized cells, which differ from the classical Fc receptor mechanism. Since phagocytosis was inhibited by D-galactose/ D-mannose in the presence of serum as well (35), it seems that sugar recognition of damaged RBCs by macrophages may play a role in vivo.

Inhibition of phagocytosis was also obtained by removal of ß-galactosyl residues from the cell surface with ß-galactosidase (Fig. 2). These results complement the previous sugar inhibition experiments and further support the role played by sugars in recognition and phagocytosis. Taken together, these data imply that during oxidation, sugar residues are exposed on RBC membrane and are recognized by the lectin-like receptors. Since phagocytosis of G6PD deficient RBCs was almost identically affected by the same treatments as Phz-oxidized cells, it seems that these cells have already been exposed to some oxidative processes in vivo, inducing ultimate membranal carbohydrate modifications similar to those obtained with Phz.

Various explanations have been reported for exposure of sugars. Among these are desialization of the RBC membrane (as in thalassemic and senescent RBCs) (2,3,14), by the presence of contaminating glycosidases and by enchanced proteolysis occurring during Phz oxidation (47,48,49). Results from previous studies do not support massive membrane desialization as the main mechanism for D-galactose/D-mannose exposure (35). Contaminating glycosidases of leukocyte origin are probably not responsible for the exposure of sugars, since RBC preparations freed from leukocytes by cellulose columns were phagocytosed to the same extents (36). Enhanced proteolysis has been reported in erythrocytes treated with hydrazines (48,49). It is conceivable that the proteolytic removal of oxidatively-damaged proteins could contribute to the exposure of cryptic sugars on the cell surface. This possibility remains to be tested.

Since phagocytosis of the oxidized and G6PD deficient RBCs was only partially inhibited by sugars, additional mechanisms that could be responsible for phagocytosis were examined. Binding of autologous antibodies to senescent RBC membranes is a well known mechanism for removal of old cells from circulation. However it is not clear whether immunoglobulins also trigger the removal of RBCs damaged by oxidative stress. Magnani et al. (25) have shown pronounced IgG binding and phagocytosis of rabbit RBCs treated with 10 mM Phz and subsequently incubated in autologous serum, suggesting that the mechanism of Phz-RBCs sequestration is mediated by autoantibodies. On the other hand, Beppu et al. (24) obtained enhanced phagocytosis of mouse RBCs treated with Phz in the absence of serum, concluding that the recognition mechanism of Phz-RBCs is not mediated by binding of autologous antibodies, but rather through alteration of membrane physicochemical properties. Similarly, G6PD deficient RBCs have been reported to be phagocytosed only in the presence of serum, although no increase in immunoglobulin binding to the cells was observed (46,50).

In the present study, we have examined the possible role of autologous immunoglobulins in recognition and phagocytosis of human Phz-oxidized and G6PD deficient RBCs.

RBCs treated with 3mM Phz were efficiently phagocytosed by macrophages in serum free medium. The addition of 50% autologous serum did not significantly affect the phagocytosis values (Fig. 1). These results indicate that binding of serum immunoglobulins to the membrane is not essential for recognition and removal of the oxidized RBCs by macrophages. Paradoxically, the phagocytosis of these cells was partially inhibited by aIg, BSA complexes and by monoclonal anti Fc receptor antibodies (Fig. 3). These results indicate that Fc macrophage receptors are indeed involved in the recognition of Phz treated RBCs, even in the absence of serum. We, therefore, examined the amount of bound immunoglobulins on the surface of the RBCs before and after oxidation. A small amount of bound immunoglobulins (28 molecules/cell) was found present on control and oxidized RBCs and was not affected By incubation of the cells with autologous serum (Fig. 5). This basal amount of bound immunoglobulins seems to be relevant in the recognition and phagocytosis of oxidized RBCs, since the addition of $F(ab')_2$ fragments of anti human IgG to the cells, prevented their phagocytosis (Tab. 1). We therefore suggest that during mild oxidation (3mM Phz), membrane bound immunoglobulins undergo configurational or topological membrane modifications, enabling recognition and phagocytosis of the cells through macrophage Fc receptors. This suggestion is further supported by Kannan et al. (51), who recently reported that most of the surface bound IgG on sickle anemic RBCs was clustered at aggregate sites on hemicromes, suggesting that membrane reorganization of immunoglobulins may be an important cause of sickle cell RBCs removal. In line with our results Bosman et al. (44) did not find enhanced binding of autologous immunoglobulins to RBCs treated with a variety of other oxidative agents.

A significant increase in the binding of serum autoantibodies was obtained only for RBCs oxidized with high concentrations of Phz (9-14 mM), as observed also in Low's reports (22,45). These results suggest that at severe oxidative conditions, in which pronounced cell lysis is observed, a subpopulation of surviving RBCs are recognized by autologous antibodies and are opsonized.

As observed with Phz-oxidized RBCs, G6PD deficient cells were also phagocytosed in the absence of serum and their recognition by macrophages was affected by Fc receptor blockers. Still, it is possible that these cells, as results of oxidation, have already been opsonized in vivo (46,50). Determination of cell-bound imunoglobulins on various G6PD patients shows average numbers of 33 molecules Ig/cell, which were not significantly different from the in vitro oxidized cells (Fig. 5). Yet, analysis of each patient shows elevated numbers of bound immunoglobulins only on RBCs from anemic patients, which are in direct correlation with erythrophagocytosis values (Tab. 2). These results support our previous hypothesis that enhanced binding of autologous immunoglobulins to damaged RBCs occurs only under severe oxidative/hemolytic conditions. It should be noted that differences observed in the levels of bound immunoglobulins between the various G6PD patients were in the range of about 15 molecules/cell, therefore suggesting the relevance of small amounts of bound antibodies on the observed pronounced phagocytosis. Lutz et al. (52) also showed a very slight increase in binding of immunoglobulins to diamide oxidized RBCs (10 moleules IgG/cell) and found that this treatment causes clustering of immunoglobulins which then efficiently bind complement.

The role of complement in phagocytosis of Phz-oxidized and G6PD deficient RBCs is being studied in our assay, by the use of monoclonal antibodies against C3c complement components. Preliminary results show enhanced levels of bound serum C3b and C3c to both mildly and severely oxidized cells (Phz, G6PD), confirming the role played by complement in opsonization of RBCs.

We conclude that Phz-oxidized and G6PD deficient RBCs are similarly opsonized and recognized by murine macrophages, suggesting Phz-oxidation as an optimal _in vitro_ model for studying some hemolytic processes.

In this study, macrophage lectin-like and Fc receptors were found to be partially involved in the phagocytosis of G6PD and Phz-RBCs since only partial inhibition of phagocytosis was observed after addition of galactose (Fig. 2) or Fc receptor blockers (Fig. 3). When either sugar and aIg were added simultaneously to the macrophages, an additive inhibitory effect of phagocytosis (90%) was obtained (Fig. 4). Phagocytosis of F(ab') treated RBCs in the presence of galactose was also additively inhibited (Tab. 1).

We conclude that there are at least two independent mechanisms involved in the recognition of oxidized RBCs by macrophages: one involving a lectin-like macrophage receptor and a second involving macrophage Fc receptors. The second mechanism is mediated by small amounts of immunoglobulin attached to the RBC surface which during oxidation undergo modifications, enabling phagocytosis.

REFERENCES

1. N. I. Berlin, P. O. Berk, The biological life of the red cell., in: The Red Blood Cell, Surgenor, MacN. eds. 2nd ed., vol. 2, New York, p. 957 (1975).
2. E. M. Alderman, H. H. Fudenberg, R. E. Lovins, Isolation and characterization of an age-related antigen present on senescent human red blood cells, Blood 58:34 (1981).
3. S. Kelm, A. K. Shukla, J. C. Paulson, R. Schauer, Reconstitution of the masking effect of sialic acid groups on sialidase treated erythrocytes by the action of sialyltransferases, Carb. Res. 149:59 (1986).
4. D. Aminoff, The role of sialoglycoconiugates in the aging and sequestration of red cells from circulation, Blood Cells 14:229 (1988).

5. J. Schleppe-Schafer, V. Kolb-Bachofen, H. Kolb, Identification of a receptor for senescent erythrocytes on liver macrophages, Biochem. Biophys. Res. Comm. 115:551 (1983).

6. J. Schlepper-Schafer, V. Kolb-Bachofen, Red cell aging results in a change of cell surface carbohydrate epitopes allowing for recognition by galactose specific receptors of rat liver macrophages, Blood Cells 14:170 (1988).

7. N. Vaysse, L. Gattegno, D. Bladier, D. Aminoff, Adhesion and erythrophagocytosis of human senescent erythrocytes by autologous monocytes and their inhibition by galactosyl RBC derivates, Proc. Natl. Acad. Sci. USA 83:1339 (1986).

8. M. M. B. Kay, Aging of cell membrane molecules leads to appearance of an aging antigen and removal of senescent cells, Gerontology, 31:215 (1985).

9. H. U. Lutz, S. Fasler, P. Stammler, F. Bussolino, P. Arese, Naturally occurring antiband 3 antibodies and complement in phagocytosis of oxidatively-stressed and in clearance of senescent red cells, Blood Cells 14:175 (1).

10. S. M. Waugh, B. M. Williardson, R. Kannan, R. J. Labtka, P. S. Low, Heinz bodies induce clustering of Band 3, glycophorin and ankirin in sickle cell erythrocytes, J. Clin. Invest, 78:1155 (1986).

11. A. J Schroit, Y. Tanaka, J. Madsen, I. J. Fidler, The recognition of red blood cells by macrophages: role of phosphatidylserine and possible implications of membrane phospholipid asymetry, Biol. Cell 51:227 (1984).

12. L. McEvoy, P. Williamson, R. A. Schlegel, Membrane phospholipid asymmetry as a determinant of erythrocyte recognition by macrophages, Proc. Natl. Acad. Sci. Usa 83:3311 (1986).

13. U. Galili, I. Flechner, A. Knyszynsky, D. Danon, E. A. Rachmilewitz, The natural antigalactosyl IgG on human normal senescent red blood cells, Br. J. Haem. 62:317 (1986).

14. A. Knyszynski, D. Danon, I. Kahane, E. A. Rachmilewitz, Phagocytosis of nucleated and mature B thalassemic red blood cells by mouse macrophages "in vitro", Br. J. Haem. 43:251 (1979).

15. U. Galli, A. Korkesh, I. Kahane, E. A. Rahmilewitz, Demonstration of a natural antigalactosyl IgG antibody on thalassemic red blood cells. Blood 61:1258 (1983).

16. N. Bashan, R. Pothashnik, R. Feozer, S. W. Moses. The effect of oxidative agents on normal and G6PD deficient red blood cell membranes, in: Advances in Red Cell Biology, D. J. Weatheral, G. Fiorelly, S. Gorini eds., New York, p. 365 (1982).

17. L. M. Snyder, N. L. Fortier, J. Trainor, J. Jacobs, L. Leb, B. Lubin, D. Chim. S. Shohet, N. Mohandas, Effect of hydrogen peroxide exposure on normal human erythrocyte deformability, morphology, surface characteristics and spectrin-hemoglobin crosslinking, J. Clin. Invest. 76:1971 (1985).

18. S. K. Jain, P. Hochstein, Generation of superoxide radicals by hydrazine. Its role in phenylhydrazine induced hemolytic anemia, Biochim. Bipohys. Acta 586:128 (1979).

19. C. C. Winterboum, Free radical production and oxidative reactions of hemoglobin, Envir. Health Persp. 64:321 (1985).

20. B. Vilsen, H. Nielsen, Reaction of phenylhydrazine with erythrocytes, Clin. Pharm. 33:2739 (1984).

21. A. Arduini, A. Stern, Spectrin degradation in intact red blood cells by phenylhydrazine, Biochem. Pharmacol. 34:4238 (1985).
22. P. S. Low, S. M. Waugh, K. Zinke, D. Dreckhahn, The role of hemoglobin denaturation and band 3 clustering in red blood cell aging, Science 227:531 (1985).
23. O. Shalev, M. N. Leida R. P. Hebbel, H. S. Jacob, J. W. Eaton, Abnormal erythrocyte calcium hemostasis in oxidant-induced hemolytic disease. Blood 58:1232 (1981).
24. M. Beppu, H. Ochiai, K. Kikugawa, Macrophage recognition of the erythrocytes modified by oxidizing agents, Biochim. Biophys Acta 930:244 (1987).
25. M. Magnani, V. Stocchi, L. Cucchiarini, L. Chiarantini, G. Fornaini, Red blood cell phagocytosis and lysis following oxidative damage by phenylhydrazine, Cell. Biochem. and Function 4:263 (1986).
26. G. S. Platt., J. F. Falcone, Membrane protein lesion in erythrocytes with Heinz bodies, J. Clin. Invest. 82:1051 (1988).
27. T. P. Flynn, G. J. Jahnson, D. W. Allen, Mechanism of decreased erythrocyte deformability and survival in glucose 6 phosphate dehydrogenase mutants, in: Recent Clinical and Experimental Advances, Alan R. Liss A. Eds., New York, Raven p. 231 (1981).
28. A. Yoshida, Hemolitic anemia and G6PD deficiency, Science 179:532 (1973).
29. S. L. Schrier, Human erythrocyte G6PD deficiency: pathophysiology, prelevance, diagnosis and management, Blood Dis. 41 (1980).
30. G. J.Johnson, D. W. Allen, S. Cadman, V. F. Fairbanks, J. G. White, B. C. Lampkin, M. E. Kaplan, Red-cell-membrane aggregates in glucose-6-phosphate dehydrogenase mutants with chronic hemolytic disease, New Engl. J. Med. 301:522 (1979).
31. T. P. Flynn, G. J. Johson, D. W. Allen, Mechanism of decreased deformability and survival in glucose-6-phosphate dehydrogenase mutants, in: Erythrocyte Membranes 2: Recent Clinical and Experimental Advances, Alan R Liss, New York (1981).
32. E. Alhanaty, M. Snyder, M. B. Sheetz, Glucose-6-phosphate dehydrogenase have an impaired shape recovery system, Blood 63(5):1198 (1984).
33. M. A. Klausner, L. J. Hirsch, P. F. Leblond, J. K. Chamberlain, M. R. Klemperer, G. B. Segel, Contrasting splenic mechanism in the blood clearance of red blood cells and colloidal particles, Blood 46(6):965 (1975).
34. S. Kyoizumi, T. Masuda, A lectin-like receptor on murine macrophage cell line cells Mm_1: involvement of sialic acid-binding sites in opsonin-independent phagocytosis for xenogenic red cells, J. Leu . Biol. 37:289 (1985).
35. S. Horn, J. Gopas, N. Bashan, A lectin-like receptor on murine macrophage is involved in the recognition and phagocytosis of human red cells oxidized by phenylhydrazine, Biochem. Pharmacol. 39(4):775 (1990).
36. E. Beuler, C. West, K. G. Blume, The removal of leukocytes and platelets from whole blood, J. Lab. Clin. Med. 88:328 (1976).
37. P. Yam, L. D. Petz, P. Spath, Detection of IgG sensitization of red cells with ^{125}I-Staphylococcal protein, Am. J. Hematol. 12:337 (1983).

38. A. Knyszynsky, J. S. Leibovich, Interaction of macrophages with "old" red blood cells from syngeneic mice in vitro and independence of the recognition process on macrophage Fc receptors, Mech. Aging Dev. 29:171 (1985).

39. G. Kaplan, T. Eskeland, R. Seljelid, Difference in the effect of immobilized ligands on the Fc and C3 receptors of mouse peritoneal macrophages in vitro, Scand. J. Immunol. 7:19 (1978).

40. E. L. Kean, N. Sharon, Inhibition of yeast binding to mouse peritoneal macrophages by wheat germ agglutinin: a novel effect of the lectin on phagocytic cells BBRC 148(3):1202 (1987).

41. A. Perry, Y. Keisari, I. Ofek, Liver and macrophage surface lectins as determinants in blood clearance and cellular attachment of bacteria, FEMS Microbiol. Lett. 27:345 (1985).

42. N. Sharon, Surface carbohydrates and surface lectins are recognition determinants in phagocytosis, Immunol. Today 5:143 (1984).

43. C. E. Smalley, E. M. Tucker, Blood group A antigen site distribution and immunoglobulin binding in relation to cell age, Br. J. Haematol. 54:209 (1983).

44. G. J. Bosman, M. M. B. Kay, Erythrocyte aging: a comparison of model systems for stimulating cellular aging in vitro, Blood Cells. 14(1):19 (1988).

45. S. Horn, N. Bashan, J. Gopas, Phagocytosis of phenyhydrazine oxidized erythrocytes: the role of cell-bound immunoglobulins, submitted.

46. F. Bussolino, F. Turrini, P. Arese, Measurements of phagocytosis utilizing $[^{14}C]$ cyanate-labelled human red cells and monocytes, Br. J. Haem. 66:271 (1986).

47. A. Brovelli, C. Seppi, A. Bardoni, C. Balduini, H. U. Lutz, Re-evaluation of the structural integrity of red-cell glycoproteins during aging in vivo and nutrient deprivation, Biochem. J. 242:115 (1987).

48. D. Cola, P. Sacchetta, P. Battista, Proteolysis in human erythrocytes is triggered only by selected oxidative stressing agents, Ital. J. Biochem. 37(3):129 (1988).

49. M. A. Runge-Morris, S. Jacob, R. F. Novak, Characterization of hydrazine-stimulated proteolysis in human erythrocytes Toxicol. Appl. Pharmacol. 94:414 (1988).

50. M. M. B. Kay, G. J. C. G. M. Bosman, G. J. Johnson, A. H. Beth, band-3-polymers and aggregates, and hemoglobin precipitates in red cell aging, Blood Cells. 14(1):275 (1988).

51. R. Kannan, R. Laboyka, P. S. Low, Isolation and characterization of the hemichrome- stabilized membrane protein aggregates from sickle erythrocytes. Major sites of autologous antibody binding, J. Biol. Chem. 263(27):13766 (1988).

52. H. U. Lutz, F. Bussolino, R. Flepp, S. Fasler, P. Stammler, M. D. Kazatchkine, P. Arese, Naturally occurring anti-band-3 antibodies and complement together mediate phagocytosis of oxidatively stressed erythrocytes, Proc. Natl. Acad. Sci. USA 84:7368 (1987).

THE SENESCENT RED CELL AND ITS REMOVAL

MOLECULAR MAPPING OF THE ACTIVE SITE OF AN AGING ANTIGEN

Marguerite M. B. Kay and John J. Marchalonis

University of Arizona Health Sciences Center
Dept. of Microbiology and Immunology
University of Arizona
Tucson USA

INTRODUCTION

An aging antigen, senescent cell antigen, resides on the 911 amino acid membrane protein band 3. It marks cells for removal by initiating specific IgG autoantibody binding (1-22). This appears to be a general physiologic process for removing senescent and damaged cells in mammals and other vertebrates (4). Although the initial studies were done using human erythrocytes as a model, senescent cell antigen occurs on all cells examined (4). The aging antigen itself is generated by the degradation of an important structural and transport membrane molecule, protein band 3 (5). Besides its role in the removal of senescent and damaged cells, senescent cell antigen also appears to be involved in the removal of erythrocytes in clinical hemolytic anemias (7,8), and the removal of malaria-infected erythrocytes (23,24). Oxidation generates senescent cell antigen _in situ_ (6). Neither cross-linking nor hemoglobin appear to play a role. Although storage is the only _in vitro_ model that mimics cellular aging _in situ_, we have discovered three alterations/mutations of band 3 that permit insight into aging _in situ_. One mutation with an addition to band 3 has normal or decelerated red cell aging. In contrast, another band 3 alteration with a suspected deletion or substitution that renders band 3 more susceptible to proteolysis, shows accelerated aging. The third alteration which is also more susceptible to proteolysis is associated with neurologic defects.

Band 3 is ubiquitous membrane transport protein found in the plasma membrane of diverse cell types and tissues, and in nuclear, mitochondrial, and Golgi membranes (25-33). Band 3 in tissues such as brain performs the same functions as it does in red cells. Senescent cell antigen is generated on brain membranes. Oxidation is a mechanism for generating senescent cell antigen. Band 3-like proteins in nucleated cells participate in band 3 antibody induced cell surface patching and capping (25). Band 3 maintains acid-base balance by mediating the exchange of anions (eg. chloride, bicarbonate) (34-36). Because of its central role in respiration of CO_2,

Red Blood Cell Aging, Edited by M. Magnani and
A. De Flora, Plenum Press, New York, 1991

band 3 is the most heavily used ion transport system in vertebrate animals.
Band 3 is a major transmembrane structural protein (37) which attaches the
plasma membrane to the internal cell cytoskeleton by binding to band 2.1
(ankyrin) (38,39). The transport and the cytoskeletal domains can be
separated by proteolysis with trypsin. Digestion with trypsin yields a
52,000 Da membrane bound transport domain and a 40,000 Da water-soluble
cytoplasmic domain that binds cytoskeletal proteins. The transport domain of
band 3 is highly conserved evolutionarily and no polymorphisms of it have
been found.Senescent cell antigen is generated on this transport domain (5).

The complete sequence of mouse (28), chicken (40), and human (41)
erythrocyte, and human K562 cell line (27), and kidney (29,30) band 3 anion
transport protein have been deduced from the cDNA sequence. These band 3
cDNAs are closely related. For example, even under high stringency
conditions, kidney band 3 mRNA hybridizes with cDNA probes from the 3'
untranslated sequence, sequences encoding parts of the cytoplasmic domain,
and sequences encoding the membrane spanning domains of erythrocyte band 3
(30).

We have used synthetic peptides to identify antigenic sites on band 3
recognized by the IgG that binds to old cells. Results indicate that: a) the
active antigenic sites of the aging antigen reside on membrane protein band
3 residues which are extracellular regions implicated in anion transport; b)
a putative ankyrin binding region peptide is not involved in senescent cell
antigen activity; and c) carbohydrate moieties are not required for the
antigenicity or recognition of senescent cell antigen since synthetic
peptides alone abolish binding of senescent cell IgG to erythrocytes. One of
the putative transport sites that contributes to the aging antigen is
located toward the carboxyl terminus.

MATERIALS AND METHODS

Cell Separation: Red blood cells were separated into populations of
different ages on Percoll gradients as previously described (3). Middle-aged
cells were stored in Alsever's for 5 wks at 4°C.

Sodium dodecyl sulfate (NaDodSO₄) polyacrylamide gel electrophoresis
and immunostaining of peptides. Peptides are analyzed on 6-25% or 12-25%
linear gels (6,7,20,42). Immunoautoradiography was performed by the
immunoblotting technique of Towbin et al. (43) with the modifications
described previously (6,7,20,44,45). Immunoblots are exposed to Kodak Ortho
G (OG-1) film at -70°C using our standard procedures (44,45). Rabbit
antibodies to band 3 were prepared as previously described (6,7,20,25,
44,45).

IgG binding and inhibition assay: IgG was isolated from senescent RBC
from 50 liters of blood and purified as previously described (2,4). IgG
eluted from senescent cells, rather than serum IgG, was used because normal
serum contains antibodies to spectrin, actin, 2.1, etc. (44).
Competitive inhibition studies were performed using synthetic peptides to
absorb the IgG isolated from senescent erythrocytes. This is the same IgG
that initiates phagocytosis in situ. The Fc portion IgG is required for

binding and phagocytosis of cells by macrophages (1,4,7). Fab fragments were not used because we were simulating the physiological situation. Intact dimeric, senescent cell IgG containing the Fc portion binds to senescent cells in situ and initiates their removal (1-7,46). Only IgG isolated from aged erythrocytes binds specifically to senescent cells. For example, IgG eluted from young control erythrocytes did not bind to senescent cells (2,44). Moreover, the specific binding capacity of the autoantibody was eliminated to absorption with purified senescent cell antigen (7). SCIgG (3µg) is absorbed with synthetic peptides at the concentrations indicated or purified SCA, as a control, for 60 min at room temperature and incubated with stored red cells for 60 min at room temperature (1,2,4,46). Storage mimics normal aging in situ immunologically and biochemically (1-7,46). After incubation with absorbed IgG, cells are washed four times with 40-50 volumes of phosphate buffered saline containing 0.2% bovine serum albumin (BSA, fraction V, Sigma, St. Louis, MO) and 0.5% glucose. Washed cells are transferred to BSA-coated tubes (5×10^7 cells/50 µl) and incubated for 30 min. at 37°C with ^{125}I-Protein A (Amersham, Arlington Heights, IL 30-38 mCi/mg, 10-15 ng/tube). Cells are then washed four times and transferred to new tubes before counting in a gamma scintillation counter (Beckman, Gamma 5,500). The number of red cell-bound IgG molecules is quantitated before and after absorption using equilibrium binding kinetics (7,47). Scatchard analysis is performed (48). Percent inhibition is calculated from the following formula: $100 [1-(x-b/T-b)]$ where x=molecules of IgG autoantibody bound per cell; T=total number of IgG antibody molecules bound in the absence of inhibitor; b=background Protein A binding.

Peptides: Peptides are prepared by solid phase synthesis using an Applied Biosystems 430A automatic peptide synthesizer. They were analyzed by amino acid analysis, HPLC, sequencing, and/or FABS to determine purity. Amino acids are referred to by the standard single code.

Peptides that are not soluble in phosphate buffered saline are dissolved in 1-2% NaDodSO4. Others are solublized in acetic acid or trifluoroacetic acid, and sodium dodecyl sulfate (1-2%) is added. Once the peptides are in solution, they are conjugated to ovalbumin (5X crystallized) or BSA with 0.02% glutaraldehyde (49). The reaction is allowed to procede for 30-60 min, blocked with ethanolamine, and the conjugate dialysed against PBS in tubing with a molecular weight cut off of 1000. NaDodSO$_4$ does not interfer with the glutareldehyde coupling reaction. Since our assays include bovine serum albumin (BSA), the inclusion of albumin does not alter the assay.

Computer analysis: Sequence and protein structural analyses were performed using programs of the Genetics Computer Group (GCG), University of Wisconsin, Sequence Analysis Software Package (50).

RESULTS AND DISCUSSION

Identification of the aging antigenic site(s): We concluded from previous studies that senescent cell antigen is a degradation product of band 3 that includes most of the 35,000 Da carboxyl terminal segment and the 17,000 Da anion transport region (5). Both immunoblotting studies with

IgG isolated from senescent cells and peptide mapping studies of senescent cell antigen indicated that senescent cell antigen lacks a 40,000 molecular weight cytoplasmic segment which contains the amino terminus and, possibly, additional peptides of band 3 (5-7). Peptide mapping studies and anion transport studies suggested that a cleavage of band 3 occurs in the anion transport region (5). Furthermore, breakdown products of band 3 are observed in the oldest cell fractions but not in young or middle-aged cell fractions, and anion transport is impaired in old cells (5-7,46).

Based on structural, biochemical, and immunological data (1,2,4-7), we deduced that cleavage of old band 3 occurred approximately a third of the way into the transmembrane anion transport region from the carboxyl terminus end. Therefore, we synthesized peptides of erythrocyte band 3 from the anion transport region or with suspected anion transport activity. We selected one anion transport segment which appears to be exposed to the outside of the cell and one that is further along the molecule towards the amino terminus and outside the region that we speculated was included in senescent cell antigen (5). The first anion transport peptide includes two important amino acids (pep-ANION 1, position 538-554: SKLIKIFQDHPLQKTYN). The lysine at 539 is a covalent binding site for the anion transport inhibitor, diisothiocyano dihydrostilbene disulphonate (DIDS) (40), and the tyrosine at residue 553 is radio-iodinate by extracellular lactoperoxidase (41). The second anion transport region peptide is situated toward the end of the region and is probably intracellular because the potential N-glycosylation site at ASN-593 is not glycosylated (41) (pep-ANION 2, 588-602: LRKFKNSSYFPGKLR). This peptide would be predicted to lack inhibitory activity because it would not be presented as an antigen on native band 3. The last peptide from the carboxyl terminus region contains both hydrophobic and hydrophylic regions (pep-COOH, 812-827: LFKPPKYHPDVPYVKR). The lysines found in this region may comprise another binding site for DIDS (35). As a control, we used a peptide from the cytoplasmic segment of band 3 within the region of the putative ankyrin binding site (41) (pep-CYTO, 129-144: AGVANQLLDRFIFEDQ). As an additional control, a peptide containing the N-glycosylation site that is glycosylated (pep-GLYCOS, 630-648: QKLSVPDGFKVSNSSARGW) was included because it is extracellular. Peptides were synthesized based on the human sequence data from the paper by Tanner et al. (41).

Competitive inhibition studies were performed using synthetic peptides to absorb the IgG isolated from senescent erythrocytes. IgG binding and inhibition were determined with a protein A binding assay (46-47). This biological assay measures the fate of erythrocytes in vitro and in vivo (1,2,4). Results of these studies suggest that senescent cell IgG recognizes antigenic determinants that lie within the anion transport region 538-554 and putative transport site containing a cluster of lysines toward the carboxyl terminus, 812-827 (53). Pep-ANION 2 is only weakly inhibitory and pep-CYTO does not inhibit. The competitive inhibition data illustrated in Figure 1 show that peptides pep-ANION 1 and pep-COOH are inhibitory over a range of 3 to 100 µg (Figure 1), whereas the internal peptide from the anion transport region (pep-ANION 2) is only weakly inhibitory, and the putative ankyrin binding peptide does not react with the antibody (Figure 1).

We then decided to mix the two inhibitory peptides to determine whether they acted synergistically for two reasons. First, both are inhibitory, but the inhibition is not complete ever at 300 µg. Second, the results of our early peptide mapping studies with topographically defined segments of band 3 suggested that senescent cell antigen was composed of peptides from both the anion transport transmembrane region and the 35,000-38,000 Da carboxyl terminal segment (5). Mixing of these two regions produced a peptide map that closely resembled senescent cell antigen even though it contained more peptides. The mixture of these two peptides produced ∼50% inhibition at 0.1 µg (i.e. 0.05 µg of each peptide) indicating that pep-ANION 1 and pep-COOH interact together to form a three dimensional structure that functions as an aging antigen (Figure 1). Pep-COOH hexamer (N6) consisting of 6 amino acids on the amino side of pep-COOH gave significant inhibition (∼50% at 10 µg) but did not synergize as well with pep-ANION 1 as did pep-COOH itself since a ∼10 fold increase in peptide is required to obtain inhibition (∼50% inhibition at 1µg), pep-COOH decamer (C10) consisting of 10 amino acids on the carboxyl side of pep-COOH gave 54±3% inhibition at 30 µg. A. mixture of pep-ANION 2 and pep-COOH did not exhibit synergy (inhibition: 18±4% at 30 µg).

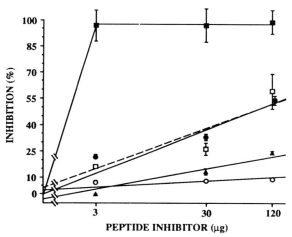

Fig. 1. <u>Inhibition of senescent cell IgG binding to erythrocytes by synthetic peptides or synthetic peptide mixtures.</u> Competitive inhibition studies were performed as described in materials and methods at the concentrations indicated on the graph. For peptide mixtures, the total peptide used was the amount indicated with each peptide constituting half of that amount. (o) pep-CYTO (R139-159); (□) pep-ANION 1 (R538-554); (▲) pep-ANION 2 (R588-602); (●) pep-COOH (R812-827); (■) pep-ANION 1 and pep-COOH mixture. Experiments considered of replicative determinations. Lines were fitted by the method of least squares. The data of pep-ANION 1 and pep-COOH could be fitted by a single line. Other pep-COOH subpeptides gave the following percent inhibition: R800-818 (10µ), 45±9; R818-827 (17 µM, 12.2 µg), 54±7; R818-823 (17µM, 6.9 µg), 27±4; R822-827 (70.6µM, 30 µg), 30±2. Reproduced from reference 52.

Synthetic peptides would not be expected to be as effective as the native band 3 molecule itself because the short peptide segments do not assume the same tertiary configuration as that of the 911 amino acid band 3 molecule. The synergism of peptides pep-ANION 1 and pep-COOH suggest that the conformation of the determinants of these two peptides interacting with each Other is similar to that of the intact aging antigen. These results, together with data indicating that DIDS crosslinks these two regions (52), suggests that these peptides lie in close spatial proximity in native, aged band 3. This is consistent with other data indicating that these two sites are in close proximity in native band 3 (36,51,57).

Binding of senescent cell IgG to pep-ANION 1 and peptide pep-COOH which is close to the carboxyl terminus of band 3 suggests that these segments are extracellular because the IgG molecule is 150,000 molecular weight and too large to enter a cell. This is consistent with the data of Jennings (34) which indicates that these regions are extracellular and that a segment located carboxyl to the trypsin cleavage site at lysine 743 and an S-cyanylation cleavage site ∼7000 Da from the C-terminus in crosslinked by extracellular DIDS (34).

Results of these studies with synthetic peptides are consistent with the physiological data demonstrating that old erythrocytes have impaired anion transport (6,7,46), the biochemical and immunological data indicating that band 3 undergoes degradation with loss of a cytoplasmic segment during the aging process (5-7, 46), and the data derived from alterations and/or mutations of band 3 indicating that changes associated with accelerated aging involve changes in the anion transport region of band 3 (6,7,46).

Model of the membrane associated region of human band 3. Based on the above and other considerations, we developed a working model for the membrane associated region of human band 3 protein, approximate residues (R) 400-870. The model was constructed using the program PEPPLOT of the GCG package to identify membrane spanning nonpolar helices (53) and intervening hydrophilic loops (54). The location of the hydrophilic loops as extracellular or intracellular was predicted on the basis of established chemical or biological markers, e.g., the demonstration that residues 814-829 contain a DIDS binding site (41) or the availability of Y553 for external radioiodination (41). Table 1 defines these regions and indicates the number of residues in each. These regions are assembled into a model in (52). Our present results show that close steric association must be maintained by external loops O2 and O4. If these regions are associated on the same band 3 monomer then band 3 loops back upon itself so that these regions are contiguous. Alternatively, the functional assembly may be dimers in which close associations of O2 and O4 form between separate molecules. Our model for the membranes-associated portion of human band 3 is largely congruent with the model of Jay and Cantley (54) for the murine homolog. Our external segments O1, O1a, O2, and O3 correspond completely, or to a major degree with external segments identified by those workers. Our incorporation of recent human data (41) with the antigenic results presented here allows refinement and development of the model for the segment of band 3 designed O4 that contains the peptides pep-COOH and its component N6 which are potent inhibitors of the binding of autoantibodies to the aging antigen to red cells. All of the data support the external presentation of this. This segment occurs in the murine sequence (residues 825–853) but a prior model indicates that its location is uncertain. The short bilayer spanning segments may represent segments that enter the bilayer and exit again, looping back on themselves, without transversing the bilayer to the other side. These projections are based on hydrophobicity plots and may or may not exist in the membrane.

Amino acids that are exposed to the outside could reside on a hydrophobic helix within a membrane pore and still be accessible to the outside even though they are not on the outer most membrane. This probably applies to some band 3 sites designated as external although it is not reflected in the model. We suspect that the tertiary structure of band 3, when finally elucidated, will turn out to be a ring.

Table 1. Bilayer spanning helices and intervening hydrophilic loops (approximate) predicted by the model[a].

BILAYER SPANNING NONPOLAR HELICES	INTERVENING HYDROPHILIC LOOP	
	EXTRACELLULAR	INTRACELLULAR
1. 400–425 (25)		
	0.1: 426–439 (13)	
2. 440–475 (35)		
	0.1a: 476–489 (13)	
3. 490–510 (20)		
4. 525–536 (11)		
	0.2: 537–569 (32)	
5. 570–588 (18)		
		IN-1: 589–609 (20)
6. 610–620 (10)		
	0.3: 621–654 (33)	
7. 655–683 (28)		
		IN-2: 684–704 (20)
8. 705–732 (27)		
		IN-3: 733–769 (36)
9. 770–806 (36)		
	0.4: 807–835 (35)	
10. 836–861 (25)		
11. 862–870 (8)		
		IN-4: 871–911 (40)

a. () indicate approximate number of amino acids in segment Model shown in ref. 52.

Computer search for regions of internal homology within band 3 was performed using programs COMPARE and DOTPLOT with a window of 30 residues and stringencies of 10.0 (low) and 15.0 (high). Homologies were disclosed among the membrane spanning nonpolar helical regions. For example, a segment of bilayer spanning helix adjacent to extracellular loop 02 had ~40% identity to a corresponding segment of bilayer spanning segment between international loops 2 and 3; viz:

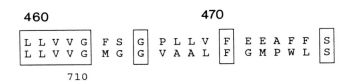

Such relationships were not obvious among the hydrophilic regions, e.g. external hydrophilic 02 and 04 did not show significant sequence homology.

Molecular "walking" of band 3 to define the antigenic site. Our previous experiments suggest that the active antigenic sites of senescent cell antigen reside on the peptides which we have designated pep-ANION 1 and pep-COOH. In the series of experiments, we attempted to define the active antigenic site by "walking" the anion transport domain of band 3 molecule which our studies indicate contain the antigenic determinants.

By "walking" we mean the antigenic analysis of a series of synthetic overlapping peptides that encompass the entire polypeptide chain adjacent to the active sites we have identified and include all predicted extracellular segments of band 3. The synthetic peptides are 17–19 mer and overlap their adjacent neighboring peptides by 6 residues in the overlap regions in order to optimize the feasibility of synthesis and to expect reasonable resolution of individual antigenic sites. In addition, we synthesized 2 predicted external sequences (R 426–440, LLGEKTRNQMGVSEL; 645–659, ARGWVIHPLGLRSEF). The former sequence is not a putative transport region. This was done to complete the testing of all predicted external band 3 sequences for senescent cell antigen since the antigenic determinants must be external to be accessible to the 150,000 molecular weight IgG molecule.

Peptides used for "walking " the molecule are:
A. pep-CYTO, 129–144: AGVANQLLDRFIFEDQ;   B. 426–440, LLGEKTRNQMGVSEL;
C. 515–531, FISRYTQEIFSFLISLI;   D. 526–541, FLISLIFIYETFSLI;
E. pep-ANION 1, 538–554, SKLIKIFQDHPLQKTYN; F. 549–566, LQKTYNYNVLMVPKPQGP;
G. 561–578, PKPQGPLPNTALLSLVLM;   H. 573–591, LSLVLMAGTFFFAMMLRKF;
I. pep-ANION 2, 588–602, LRKFKNSSYFPGKLR;   J. 597–614, FPGKLRRVIGDFGVPISI;
K. 609–626, GVPISILIMVLVDFFIQD;   L. 620–637, VDFFIQDTYTQKLSVPD;
M. pep-GLYCOS,630–648, QKLSVPDGFKVSNSSARGW; N. 645–659, ARGWVIHPLGLRSEF;
O. 776–793, MEPILSRIPLAVLFGIFL;   P. 788–805, FGIFLYMGVTSLSGIQL;
Q. 800–818, LSGIQLFDRILLLFKPPKY;   R. pep-COOH, 812–827,LFKPPKYHPDVPYVKR;
S. 822–839, VPYVKRVKTWRMHLFTGI.

We used immunoblotting of the peptides followed by reaction with IgG eluted from senescent red cells (SCIgG) to localize the aging antigen site (Figure 1). We also used the IgG binding and inhibition assay at a single concentration of peptide and performed Scatchard analysis (Table 2). We selected 30 μg for this "single concentration" assay because our two synergistic peptides, pep-ANION 1 and pep-COOH, give 95% inhibition at this concentration; whereas, "non-inhibitry" peptides give 20% inhibition. Peptides that were negative by immunoblotting were also negative in the inhibition assay (Figure 2 & Table 2). We relied on immunoblotting with senescent cell IgG for peptides that were not completely soluble under physiologic conditions (Figure 2).

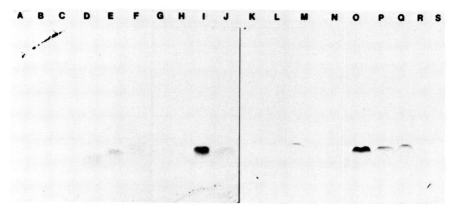

Fig. 2. "Walking" the band 3 molecule to localize the active aging antigenic site using binding of SCIgG to synthetic peptides as determined by immune labeling of electrophoretic blots. Autoradiograph of immunoblot incubated with IgG autoantibody from senescent erythrocytes. The peptides with residue number are indicated in the text.

Immunoblotting studies with senescent cell IgG showed binding to R 788–805, 800–818, in addition to pep-COOH (R812–827). Peptides R788–805, 800–818 are to the amino side of pep-COOH. R822–839, which has no labelling, is to the carboxyl side. R800–817 and 822–839 have 6 amino acids in common with pep-COOH. Binding of antibody to R630–648 and trace binding to R645–659 was observed.

However, the competitive inhibition assay shows that significant inhibition of senescent cell IgG binding is obtained only with peptide R822–839 which is on the carboxyl side of pep-COOH and contains 6 amino acids of pep-COOH. Thus, its inhibitory effect may be due to the 6 common amino acids. Peptides 776–793, 788–805, and 800–839 were not completely soluble and were not tested in the cellular binding assay.

In the pep-ANION 1 and pep-ANION 2 peptide series, trace binding was observed in immunoblots to pep-ANION 1 (R538–554) and the peptides to its amino (R526–541) and carboxyl (R549–566) side which have 6 amino acids overlap with that peptide. Senescent cell IgG bound to pep-ANION 2 (R588–602) and, faintly, to the peptide to its carboxyl side (R597–614) which has 6 amino acids overlap.

The competitive inhibition assay shows that inhibition of 52±4% is obtained with R597–614. R526–543 and R549–566 could not be tested because of solubility. The peptide from external loop R426–440 is negative.

Table 2. "Walking" of band 3 protein to define the antigenic site: inhibition of senescent cell IgG binding to erythrocytes by synthetic peptides of band 3.[a]

SYNTHETIC PEPTIDE RESIDUE	INHIBITION
426–440	11±1
515–531	15±1
526–541	NT
549–566	NT
561–578	0
573–591	NT
597–614	52±4
609–626	12±1
620–637	NT
630–648	9±0
645–659	NT
776–793	NT
788–805	NT
800–818	NT
822–839	35±2

a. Data are presented as the percentage inhibition ± standard deviation of the binding of human autoantibody to senescent cell antigen. All peptides were tested at 30µg. NT, not tested in the competitive inhibition assay because the peptide was not completely soluble in physiologic solutions even when coupled to BSA.
Reproduced from reference 52.

We used two approaches to determine the peptides with which autoantibody to senescent cell antigen reacts. The direct binding in Western blot analysis shows that the antibody bound to a number of peptides, including some that are internal and do not block in the cellular inhibition assay. This probably results from the fact that aged cells that bind autoantibody are phagocytosed by macrophages and antigenic peptides are generated for presentation to T cells. These peptides are not exposed in the normal presentation of band 3 and senescent cell antigen on the erythrocyte surface, however. Our competitive inhibition studies show that the native determinant is formed by interaction of two exposed peptide regions contributed by pep-ANION 1 and pep-COOH.

Thus, senescent cell antigen is localized to a region within residues 538–554 and 788–827. Even though an antigen binding site is only 6 amino acids in size, these amino acids are probably not adjacent in primary structure. A minimal number of amino acids is probably required to generate the twists and turns necessary for the active three dimensional structure. The synergy between pep-ANION 1 and pep-COOH supports this. The data from the N terminus 6 amino acids of pep-COOH indicate that even though significant inhibition can be obtained with a six met peptide, larger quantities of the peptide are required, and synergy is impaired because 95% inhibition cannot be obtained at concentrations up to 100 µg. In contrast, 95% inhibition can be obtained with the mixture of pep-ANION 1 and pep-COOH at 10 µg, and, in some experiments, at 3 µg.

Studies indicate that the transport region of band 3 is highly conserved across tissues, individuals, and species (29,30,40,41,55). Since senescent cell antigen is generated on all cell types, tissues,and species examined (4), the regions that comprise senescent cell antigen must also be highly conserved both evolutionarily and in various tissues. In this study, we show that active antigenic determinants of senescent cell antigen reside on peptides pep-ANION 1 and pep-COOH within residues 538-554 and 788-827. These peptides reside in highly conserved regions (41,52,55).

We have localized senescent cell antigen, an aging antigen, in terms of primary structure. Localization of the active antigenic site of senescent cell antigen facilitates the next logical step, namely, definition of the molecular changes occurring during aging that imitate molecular as well as cellular degeneration and regulation of cellular lifespan.

ACKNOWLEDGEMENTS

We are grateful to Robert Poff and Gordon Purser for assistance with the figures. This work was supported by the International Foundation for Biomedical Aging Research.

REFERENCES

1. M. M. B.Kay, Mechanism of removal senescent cells by human macrophages in situ, Proc. Natl. Acad. Sci 72:3521 (1975).
2. M. M. B. Kay , Role of physiologic autoantibody in the removal of senescent human red cells, J. Supramol. Struct. 9:555 (1978).
3. G. D. Bennett and M. M. B. Kay, Homeostatic removal of senescent murine erythrocytes by splenic macrophages, Exp. Haematol. 9:297 (1981).
4. M. M. B. Kay, Isolation of the phagocytosis inducing IgG-binding antigen of senescent somatic cells, Nature 289:491 (1981).
5. M. M. B. Kay, Localization of senescent cell antigen on band 3, Proc. Natl. Acad. Sci. 81:5753 (1984).
6. M. M. B. Kay, G. J. C. G. M. Bosman, S. S. Shapiro, A. Bendich and P.S. Bassel, Oxidation as a possible mechanism of cellular aging: Vitamin E deficiency causes premature aging and IgG binding to erythrocytes, Proc. Natl. Acad. Sci. USA 83:2463 (1986).
7. M. M. B. Kay, N. Flowers, J. Goodman and G.J. Bosman, Alteration in membrane protein band 3 associated with accelerated erythrocyte aging, Proc. Natl. Acad. Sci. 86:5834 (1989).

8. R. P. Hebbel and W. J. Miller, Phagocytosis of sickle erythrocytes. Immunologic and oxidative determinants of hemolytic anemia, Blood 64:733 (1984).

9. J. A. Singer, L. K. Jennings, C. Jackson, M.E. Doctker, M. Morrison and W.S. Walker, Erythrocyte homeostasis: Antibody-mediated recognition of the senescent state by macrophages, Proc. Natl. Acad. Sci. USA 83:5498 (1986).

10. G. A. Glass, H. Gershon and D. Gershon, The effect of donor and cell age on several characteristics of rat erythrocytes, Exp. Hematol. 11:987 (1983).

11. G. A.Glass, D. Gershon and H. Gershon, Some characteristics of the human erythrocyte as a function of donor and cell age, Exp. Hematol. 13:1122 (1985).

12. G. Bartosz, M. Sosynski and J. Kredziona, Aging of the erythrocyte. VI. Accelerated red cell membrane aging in Down's syndrome, Cell Biol. Int. Rep. 6:73 (1982).

13. G. Bartosz, M. Sosynski and A. Wasilewski, Aging of the erythrocyte. XVII. Binding of autologous immunoglobulin, Mech. Aging Dev. 20:223 (1982).

14. N. Khansari, G. F. Springer, E. Merler and H.H. Fudengerg, Mechanisms for the removal of senescent human erythrocytes from circulation: specificity of the membrane-bound immunoglobulin G., J. Mech. Aging Dev. 21:49 (1983).

15. N. Khansari and H. H. Fudenberg, Immune elimination of autologous senescent erythrocytes by Kupffer cells in vivo, Immunol. 80:426 (1983).

16. E. M. Alderman, H. H. Fudenberg and R. E. Lovins, Binding of immunoglobulin classes to subpopulations of human red blood cells separated bydensity-gradient centrifugation, Blood 55:817 (1980).

17. C. H. Tannert, Untersuchungen zum altern roter blutzellen, Ph.D. Thesis, Humboldt University, Berlin, GDR (1978).

18. G. Wegner, C. H. Tannert, D. Maretzki, W. Schossler and D. Stanss, IgG Binding to glucose depleted and preserved erythrocytes, 9th Int. Symp. Struct. Function Erythroid Cells, Berlin GDR 57 (1980).

19. W. S. Walker, J. A. Singer, M. Morrison and C.W. Jackson, Preferential phagocytosis of in vivo aged murine red blood cells by a macrophage-like cell line, Br. J. Haematol. 58:259 (1984).

20. M. M. B. Kay, S. Goodman, K. Sorensen, C. Whitfield, P. Wong, L. Zaki, V. Rudloff, The senescent cell antigen is immunologically related to band 3, Proc. Natl. Acad. Sci. 80: 1631 (1983).

21. H. U. Lutz, R. Flepp and G. Stringaro-Wipf, Naturally occurring autoantibodies to exoplasmic and cryptic regions of band 3 protein, the major integral membrane protein of human red blood cells, J. Immunol. 133:2160 (1984).

22. H. Miller and H.U. Lutz, Binding of autologous IgG to human red blood cells before and after ATP-depletion. Selective exposure of binding sites (autoantigens) on spectrin-free vesicles, Biochim. Biophys. Acta 729:249 (1983).

23. M. J. Friedman, M. Fukuda and R. A. Laine, Evidence for a malarial parasite interaction site on the major transmembrane protein of the human erythrocyte, Science 228:75 (1985).

24. A. R. Dluzewski, K. Rangachari, M. J. Tanner, et al., Inhibition of malarial invasion by intracellular antibodies against intrinsic membrane proteins in the red cell, Parasitology 93:427 (1986).

25. M. M. B. Kay, G. Bosman, M. Notter, et al., Life and death of neurons: The role of senescent cell antigen, Ann. N.Y. Acad. Sci. 521:155 (1988).

26. M. M. B. Kay, C. M. Tracey, J. R. Goodman, et al., Polypeptides immunologically related to erythrocyte band 3 are present in nucleated somatic cells, Proc. Natl. Acad. Sci. 80:6882 (1983).

27. D. R. Demuth, L. C. Showe, M. Ballantine, A. Palumbo, P.J. Fraser, L. Cioe, G. Rovera and P.J. Curtis, Cloning and structural characterization of a human non-erythroid band 3-like protein. EMBO Journal 5:1205 (1986).

28. R. R. Kopito and H. F. Lodish, Structure of the murine anion exchange protein, J. Cell. Biochem. 29:1 (1985).

29. K. E. Kudrycki and G. E. Shull, Primary structure of the rat kidney band 3 anion exchange protein deduced from a cDNA, J. Biol. Chem. 264:8185 (1989).

30. S. L. Alper, R. R. Kopito, S. M. Libresco and H. Lodish, Cloning and characterization of a murine band 3-related cDNA from kidney and from a lymphoid cell line, J. Biol. Chem. 263:17092 (1988).

31. D. J. Hazen-Martin, G. Pasternack, S. S. Spicer and D.A. Sens, Immunolocalization of band 3 protein in normal and cystic fibrosis skin, J. Histochem. Cytochem. 34:823 (1986).

32. V. L.Schuster, S. M. Bonsib and M. L. Jennings, Two types of collecting duct mitochondria-rich (intercalated) cells: lectin and band 3 cytochemistry, Am. J. Physiol. 251:C347 (1986).

33. S. Kellokumpu, L. Neff, S. Jamsa-Kellokumpu, R. Kopito and R. Baron, A 115-kD polypeptide immunologically related to erythrocyte band 3 is present in Golgi membranes, Science 242:1308 (1988).

34. M. L. Jennings, M. P. Anderson and R. Monaghan, Monoclonal antibodies against human erythrocyte band 3 protein. Localization of proteolytic cleavage sites and stilbenedisulfonate-binding lysine residues, J. Biol. Chem. 261:9002 (1986).

35. J. Falke, K. J.Kanes and S. I. Chan, The minimal structure containing the band 3 anion transport site. A 35CI NMR study, J. Biol. Chem. 260:13294 (1985).

36. P. J. Bjerrum, J. O. Wieth and S. Minakami, Selective phenylglyoxalation of functionally essential arginyl residues in the erythrocyte anion transport protein, J. Gen. Physiol. 81:453 (1983).

37. T. L. Steck, The organization of proteins in human red blood cell membranes, J. Cell. Biol. 62:1 (1974).

38. S. R. Goodman and K. Shiffer, The spectrin membrane skeleton of normal and abnormal human erythrocytes: A review, Am. J. Physiol. 244:C121 (1983).

39. V. Bennett, Immunoreactive forms of human erythrocyte ankyrin are present in diverse cells, Nature, Lond. 281:597 (1979).

40. J. V. Cox and E. Lazarides, Alternative primary structures in the transmembrane domain of the chicken erythroid anion transporter, Mol. Cell. Biol. 8:1327 (1988).

41. M. J. A. Tanner, P. G. Martin and S. High, The complete amino acid sequence of the human erythrocyte membrane anion-transport protein

deduced from the cDNA sequence, Biochem. J. 256:703 (1988).

42. U. K. Laemmli, Cleavage of structural proteins during the assembly of the head of bacteriophage T4, Nature, Lond. 227:680 (1970).

43. H. Towbin, T. Staehelin and J. Gordon, Electrophoretic transfer of proteins from polyacrylamide gels to nitrocellulose sheets: procedure and some applications, Proc. Natl. Acad. Sci. 76:4350 (1979).

44. M. M. B. Kay, K. Sorensen, P. Wong and P. Bolton, Antigenicity, storage & aging: Physiological autoantibodies to cell membrane and serum proteins and the senescent cell antigen, Mol. Cell. Biochem. 49:65 (1982).

45. M. M. B. Kay, Immunologic techniques for analyzing red cell membrane proteins, in: "Methods in Hematology: Red Cell Membranes", S. Shohet and N. Mohandas, eds. Churchill Livingston, Inc. New York (1988).

46. M. M. B. Kay, Red cell aging: Senescent cell antigen, band 3, and band 3 mutations associated with cellular dysfunction, Proc. Clin. Biol. Res. 318:199 (1989).

47. P. Yam, L. D. Petz and P. Spath, Detection of IgG sensitization of red cells with 125I staphycoccal protein A, Am. J. Hematol. 12:337 (1982).

48. G. Scatchard, The attraction of proteins for small molecules and ions, Ann. N.Y. Acad. Sci. 51:660 (1949).

49. M. M. B Kay, Multiple labeling technique used for kinetic studies of activated human B lymphocytes, Nature, Lond. 254:424 (1975).

50. J. Devereux, P. Haeberli and O. Smithies, A comprehensive set of sequence analysis programs for the VAX, Nucleic Acid Res. 12:387 (1984).

51. L. Zaki, Anion transport in red blood cells and arginine specific reagents. (1) Effect of chloride and sulfate ions on phenylglyoxal sensitive sites in the red blood cell membrane, Biochem. and Biophys. Res. Comm. 110:616 (1983).

52. M. M. B. Kay, J. J. Marchalonis, J. Hughes, J. Hughes, L. Watanabe and S.F. Schluter, Definition of a physiologic aging autoantigen using synthetic peptides of membrane protein band 3: Localization of the active antigenic sites, Proc. Natl. Acad. Sci. USA, in press (1990).

53. D. M. Engelman, T. A. Steitz and A. Goodman, Identifying nonpolar transbilayer helices in amino acid sequences of membrane proteins, Ann. Rev. Biophys. Com. 15:321 (1986).

54. J. Kite and R. F. Doolittle, A simple method of displaying the hydropathic character of a protein, J. Mol. Biol. 157:105 (1982).

55. M. M. B. Kay, F. Lin, G. Bosman, J. Marchalonis and S. Schluter, Human erythrocyte aging: Cellular and molecular biology, Trans. Med. Revs., in press (1990).

RECOGNITION SIGNALS FOR PHAGOCYTIC REMOVAL OF FAVIC

MALARIA-INFECTED AND SICKLED ERYTHROCYTES

Paolo Arese[1], Franco Turrini[1], Federico Bussolino[1],
Hans U. Lutz[2], Danny Chiu[4], Lynn Zuo[4]
Frans Kuypers[4] and Hagai Ginsburg[3]

[1]Dipartimento di Genetica, Biologia e
Chimica Medica, Università di Torino
[2]Laboratorium für Biochemie
Swiss Federal Institute of Technology, Zurich
[3]Department of Biological Chemistry
The Hebrew University, Jerusalem, Israel
[4]Children's Hospital of Northern California
Medical Center, Oakland, CA

According to a recently published concept (1), immunologically-mediated erythrocyte (RBC) removal involves binding of autologous, naturally circulating antibodies directed against normally hidden epitopes of integral RBC membrane proteins. Prerequisite of antibody binding are modifications of specific RBC proteins, presumably band 3 and the glycophorins. The exact nature of the modifications resulting in exposure of hidden epitopes or generation of neoantigenic sites, is still unknown. Evidence that oxidant-elicited oligomerization of band 3 is a sufficient change to enhance anti-band 3 binding has been provided (2,3). The precise molecular composition of the opsonin complex, and the nature of the bonds which keep its components together, are poorly defined as yet. An essential characteristic of the anti-band 3-mediated removal is the precipitation of complement (1-3). Indeed, conditions that inactivate complement convertases and abrogate the formation of the active complement component C3b also extensively inhibit phagocytic removal of variously damaged RBC. The same inhibition of phagocytosis occurs after blockage of the C3b receptor (CR1, complement receptor type one) on the macrophage surface (3,4). Lutz et al.(2,3-5) hypothesized that anti-band 3 antibodies bivalently bound to aggregated band 3 stimulate generation and deposition of activated C3 on the RBC surface.

The details of the specific enhancement of complement deposition by aggregated anti-band 3 antibodies are not known yet. The sequence outlined before, and summarized in Fig. 1 and Fig. 2 has found experimental

support in artificially oxidatively stressed human RBC (2,3) and in
physiologically senescent human RBC (5). Involvement of anti-band 3 binding
has also been demonstrated in sickle RBC, malaria-infected RBC, and in
RBC with unstable hemoglobins (6-8). In none of these pathophysiological
conditions that are accompanied by accelerated removal of RBC has complement
yet been shown to play a role.

   Here we will consider three important conditions leading to, or
accompanied by, massive extravascular - i.e. phagocytic- RBC removal. The
conditions under scrutiny are: favic hemolysis, P. falciparum infection and

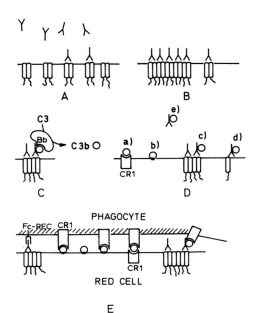

Fig. 1. Possible sequence of events after oxidant damage of RBC leading to
       IgG binding, complement deposition and their recognition and
       removal by the macrophage. A) Normal, non-oxidized RBC. Band 3 is
       not aggregated or clustered, and only a few anti-band 3 IgG
       molecules are bound. B) Oxidant insult causes aggregation or
       clustering of band 3, which in turn enhances binding of anti-band
       3 IgG. Anti-band 3 has a special affinity for spontaneously
       generated C3b. C) Binding of C3b to IgG prevents inactivation of C3b
       and allows formation of the alternative pathway convertase C3bBb.
       An amplification loop is started which generates more C3b molecules.
       D) Nascent C3b can bind to complement receptor type 1 (CR1) (a), can
       covalently bind to surface components of RBC, like oligosaccharides
       or aminophospholipids (b), to anti-band 3 clusters (c) and to
       monovalently bound anti-band 3 (d). Binding modes (c) and (d) are
       predominant. E) Bound C3b is recognized predominantly by CR1 of the
       macrophage. Fc-mediated recognition of bound anti-band 3 IgG is also
       possible (10, reprinted with permission).

318

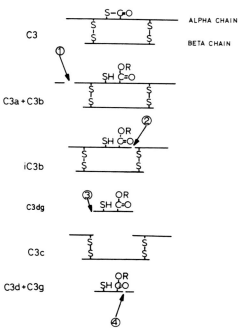

Fig. 2. Schematic representation of complement component C3 modifications. C3 is composed of two disulfide-linked chains. Native C3 in plasma undergoes continuous slow-rate hydrolysis of the internal thioester bond (see below). The product (C3i) is a binding site for factor B. Factor B bound C3i is split into Ba and Bd by factor D. C3iBb is a C3 convertase which actively cleaves C3 to C3a and C3b. Some C3b bound to the RBC surface acts as a binding site for more factor B and initiates an amplification loop. The splitting of the small peptide C3a from the alpha chain of C3 by the C3iBb convertase causes a conformational change and exposure of the internal thioester bond (step 1). The bond is metastable and the electrophilic carbonyl group is susceptible to attack by adjacent nucleophilic (electron donor) groups. If the nucleophile is a carbohydrate hydroxyl group, a stable ester bond is formed, while in case of amino group, formation of amide bond ensues (step 1). Step 2 is mediated by the serine protease factor I resulting in the formation of iC3B. Subsequent cleavage (step 3) generates fragments C3dg and C3c. Final cleavage (step 4) of C3dg into C3d and C3g is effected by plasmin, trypsin and leukocyte elastase. C3b and C3c bind exclusively to CR1. iC3b binds to CR1 but also to other complement receptors such as CR2, CR3 and CR4 (9).

sickle cell anemia. The following RBC or RBC-associated parameters have been studied by homogeneous techniques and criteria: extent of binding of autologous IgG to RBC; deposition of complement components; phagocytosis and its dependence on complement activation; isolation and composition of the high-molecular weight aggregates; specificity of bound autologous IgG.

As far as the macrophage was concerned, modulation of phagocytic activity by selective inhibitors of Fc or complement receptors has been analyzed. Under all conditions studied, complement was found to play an important role, and in all cases, except in RBC infected by late stage forms of P. falciparum, abrogation of complement activation or deposition by a variety of means led to substantial reduction of erythrophagocytosis.

FAVISM

Despite massive hemoglobinuria and increased hemoglobinemia commonly observed during favism, circumstantial evidence indicates that removal of damaged RBC is predominantly extravascular (10). This means that affected RBC do not lyse in the bloodstream due to osmotic or mechanical reasons but are removed phagocytically by macrophages. Evidence (for a review, see ref. 10) supporting this contention is based on: a) balance studies between total blood disappearance, excretion of hemoglobin in urine, and other routes of hemoglobin disposal. Those studies demonstrated a low share of intravascular hemolysis in favism, not exceeding 20 percent of total hemolysis as a mean value; b) RBC isolated during early phases of the favic crisis were actively phagocytosed by adherent, monocyte-derived macrophages in vitro, while RBC isolated from late crisis patients were not phagocytosed; c) phagocytes isolated from early crisis blood or bone marrow were engulfed with ingested RBC, and showed enhanced eigenluminescence; d) the total number of phagocytes present in the human body and their erythrophagocytic capacity are compatible with the number of RBC disappearing during favic crisis. Studies (10) performed on RBC isolated during the early phases of the crisis and bound to be eliminated, support the idea that the favic crisis can be considered as sharply accelerated senescence affecting a large proportion of the RBC population. Membrane alterations leading to RBC recognition, the specificity of deposited autologous IgG, and the complement-dependence of phagocytic removal are comparable to, and compatible with, observations performed on normally senescent RBC. As shown in Fig. 3, RBC membranes isolated from favic patients contained elevated amounts of C3b-IgG complexes. The relative amount of complexes correlated with the severity of the crisis and decreased to normal values within 4 to 5 days after fava bean ingestion. Cleavage of C3b-IgG complexes with hydroxylamine, which hydrolyzes primarily ester bonds, liberated roughly half of the bound IgGs. These IgGs were purified on Protein A-Sepharose and radioiodinated. When applied to electrophoretically spread and blotted RBC membrane proteins, IgG bound exclusively to band 3 protein. The same specificity was detected with IgG liberated from C3b-IgG complexes of normal senescent RBC (1,5). In addition, the importance of complement in favism is also implied from: a) increased, complement-dependent phagocytosis of damaged RBC isolated during the crisis; abrogation of phagocytosis by complement convertase inactivators; b) presence of covalently bound complexes between (predominantly) anti-band 3 IgG and complement on the damaged RBC membrane. Complement fragments were in excess over IgG in these complexes; c) increased amounts of the stable complement fragment C3c bound to damaged RBC; d) correlation between clinical time-course of the crisis and disappearance of RBC bearing complement fragments in periferal blood.

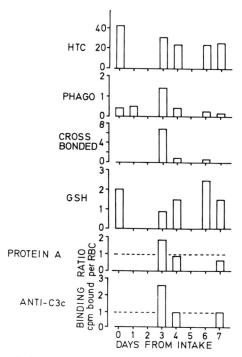

Fig. 3. Binding of autologous IgG and deposition of complement fragments in
G6PD-deficient RBC bound to be removed during hemolytic crisis due
to fava bean ingestion (favism). Data refer to a female heterozygous
patient who ingested 80-100 g fresh fava beans three days before
admission. RBC bearing increased amounts of IgG and complement
fragments are removed from circulation within four days after fava
bean ingestion. Phagocytosis by monocyte-derived human macrophage
is expressed as RBC ingested per monocyte. Binding ratio of
radioiodinated protein A or radioiodinated anti-C3c antibodies
is referred to normal, non-G6PD-deficient controls. Bound IgG has
exclusive anti-band 3 specificity (H. U. Lutz, F. Turrini, S.
Fasler, D. Alessi, G. Giribaldi and P. Arese, unpublished results).

P. FALCIPARUM -INFECTED RBC

     Malaria-infected RBC are intensely phagocytosed in vivo and in vitro
(11,12). In vivo, circulating and resident macrophages loaded with
infected RBC are commonly observed in falciparum malaria patients. In
vitro, cultured human macrophages phagocytose infected RBC opsonized with
naive or immune sera. We have investigated the phagocytosis of infected RBC
(IRBC) at various stages of parasite development, dissected the recognition
signals, quantitated and characterized their relative role and importance.
Asynchronous cultures of P. falciparum infected RBC were fractionated into
different development stages (ring-forms, RIRBC; trophozoite, TIRBC; and
schizonts, SIRBC) on a Percoll-sorbitol gradient (13). As shown in Fig. 4,
infected RBC were phagocytosed more intensely than normal controls. An
average of 3 RIRBC and 8-9 TIRBC or SIRBC were ingested per monocyte, in

comparison to average 4-5 diamide stressed RBC and 0.8 normal non- infected controls. Opsonization with serum in which the complement convertases were inactivated, inhibited phagocytosis of RIRBC by 78-100 percent, and that of TIRBC by 34-45 percent. Blockage of macrophage CR1 by anti CR1 monoclonal antibodies reduced phagocytosis of RIRBC by 91 percent and that of TIRBC by 25 percent. Blockage of the Fc portion of IgG bound to the RBC by F(ab)2 fragment of anti-Fc polyclonal antibodies reduced phagocytosis by about 14 percent in both RIRBC and TIRBC. Residual, non-complement, non-IgG and non-phosphatidylserine-dependent (see below) phagocytosis amounted to approx. 10 percent in RIRBC and to approx. 20 percent in TIRBC, respectively (F. Turrini, H. Ginsburg, M. V. Serra and P. Arese, submitted). The intensity and mode of phagocytic removal of oxidatively stressed normal RBC and RIRBC are superimposable, while in TIRBC a higher portion of phagocytic susceptibility appears to be non-complement dependent and suppressible by preincubation of macrophages with phosphotidylserine-containing liposomes. Determination of bound IgG by radioiodinated protein A and bound C3c by radioiodinated anti-C3c antibodies showed that IgG binding was increased 2.6 times in RIRBC, 21 times in TIRBC and 24 times in SIRBC compared to non-infected controls. C3c bound 2.8 times more to RIRBC, 6.5 times more to TIRBC and 6.7 times more to SIRBC, compared to C3c bound to non-infected controls. In summary, our observations show that recognition signals generated by the interaction of the parasite with the host RBC seem to be similar to those present in oxidatively stressed RBC in the early stages of infection. In late stages, additional, still undefined signals are produced, which induce dramatically higher and clearly distinct phagocytic susceptibility.

SICKLE CELL ANEMIA

We were not able to show increased phagocytic susceptibility of sickle RBC subjected to 2-3 oxygenation-deoxygenation cycles, and having only a limited percentage of irreversibly sickled RBC. On the other hand, increases in adherence to endothelial cells and in phagocytic susceptibility are characteristic of irreversibly sickled RBC (14-16).

Recently, it has been shown (F. Turrini, P. Arese, J. Yuan and P. S. Low, submitted) that clustering of band 3 induced by non-oxidative membrane perturbations (such as treatment with zinc, acridine orange or melittin) strongly enhanced complement-mediated phagocytosis, provided that the perturbations were "fixed" by crosslinkers. The experimental design used to provide the positive evidence and some of the controls are depicted in Fig. 5. The same design was used with sickle cells subjected to a limited number (2,3) of oxygenation-deoxygenation cycles. Our reasoning was that sickling could possibly bring about topical redistribution of integral proteins leading to clustering of band as long as the deoxy- conformation lasted. Re-oxygenation would change back to the homogenous distribution, unless the redistribution process is impeded by "freezing" the clustered distribution with crosslinkers.

The results obtained with RBC isolated from a homozygous, non transfused sickle cell anemia patient are shown in Fig. 6. Sickling alone,

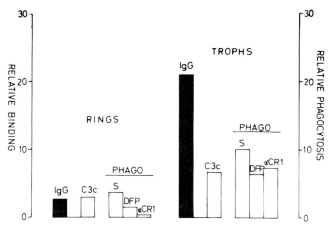

Fig. 4. Binding of autologous IgG, deposition of complement fragments and phagocytosis by monocyte-derived human macrophages of malaria-infected RBC (IRBC). Stage-dependent separation of IRBC was performed on a Percoll-sorbitol gradient (13). After opsonization, binding of radioiodinated protein A and radioiodinated anti- human C3c antibodies was determined. Phagocytosis was assayed by measurement of luminescence elicited by the macrophage-ingested hemoglobin. In this assay, electrons were transferred from t-butylhydroperoxide to luminol in a heme-dependent, heme-proportional, photon-producing reaction. Photons were counted with a bioluminescence counter. Abbreviations: IgG, bound autologous IgG, expressed as radioiodinated protein A bound by treated IRBC relative to controls; C3c, bound C3c, expressed as amount of radioiodinated anti-human C3c antibodies bound by treated IRBC relative to controls; PHAGO, phagocytosis, expressed as the number of ingested infected RBC per monocyte relative to the number of ingested normal RBC; S, native fresh autologous serum. DFP, siisopropylfluorophosphate-treated serum, used to inactivate complement convertases (see legend to Fig. 6); α CR1, anti-CR1 monoclonal antibody (see legend to Fig. 6), added to adherent macrophages in order to block CR1-dependent recognition.

followed by opsonization with serum but without subsequent crosslinking treatment resulted in only a minor increase in autologous IgG binding, no C3c deposition and no phagocytosis (Fig. 6, lane B). When sickling and opsonization were followed by treatment with the crosslinking agent DSP, high IgG binding, high C3 deposition and increased phagocytosis were observed (Fig. 6, lane C). The presence of active complement convertases was essential for phagocytosis. As shown in Fig. 6, lane D, despite high IgG binding, opsonization with DFP-treated serum almost totally abrogated complement deposition and phagocytosis. Availability of complement receptor CR1 on the phagocyte was also essential for phagocytosis.

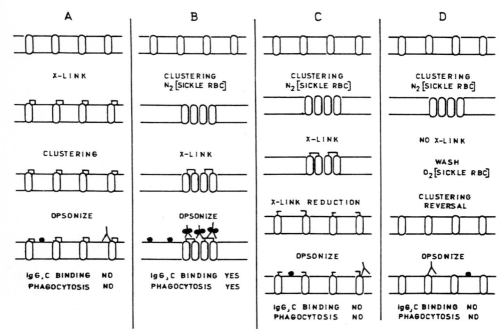

Fig. 5. Schematic sequence of treatments showing the role of band 3 clustering in eliciting recognition and phagocytosis of normal and sickle RBC.

Lane A: Normal RBC were treated first with 1 mM DSP (dithiobis succinimidyl propionate), a reducible, membrane-penetrating crosslinker. Band 3 and other integral proteins were "frozen" in their native position and conformation. Successive treatment with band 3 clustering agents such as 1 mM zinc or 1 mM acridine orange was unable to cluster band 3. After clustering treatment, a second crosslinking step with 1 mM DSP was performed. Finally, RBC were opsonized by incubation at 37°C for 60 min in fresh autologous serum. Pre-crosslinked RBC bound no IgG and C3c, and phagocytosis was not increased. Lane B: Normal and sickle RBC were treated first with clustering agents or with repeated oxygenation-deoxygenation cycles, respectively. Cells were crosslinked by 1 mM DSP and opsonized. Both treated normal and sickle RBC showed increased IgG and C3c binding, and increased phagocytosis. Lane C: Normal and sickle RBC were treated as detailed in Lane B. After the crosslinking treatment, incubation in the presence of 10 mM beta-mercaptoethanol completely reduced DSP and reversed clustering of band 3. Opsonization with fresh autologous serum did not result in deposition of autologous IgG and C3c, or increased phagocytosis. Lane D: Normal and sickle RBC were treated as detailed in Lane B. After the clustering treatment or the oxygenation-deoxygenation cycles, no crosslinking treatment was performed. Clustering of band 3 was reversed, and opsonization with fresh autologous serum did not induce deposition of autologous IgG and C3c or increased phagocytosis.

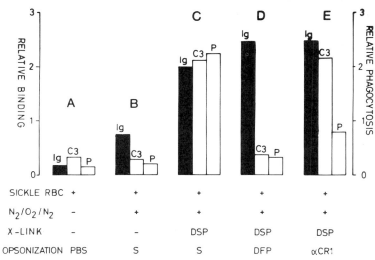

Fig. 6. Effect of sickling on binding of autologous IgG, deposition of
complement fragments, and phagocytosis by monocyte-derived human
macrophages. As outlined in Fig. 5, washed sickle RBC were subjected
to oxygenation-deoxygenation cycles by successive incubation in N2
and air atmosphere, according to the scheme: 1) Flushing with N2
during 30 min at 37°C (a thin layer of continuously shaken RBC
suspended at 25 percent hematocrit was flushed with a gentle stream
of humidified N2); 2) Flushing with air during 15 min at 37°C; 3)
Flushing with N2 during 30 min at 37°C (same conditions as in the
first cycle). In order to fix the sickled conformatioin, 1 mM DSP
was added to the deoxygenated RBC before opsonization with
autologous serum. Following opsonization, binding of radioiodinated
protein A or radioiodinated anti-human C3c antibodies was
determined. Abbreviations: Ig, bound autologous IgG, expressed as
radioiodinated protein A bound by treated RBC relative to controls;
C3, bound C3c, expressed as amount of radioiodinated anti-human C3c
antibodies bound by treated RBC relative to controls; P,
phagocytosis, expressed as number of ingested treated RBC relative
to untreated controls; PBS, phosphate- buffered saline; S, native
fresh autologous serum; DFP, diisopropylfluorophosphate-treated
serum, 5 mM DFP (inhibitor of C3 convertases) was added to S, and
extensively dialyzed against PBS after 30 min reaction at room
temperature αCR1, anti- CR1 monoclonal antibody J3D3 (kindly
provided by Dr. M. D. Kazatchkine, Paris) was added to adherent
macrophages in order to block CR1-dependent recognition mechanisms
(4).

Phagocytosis was abrogated if CR1 on the macrophages was blocked by previous treatment with anti–CR1 antibodies (Fig. 6, lane E). In parallel studies (S. Claster, unpublished results) ghosts isolated from sickled, crosslinked and serum–opsonized RBC were rich in high–molecular weight aggregates that reacted positive with anti–band 3 antibodies. Those aggregates were absent when sickled RBC were not crosslinked after 2–3 oxygenation– deoxygenation cycles. These data demonstrate that a small number of oxygenation– deoxygenation cycles resulting in reversible sickling is accompanied by reversible re– distribution of integral membrane proteins similar to the re-distribution caused by band 3 clustering agents. "Freezing" the change by treating the sickled RBC with crosslinkers generated a very efficient signal for IgG and complement deposition, and phagocytosis. It is conceivable that the same sequence of events may also occur *in vivo*, whereby aldehydes such as malonaldehyde, known to be generated in increased amounts in sickled RBC (17) may function as endogenous crosslinker of reversibly clustered integral proteins

ACKNOWLEDGEMENTS

Work supported by CNR, Target Project on Biotechnology and Bioinstrumentation, by the WHO/UNDP/World Bank Special Programme for Research and Training in Tropical Diseases, and by a NATO Collaborative Research Grant to P.A. and D.C.

REFERENCES

1. H. U. Lutz, Erythrocyte clearance, in: "Blood Cell Biochemistry, Vol. 1, Erythroid Cells", J. R. Harris, ed., Plenum Press, New York (1990).
2. H. U. Lutz, F. Bussolino, R. Flepp, P. Stammler, M. D. Kazatchkine and P. Arese, Naturally occurring anti–band 3 antibodies and complement together mediate phagocytosis of oxidatively stressed human red blood cells, Proc. Natl. Acad. Sci. USA 84:7368 (1987).
3. H. U. Lutz, S. Fasler, P. Stammler, F. Bussolino and P. Arese, Naturally occurring anti–band 3 antibody and complement in phagocytosis of oxidatively-stressed and in clearance of senescent red cells, Blood Cells 14:175 (1988).
4. F. Bussolino, E. Fischer, F. Turrini, M. D. Kazatchkine and P. Arese, Platelet–activating factor enhances complement-dependent phagocytosis of diamide-treated erythrocytes by human monocytes through activation of protein kinase C and phosphorylation of complement receptor type one (CR1), J. Biol. Chem. 264:21711 (1989).
5. H. U. Lutz, P. Stammler, C. Furter and S. Fasler, Anti–Bande 3 – Antikörper aktivieren komplement über den alternativen Weg, Schweiz. Med. Wochenschr. 117:1821 (1987).
6. R. Kannan and P. S. Low, Isolation and characterization of the hemichrome– stabilized membrane protein aggregates from sickle erythrocytes, J. Biol. Chem. 263:13766 (1988).
7. E. Winograd, J. R. T. Greenan and I. W. Sherman, Expression of senescent antigen on erythrocytes infected with a knobby variant of the human

malaria parasite <u>Plasmodium falciparum</u>, <u>Proc. Natl. Acad. Sci. USA</u> 84:1931 (1987).

8. K. Schlüter and D. Drenckhahn, Co-clustering of denatured hemoglobin with band 3: Its role in binding of autoantibodies against band 3 to abnormal and aged erythrocytes, <u>Proc. Natl. Acad. Sci. USA</u> 83:6137 (1986).

9. D. T. Fearon and K. F. Austen, The alternative pathway of complement. A system of host resistance to microbial infections, <u>New Engl. J. Med.</u> 303:259 (1980).

10. P. Arese and A. De Flora, Pathophysiology of hemolysis in glucose- 6-phosphate dehydrogenase deficiency, <u>Sem. Hematol.</u> 27:1 (1990).

11. A. Celada, A. Cruchaud and L. H. Perrin, Opsonic activity of human immune serum on in vitro phagocytosis of <u>Plasmodium falciparum</u> infected red blood cells by monocytes, <u>Clin. Exp. Immunol.</u> 47:63 (1982).

12. A. Vernes, Phagocytosis of <u>P. falciparum</u> parasitized erythrocytes by periferal monocytes, <u>Lancet</u> ii:1297 (1980).

13. S. Kutner, W. V. Breuer, H. Ginsburg, S. B. Aley and Z. I. Cabantchik, Characterization of permeation pathways in the plasma membrane of human erythrocytes infected with early stages of <u>Plasmodium falciparum</u>: Association with parasite development, <u>J. Cell. Physiol.</u> 125:521 (1985).

14. G. A. Barabino, L. V. McIntire, S. G. Eskin, D. A. Sears and M. Udden, Endothelial cell interactions with sickle cell, sickle cell trait, mechanically injured, and normal erythrocytes under controlled flow, <u>Blood</u> 70:152 (1987).

15. N. Mohandas, E. Evans, Adherence of sickle erythrocytes to vascular endothelial cells: Requirements for both cell membrane changes and plasma factors, <u>Blood</u> 64:282 (1984).

16. R. P. Hebbel and W. J. Miller, Phagocytosis of sickle erythrocytes: Immunologic and oxidative determinants of hemolytic anemia, <u>Blood</u> 64:733 (1984).

17. S. K. Das and R. C. Nair, Superoxide dismutase, glutathione peroxidase, catalase and lipid peroxidation of normal and sickle erythrocytes, <u>Br. J. Haematol.</u> 44:87 (1980).

INHIBITION BY CARBOHYDRATES AND MONOCLONAL ANTICOMPLEMENT

RECEPTOR TYPE 1, ON INTERACTIONS BETWEEN SENESCENT HUMAN

RED BLOOD CELLS AND MONOCYTIC MACROPHAGIC CELLS

Liliane Gattegno[1], Dominique Bladier[2],
Jenny Vaysse[1] and Line Saffar[1]

[1]Laboratoire de Biologie Cellulaire et
[2]Laboratoire de Biochimie-Faculté de Medicine
Paris – Nord, France

INTRODUCTION

In order to be recognized and phagocytosed by Kupffer cells in the liver or by splenic or bone marrow macrophages, the human senescent red blood cells (Sen-RBC) must carry specific surface alterations that are recognized by the phagocytic cell. Previous works have shown that after injection into rabbits, neuraminidase-treated, sialic acid less autologous red blood cells (RBC) are promptly removed from the circulation; intact RBC previously incubated under the same conditions, but without neuraminidase, are removed from the circulation, after a significantly longer period (1–3). These results have suggested a role of sialic acid in the survival time of RBC in the circulation. However, treatment of RBC with sialidase did not result in the rapid clearance of young desialylated cells as might have been expected if loss of sialic acid were the primary signal for cell removal: the half-life of intact Sen-RBC was found to be significantly shorter than that of sialidase-treated young red blood cells (Y-RBC) with a similar sialic acid content (4).

Moreover, in humans, it was clearly demonstrated that neuraminidase treated, sialic less RBC do not constitute a model for studying in vivo aged RBC, since, whereas in vivo aged, untreated RBC give an autorosette forming cells percentage with lymphocytes (A-RFC) lower as compared with the young ones, young and in vivo aged RBC treated with neuraminidase show a parallel increase of A-RFC as compared with untreated cells (5). In addition, we have observed that the contents of fucose, mannose, galactose, glucose, N-acetylglucosamine, N-acetylgalactosamine and sialic acid are slightly but significantly lower in the membranes of in vivo aged human RBC than in the membranes of young ones; when the results are expressed as nmol per mg of

Red Blood Cell Aging, Edited by M. Magnani and
A. De Flora, Plenum Press, New York, 1991

membrane dry weight. However, no significant difference is found between young and old RBC membranes when these values are expressed as residues per one hundred carbohydrate residues (6). These results demonstrate that during human RBC _in vivo_ ageing, a slight but homogeneous and significant decrease of the carbohydrate moieties occur, and that sialic acid is not the only membrane carbohydrate composant which significantly decreases during human RBC _in vivo_ ageing.

On another hand, several authors (7-9) have shown evidence that the interactions between Sen-RBC and macrophages are mediated by immunoglobulins-G bound to specific neoantigens that appear on the Sen-RBC membrane. Moreover, the recognition of particles coated with IgG and/or C3b by the Fc and the C3b receptors of macrophages is the main mechanism responsable for immune phagocytosis (10). Macrophages also exhibit lectin-like receptors (11). M. Malaise et al (12) have recently shown that the ability of normal monocytes to phagocytose IgG-coated RBC is related to the number of accessible galactosyl and mannosyl residues in the Fc domain of the anti-RBC IgG antibody molecules. These data suggest that during immune phagocytosis, lectin-carbohydrate interactions between galactosyl or mannosyl residues of the Fc domain of IgG and macrophagic carbohydrate receptors may also occur. The aim of the present report was thus to investigate whether lectin-carbohydrate interactions and opsonins might play a synergic role in the recognition of events leading to the specific adhesion and phagocytosis of Sen-RBC by macrophages.

MATERIAL AND METHODS

Human blood was drawn from healthy donors of the Seine-Saint-Denis blood Transfusion Center and was collected into sterile heparinized tubes. Sen-RBC were separated from young RBC (Y-RBC) within 15 minutes after venipuncture on the basis of differences in density in autologous plasma, as previously described (6) and according to O'Connell et al (13): whole blood was spun at 2000 g for 120 min at 15° C and the plasma was removed. The top 5% and the bottom 5% RBC layers were collected by aspiration. Mononuclear cells were isolated from peripheral blood and adjusted to 2×10^6 cells/mL in RPMI 1640 medium (Eurobio, Paris, FRANCE) supplemented with 10% heat-inactivated human AB serum or foetal calf serum. One mL aliquots were layered into 16 mm-diameter flat-bottomed wells (Costar, Cambridge, MA) containing 12 mm sterile glass coverslips and incubated for 2h at 37°C in humidified air with 5% CO_2. Non adherent cells were removed by four washings with RPMI 1640 medium. 95% of the adherent cells were identified as monocytes, as judged by phagocytosis of latex particles or by morphological criteria after May-Grünwald Giemsa staining; $94 \pm 1\%$ of the monocytes were able to inernalize zymosan particles opsonized with plasma.

Adherent monocytes were either used at the first day of _in vitro_ cultivation for erythrophagocytosis assays or cultured at 37°C in RPMI 1640 medium supplemented by 2 μ moL/ml glutamine, 100 U/mL penicillin, 100 μg/mL streptomycin and 20% heat-inactivated AB serum or fetal calf serum. Monocyte-derived-macrophages (MDM) obtained after 7 days of culture – the medium being changed at day 3,5,7 – were also used for erythrophagocytosis

330

assays. More than 95% of the cells were viable as determined by Trypan Blue exclusion; 87 ± 10% (n=7) of the MDM could ingest unopsonized zymosan particles and 93 ± 6% (n=6) could internalize zymosan particles opsonized with plasma. Before, the erythrophagocytosis assays, the monocyte or MDM monolayers were washed three times with serum-free RPMI; then, 0.5 mL medium was added to each well before the addition of 10 autologous or allogeneic Y-RBC or Sen-RBC, freshly drawn, unwashed and just suspended in 0.1 mL serum-free RPMI medium.

Cells were coincubated for 2 h at 37°C under 5% CO_2 humidified atmosphere. The glass coverslips were then washed twice with Hank's medium (Eurobio) to remove unbound RBC, then fixed with methanol, dried and stained with May-Grünwald Giemsa reagent. In duplicate experiments, RBC that had been bound but not internalized were lyzed with TRIS-ClNH buffer pH 7.4 just before fixing. Random fields of the coverslips were examined by light microscopy at 1000 x magnification under oil; 1000 monocytes or MDM were scored for RBC binding and internalization. The results are expressed as the average of different experiments carried out with the cells from different donors. For inhibition experiments by carbohydrates or carbohydrate derivatives, MDM were preincubated for 5 minutes before addition of the Sen-RBC with one of the following compounds: Me-α-Glc, Me-ß-Gal, Me-α-Gal, Me-α-Man, GlcNac, mannan, fetuin, asialofetuin, glucocerebroside (SIGMA Chemical Co, St Louis, MO, USA), glycopeptides from asialofetuin, ß-D-GlcNAC47-BSA, α-L Fuc_{12}-BSA, α-D-Man_{23}-BSA- prepared according to Westphal and Schmidt (14).

These compounds were then coincubated with the MDM and the Sen-RBC during the erythophagocytosis assays. For inhibition experiments by monoclonal (Mab) anticomplement receptors, the monocytes or the MDM were preincubated for 1 h at 20°C before the addition of the Sen-RBC, with Mab anti human complement receptor type I (Mab anti-CRI), type 2 (Mab anti CR2), type 3 (Mab anti Leu 15) (Becton - Dickinson, Grenoble, FRANCE) or OKMI (Ortho Pharmaceutical Corp. Ramitan N.J.).

Mab anti-Leu 15 and OKMI are specific for two different epitopes of the alpha chain of CR3. Mab anti-CR2 was only tested with MDM, since CR2 is only expressed on MDM after in vitro cultivation, but not on freshly drawn monocytes. Controls were performed with Mab carrying the same isotypes as the Mab but with different specificities: Mab OKM5 and anti T3 (Ortho Pharmaceutical). After the preincubation period, the monocytes or MDM monolayers were washed four times with serum-free RPMI and the erythrophagocytosis assays were performed. In some control experiments, RBC opsonized with subagglutinating concentrations of polyclonal or monoclonal IgG anti-D (CNTS, Paris, FRANCE) were tested for their ability to bind to and to be internalized by monocytes preincubated with or without Mab anti-CR1, anti-Leu 15 and OKM1.

RESULTS AND DISCUSSION

The present experiments were undertaken with freshly drawn, unwashed human RBC which were coincubated for 2 h with freshly drawn monocytes, or

Table 1. Light Microscope Evaluation of Adhesion and
Phagocytosis of Human Allogeneic Senescent Red
Cells by Monocyte-Derived Macrophages

	Adhesion	Phagocytosis	Adhesion	Phagocytosis
	RBC per 1000 macrophages		‰ macrophages	
Y–RBC	60±15	8±2	40±10	8±3
Sen–RBC	250±50*	60±13*	135±30*	60±3*

One thousand macrophages were examined, and the results are given
as the means ± S.E.M. of 17 paired experiments run with the cells
from different healthy donors.
P of the paired differences Y–RBC–Sen–RBC were calculated by the
Student-Fisher's t test * < 0.01

Table 2. Inhibition by Carbohydrate Derivatives of Interactions Between
Human Senescent Red Cells and Monocyte-Derived Macrophages
(means ± SEM)

Carbohydrate	Concentration	% of inhibition		Nb of exp.
		Phagocytosis	Adhesion	
Me–α–Gal	20mM	54±9.5**	52±9[+]	8
Me–ß–Gal	20mM	53±13**	61±10**	7
Me–α–Man	18mM	66±7+	65±4+	11
Me–α–Glc	22mM	11±3.5	32±7	5
Glc Nac	22mM	28±9	32±9	5
ß–D–GlcNac–BSA	30µM	5±2	10±4	5
α–L–Fucose–BSA	30µM	38±14	33±13.5	5
α–D–Mannose–BSA	30µM	58±6**	47±9*	9
Mannan	2mg/ml	82±11**	56±8***	5
Fetuin	4µM	20±5	15±5	5
Asialofetuin	4µM	51±10**	45±9***	9
GP asialofetuin	160µg/ml	80±10[+]	62±15*	5
GM1 ganglioside	120µM	82±6.5***	20±7.5	5
Glucocerebroside	120µM	15±6.5	13±6	5

In comparison with phagocytosis or adhesion without added inhibitor,
p<:*0.05; **0.02; ***0.01; [+]0.001.

with MDM: these conditions were chosen in order to minimize the RBC changes
that might be induced in vitro by the washings, storage and incubation.

In these conditions, we have previously shown that monocytes bind and
internalize significantly more autologous Sen-RBC than Y-RBC (15,16). In the
present work, similar results were observed with MDM in an allogeneic
system: MDM bind 4.2 times more and phagocytized 7.5 times more Sen-RBC
than Y-RBC (Table 1). These interactions between Sen-RBC and MDM were
significantly inhibited by Me-α- Gal, Me-ß-Gal, asialofetuin and its
glycopeptides, and by GM1 ganglioside (Table 2).
Me-α-Glc, fetuin, GlcNAC, ß-D-GlcNac-BSA and α-L-Fuc-BSA were devoid of
significant inhibitory effect (Table 2).

These data are in agreement with those previously observed for the
interactions between Sen-RBC and monocytes, at the first day of in vitro
cultivation (16). In the present model, as in our previous ones (16), we
demonstrate that the inhibitory effect of galactosyl structures does not
depend on their anomeric conformation since Me-ß-Gal as well as Me- -Gal are
inhibitors. Furthermore, the inhibition by galactosyl derivatives occurs
whether the carrier is peptidic as in asialofetuin, or lipidic, as in GM1
ganglioside (Table 2). In addition, it is observed that Me-α-Man, (at the mM
level) mannan (2 mg/ml) and α-D-Man-BSA (at the μM level) are significant
inhibitors of interactions between Sen-RBC and MDM. These results strongly
suggest the involvement of the mannose receptor which is known not to be
expressed on freshly drawn monocytes, but which appears upon differentation
of monocytes into macrophages (17). In addition to these inhibitions
by carbohydrate derivatives, which suggest the involvement of lectin-
carbohydrate interactions in the recognition events of Sen-RBC by
macrophages, it is also observed here that Mab anti CR1 is a significant
inhibitor of interactions between Sen-RBC and monocytes or MDM (Table 3
and 4).

Mab anti CR1 significantly inhibit the binding and the phagocytosis of
unwashed autologous Sen-RBC by monocytes and of unwashed allogeneic Sen-RBC
by MDM. Murine Mab OKM5 carrying the same isotype as Mab anti-CR1, but a
different pecificity is devoid of inhibitory effect (Tables 3 and 4). In
addition, preincubation of monocytes with Mab anti-CR1 does not inhibit
their interaction with anti-D opsonized RBC i.e. FcR dependent interactions
(Table 5); moreover, opsonization of Y-RBC or Sen-RBC with Mab anti-CR1 does
not induce modification of their interactions with MDM (data not shown);
taken together, these results rule out a FcR blockage by Mab anti CR1
receptor in the specific recognition of Sen-RBC by macrophages. It should be
noticed that preincubation of monocytes with Mab anti-CR1 does not inhibit
their interaction with Y-RBC, suggesting the occurence of a structure
which only exists on the surface of Sen-RBC and contributes to a complement
deposition mechanism involved in the clarance of Sen-RBC. As murine Mab
anti-CR3 (anti-Leu15 and OKM1) inhibit in the same order of magnitude on one
hand, the FcR dependent rosetting and phagocytosis (Table 5) and on the
other hand, the Sen-RBC rosetting and phagocytosis by adherent monocytes
(Table 3), the specific involvement of the CR3 epitopes recognizedy of these
Mab in the interactions between Sen-RBC and monocytes or MDM could not
be demonstrated.

Table 3. Effect of Mab on Binding to and on Phagocytosis of Autologous Senescent Red Cells by Monocytes

Mab	Concentration (nM)	% of inhibition of		Nb of exp.
		Adhesion	Phagocytosis	
anti CR1 (IgG1)	0.1	66±7.5***	67±7.5***	5
anti leu15 (IgG2)	0.6	18±6.5*	15±6*	6
	6	37±5.5****	63±15**	4
OKM1 (IgG2)	6.6	34±12.5**	2±2*	6
OKT3 (IgG2A)	6.6	7±5*	25±17*	3
OKM5 (IgG1)	13.2	17±8.5	13±6.5*	4

Results are expressed as means ± S.E.M. of duplicate experiments performed with the cells from different healthy donors: p of the coupled differences between respectively the % of RBC bound or ingested by monocytes preincubated with the indicated Mab or preincubated with the medium alone calculated by the t-test of Student-Fisher. *NS, ** <0.05, ***<0.02, ****<0.01.

Several authors have demonstrated that immune phagocytosis mediated by IgG bound to specific neoantigens which appears on the RBC surface membrane during human RBC in vivo ageing is involved in the recognition events of Sen-RBC by macrophages (7-9). Our present result suggest in addition, the involvement in these recognition events of the macrophagic C3b receptor. It has been previously demostrated (10) that C3 and IgG have separated but synergistic roles in phagocytosis: IgG, through its Fc fragment, directly stimulates particle ingestion, but is relatively inefficient in inducing particle binding. Particle-bound C3 can lower 100 times the amount of IgG required to induce particle internalization.

Opsonization with C3 can thus be a necessary condition for phagocytosis to occur, although C3 itself does not make ingestion possible unless monocytes or macrophages are activated (10). However the present results also suggest that lectin-carbohydrate interactions involving galactosyl and mannosyl structures play a role in the recognition events of Sen-RBC by macrophages.

Table 4. Effect of Mab on Binding to and on Internalization of Allogeneic
Senescent Red Cells by Monocyte-Derived Macrophages

Mab	Concentration (nM)	% of inhibition		Nb of exp.
		Adhesion	Phagocytosis	
anti CR1 (IgG1)	3	52±8***	63±6.5***	8
	0.2	50±9***	52±5***	6
OKM1 (IgG2B)	6.5	12±7*	20±11.5*	3
	0.6	0	0	3
Anti Leu 15 (IgG2A)	1.5	44±14***	41±14**	7
Anti CR2 (IgG2A)	2	5±3*	16±19*	3
	0.8	0	19±8.5*	5
OKT3 (IgG2A)	6.5	8±4.5*	1±0.5	3
OKM5 (IgG1)	3.3	13±6.5*	20±10*	4

Results are expressed as means ±S.E.M. of duplicate experiments performed
on the cells from different healthy blood donors. p of the coupled
differences between the % of RBC bound to or ingested by monocytes-
derived macrophages preincubated with the indicated Mab or preincubated
with the medium alone, calculated by the t-test of Student-Fisher, * NS,
** <0.02, *** <0.01.

These results are in line with those of M. Malaise et al. which have
recently demonstrated that the ability of normal human monocytes to
phagocytose IgG-coated—RBC is related to the number of accessible galactosyl
and mannosyl residues in the Fc domain of the anti-RBC IgG antibody
molecules (12). It has also been demonstrated (18) that mouse macrophages
plated on mannan-coated surfaces lose the ability to phagocytose by their
FcR, which suggests some interrelationships between those receptors and the
lectin-like receptors in vitro.

In conclusion, the present data strongly suggest that the interactions
between human Sen-RBC and macrophages are opsonin dependent and carbohydrate
specific. The link between this opsonin dependency and the carbohydrate
specificity deserves further studies.

Table 5. Effect of Mab on Binding to and on Phagocytosis of
Anti-D Opsonized RBC by Monocytes

Mab	Concentration (nm)	% of inhibition	
		Adhesion	Phagocytosis
Anti CR1	1.5	0	0
	3	0	0
Anti Leu 15	1.5	45±3*	76±1**
Anti CR2	3	0	0
OKM1	6	41±3*	17±8

Results are expressed as means ± S.E.M. of three duplicate
experiments performed with the cells from different healthy
donors: p of the coupled diffeences between the % of RBC bound
to or ingested by monocytes preincubated with the indicated Mab
or preincubated with the medium alone * < 0.01 ** < 0.001.

REFERENCES

1. L. Gattegno, D. Bladier and P. Cornillot, The role of sialic acid in the
determination of survival of rabbit erythrocytes in the circulation,
Carbohyd. Res. 34:361 (1974).
2. J. Jancik, R. Schauer and H. J. Streicher, Influences of membrane-bound
N-acetylneuraminic acid on the survival of erythrocytes in man, Hoppe
Seyler Z. Physiol. Chem. 355:394 (1974).
3. D. Aaminoff, W. C. Bell, I. Fulton and N. Ingebristen, Effect of
sialidase on the viability of erythrocytes in circulation, Am. J.
Hematol. 1:233 (1976).
4. L. Gattegno, D. Bladier and P. Cornillot, Ageing in vivo and
neuraminidase treatment of rabbit erythrocytes: Influence on
half-life as assessed by ^{51}Cr labelling, Hop. Seyl. Z. Physiol. Chem.
356:391 (1975).
5. F. Fabia, L. Gattegno, J. C. Gluckman and P. Cornillot, Effects of
ageing, surface sialic acid and glycopeptides of erythrocytes on
auto-rosettes in man, Clin. Exp. Immunol., 34:295 (1978).
6. D. Bladier, L. Gattegno, F. Fabia, G. Perret and P. Cornillot,
Individual variations of the seven carbohydrate of human erythrocyte
membrane with ageing in vivo, Carbohyd. Res. 83:271 (1980).
7. U. Galili, E. Flechner, A. Knyszynski, A. Danon and E. A. Rachmelewitz,
The natural anti-alpha-galactosyl IgG on human normal senescent red
blood cells, Br. J. Haematol. 32:3171 (1986).
8. M. M. B. Kay, S. R. Goodman and K. Sorensen, Senescent cell antigen is
immunologically related to band 3, Proc. Natl. Acad. Sci. USA,
81:2351 (1983).
9. H. Lutz and G. Stringara-Wipf, Identification of a cell age specific

antigen from human red blood cells, Biomed. Biochim. Acta 425:112 (1984).

10. A. G. Ehlenberger and V. Nussenzweig, The role of membrane receptors for C3b and C3d in phagocytosis, J. Exp. Med. 145:357 (1977).

11. G. D. Ross, J. A. Cain and P. J. Lachman, Membrane complement receptor type three (CR3) has lectin-like properties analogous to bovine conglutinin and function as a receptor for zymosan and rabbit erythrocytes as well as a receptor for iC3b, J. Immunol 134:3307 (1985).

12. M. G. Malaise, P. Franchimont and P. R. Mahieu, The ability of normal monocytes to phagocytose IgG-coated red blood cells is related to the number of accessible galactosyl and mannosyl residues in the Fc domain of the anti-red blood cell IgG antibody molecule, J. Immunol. Methods 119:231 (1989).

13. D. J. O'Connell, C. J. Caruso and M. D. Sass, Separation of erythrocytes of different ages, Clin. Chem. 11:77 (1965).

14. O. Westphal and M. Schmidt, N-acetyl glucosamin ab determinante gruppe in kinstlichen antigenen Liebigs Ann. Chem. 575:84 (1951).

15. J. Vaysse, L. Gattegno, D. Bladier and D. Aminoff, Adhesion and erythrophagocytosis of human senescent erythrocytes by autologous monocytes and their inhibition by ß-galactosyl derivatives, Proc. Natl. Acad. Sci. USA83:1339 (1986).

16. L. Gattegno, M. J. Prigent, L. Saffar, D. Bladier, J. Vaysse and A. Le Floch, Carbohydrate specificity and opsonin dependency of the in vitro interaction between senescent human RBC and autologous monocytes in culture, Glycoconj. J. 3:379 (1986).

17. V. L. Shepherd, E. J. Campbell, R. M. Senior and P. D. Stahl, Characterization of the mannose/fucose receptor on human mononuclear phagocytes, J. Reticuloendothe. Soc. 32:423 (1982).

IN VITRO SEQUESTRATION OF ERYTHROCYTES FROM

HOSTS OF VARIOUS AGES

Harriet Gershon and Edwar Sheiban

Department of Immunology
Technion Faculty of Medicine
Haifa, Israel

ABSTRACT

Erythrocytes from young and old human donors were separated according to age-density on Stractan gradients. Old donors had more low age-density (young) erythrocytes than did young donors. Levels of IgG bound to old and young erythrocytes were determined by ELISA. Erythrocytes from old donors bore higher levels of IgG on their erythrocytes (123 ± 55 IgG molecules per young RBC and 196 ± 43 IgG per old RBC) than did those from young donors (58 ± 15 IgG per young RBC and 98 ± 20 IgG per old RBC). In an in vitro erythrophagocytosis assay, young and old erythrocytes from old donors and old erythrocytes from young donors were shown to be recognized and phagocytosed by lymphokine activated human peripheral blood monocytes. Young erythrocytes from young donors were not phagocytosed in this assay. The in vitro erythrophagocytosis of erythrocytes from old and young donors can be specifically blocked by ß-galactoside but not α-galactoside sugars. This phagocytosis is not blocked by Protein-G which specifically blocks Fc-gamma mediated erythrophagocytosis of Rh-D$^+$ erythrocytes coated with IgG anti-Rh-D antibodies. ß-galactoside and α-galactoside sugars have no inhibitory effect on erythrophagocytosis mediated by IgG anti-Rh-D antibodies coating Rh-D$^+$ erythrocytes. It thus appears that erythrophagocytosis of young and old erythrocytes from old donors and old erythrocytes from young donors are all mediated by a lectin-like receptor on the monocytes which recognizes ß-galactoside-like sugar moiety on the erythrocytes rather than by recognition of IgG on the erythrocyte and an Fc receptor on the macrophage. It also appears that the membrane of both young and old erythrocytes of old donors are marked for phagocytosis whereas only the old erythrocytes from young donors are so marked.

Red Blood Cell Aging, Edited by M. Magnani and
A. De Flora, Plenum Press, New York, 1991

INTRODUCTION

Erythrocytes have been used as a model for the study of cellular aging as a function of both cell and host age. Erythrocytes of various ages from both young and old hosts become more dense as they age in the circulation and can thus be separated by age on density gradients (1-5). The life span of the erythrocyte in the circulation of young adults is characteristic of each species (5). We and others have reported that red cell life span is significantly shortened in aged rats (6), mice (7,8) and rabbits (9). We have reported that age-density gradient separation of Fe-59 cohort labelled rat erythrocytes demonstrates a skewing of the erythrocytes of old individuals to the lighter, younger fractions in the rat (6). On the basis of density gradients, we have also demonstrated a similar skewing of the erythrocytes from elderly humans (>75 years old) to the less dense fractions, thus presenting circumstantial evidence for a shortened life span of erythrocytes in the elderly human (10 and Table 1).

The membrane changes which lead to the enhanced turnover of the erythrocyte in both the young and senescent host have not been fully elucidated. Sequestration of the effete erythrocyte in young hosts has been attributed to: 1) changes in membrane carbohydrates (11-13); 2) altered membrane proteins (14-16); and 3) externalization of membrane phosphatidyl serine (17). Suggested changes in membrane carbohydrates have been: desialization (18), exposure of ß-galacotsyl residues (19), or exposure of α-galactosyl residues (13). Possible alterations in membrane proteins have been attributed to cleavage (14) or aggregation of (15,16) Band 3. The above proposed membrane alterations can either be directly recognized by splenic macrophages or they can be recognized by an intermediary such as a physiologic autoantibody (with or without the additional binding of complement) which is then recognized by the phagocytic cell. It has thus been proposed that phagocytosis of the effete red cell is mediated by macrophage 1) lectin-like receptors (19) 2) recognition sites for exposed phosphatidyl-serine (17), 3) Fc-gamma receptors (13,16,20), or 4) C3b receptors (21).

Senescent erythrocytes in a young individual have IgG bound to their membranes (6,13,16,20,22). We have determined that young erythrocytes from both rats and humans have membrane bound IgG, albeit at lower levels than do old cells (6,22 and Table 2). Erythrocytes from senescent individuals carry higher levels of IgG than do cells from young individuals (6,22 and Table 2). Thus, normal erythrocytes in the circulation of healthy individuals bear a certain level of IgG (10,22,23) which increases with the age of the erythrocyte. It is important to determine whether such low levels of IgG on the erythrocyte can be responsible for removal of the red cell from the circulation. We have developed a Cellular ELISA which allows for the accurate determination of levels of erythrocyte bound IgG (24). We have also developed an in vitro phagocytosis assay and a method of blocking the Fc of the bound immunoglobulin with Protein-G which totally blocks macrophage recognition of the opsonized red cell and prevents IgG dependent phagocytosis (24). This assay has allowed for the evaluation of the role of IgG and sugar moieties in the sequestration of effete erythrocytes from young and old donors (25).

A study of the <u>in vivo</u> lifespan of the erythrocyte in agamma-globulinemic individuals would contribute relevant information about the role of antibody in the sequestration of effete erythrocytes. We have, therefore, performed a series of experiments on the rate of erythrocyte sequestration in CB-17 SCID mice which suffer from Severe Combined Immunodeficiency (26) due to a defective recombinase which does not allow for the proper formation of the variable regions of immunoglobulin molecules as well as T-cell receptors (27-30). This very specific SCID mutation, which developed in mice of a Balb/c background, plays no role in erythropoiesis nor in myelopoiesis (31). These experiments provide information as to the importance of purported physiologic autoantibodies in the normal removal of effete red cells from the circulation (32).

RESULTS

Age-density separation of erythrocytes from young and old donors

Erythrocytes from healthy, consenting young (20-35 years) and old (>75 years) donors were passaged through microcellulose:alpha-cellulose to remove leukocytes and platelets (33). The red cells were separated according to their density on discontinuous (1.090, 1.095, 1.100, 1.105 and 1.110g/ml) Stractan gradients (1). Erythrocytes from young donors separated into a normal distribution over the densities tested while the erythrocytes of the elderly donors skewed toward the lighter densities (Table 1), suggesting a younger population of cells.

Table 1. Density Distribution of Erythrocytes from
Young and Old Donors

Erythrocyte Density (gm/ml)	Distribution of Erythrocytes per Density Fraction	
	Young Donor	Old Donor
<1.090	6.8	27.9
<1.095	19.5	35.2
<1.100	48.7	17.2
<1.105	21.4	9.9
<1.110	3.5	9.8

Each fraction is expressed as the density of the Stractan upon which the erythrocytes settled after passing throughthe less dense fractions. Results are representative of over 30 performed.

These results are similar to our findings in the rat (6). In all further experiments, the least dense erythrocytes which settled on the 1.090gm/ml Stractan layer were harvested as young erythrocytes and the densest erythrocytes, those which passed through all but the 1.110gm/ml Stractan layer, were harvested as the old erythrocytes.

The Number of IgG Molecules Bound to Young and Old Erythrocytes

In order to approach the question of the mechanism of decreased lifespan of the erythrocyte in the circulation of the elderly, we determined the level of IgG on the erythrocytes of aged individuals and developed an in vitro phagocytosis assay to determine the relevance of erythrocyte bound IgG to the sequestration of the red cell. A corollary question which we asked was whether the shortened life span of erythrocytes in the elderly host is solely due to the selective phagocytosis of old, dense erythrocytes or whether the young, less dense, erythrocytes of the elderly are also marked for phagocytosis.

To determine the number of IgG molecules on young and old erythrocytes from young and old donors, a Cellular ELISA was developed (24). Thirty IgG molecules per erythrocyte are readily detected in this ELISA in which IgG anti-Rh-D coated Rh-D erythrocytes are used as a standard. Using this Cellular ELISA, we determined that low age-density and high age-density erythrocytes from donors of each age group bear IgG on their membranes (Table 2). The high age-density erythrocytes from donors of each age group carry approximately twice as many IgG as do the low age-density cells. The erythrocytes of old donors have about twice as many IgG per red cell as do the erythrocytes from young donors in comparable fractions, thus confirming our previous studies on rat and human erythrocytes using fluorescent antibody technology (6,10).

Table 2. Molecules of IgG per Erythrocyte

Erythrocyte Age	Density (gm/dl)	Young Donors	Old Donors
Young	<1.090	58±15	123±55
Old	<1.110	98±20	196±43

Results are the average ± SD of 8 individuals in each age group.

Phagocytosis of Density Separated Erythrocytes from Young and Old Donors

An in vitro phagocytosis assay was developed to evaluate the role of cell bound IgG in the sequestration of erythrocytes from young and old donors (24,25). In this assay, Rh-D erythrocytes coated with known levels of IgG anti-Rh-D per cell are phagocytosed by human peripheral blood monocytes from young adult donors in proportion to the levels of IgG bound to the red cell. This Fc-gamma dependent phagocytosis can be specifically inhibited by pretreatment of the IgG anti-Rh-D coated erythrocytes with Protein-G (24). Treatment of the erythrocytes with Protein-G prior to their exposure to the IgG anti-Rh-D has no detrimental effect on the subsequent ability of the monocytes to bind to the Fc of the IgG and to phagocytose the IgG coated erythrocytes. This Fc dependent phagocytosis of IgG anti Rh-D coated erythrocytes is not inhibited by the addition of 25mM concentrations of various α-glycosidic sugars (Methyl- α -D-galacto-pyranoside [α-gal], N-acetyl-glucose-amine [N-glu], α-methyl-mannoside [α-man]), nor by various ß-glycosid sugars (Methyl-ß-D-galacto-pyranoside [ß-gal], N-acetyl-galctose-amine [N-gal], Lactose [lac]).

When age-density separated erythrocytes are assayed for sequestration in this short term (4hrs) in vitro phagocytosis assay, old erythrocytes from both young and old donors are phagocytosed. Young erythrocytes from young donors are not phagocytosed in this assay, whereas young erythrocytes from old donors undergo phagocytosis to almost the same degree as do the old (Figure 1). When this phagocytosis is compared to that of IgG anti-Rh-D coated control erythrocytes, the level of phagocytosis of these age-density separated cells from young and old donors is significantly higher than would be predicted by the levels of IgG that they bear. Whereas, on the average, the high age-density erythrocytes from young individuals bear 98±20 molecules of IgG and those from old donors 196±43 molecules of IgG, the degree to which they are sequestered in vitro is comparable to the phagocytosis of anti-Rh-D coated erythrocytes bearing approximately 2400 IgG per red cell. The degree of phagocytosis of low age-density erythrocytes from old donors which bear on the average 123±55 IgG is comparable to the phagocytosis of anti-Rh-D coated erythrocytes bearing 1200 IgG per red cell.

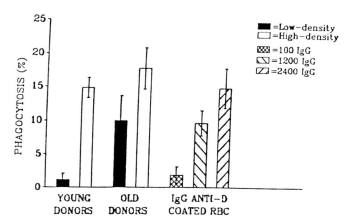

Fig 1. Phagocytosis of young and old erythrocytes from young and old donors and erythrocytes coated with various levels of IgG anti-Rh-D. Results are the average ±SD of 8 experiments

Phagocytosis of density separated erythrocytes from young and old donors:
inhibition by protein-G and various sugars

Since the experiments reported above do not show a satisfactory
correlation between the levels of IgG detected on the erythrocytes of young
and old donors and the degree of phagocytosis of these cells when compared
to the levels of IgG dependent phagocytosis obtained with IgG anti-Rh-D
coated erythrocytes, we examined the ability of Protein-G and various sugars
to inhibit the phagocytosis of these erythrocytes. Protein-G, which
completely inhibits the phagocytosis of erythrocytes coated with 2400
molecules of IgG anti-Rh-D, has no effect on the phagocytosis of old
erythrocytes from young individuals nor on that of old or young
erythrocytes from old individuals (Table 3).

We also determined that the phagocytosis of these cells was not
inhibited by the following α-glycosidic sugars: methyl-α-D-galacto-
pyranoside (α-gal), N-acetyl-glucose-amine (N-glu), norα-methyl-
mannoside (α-man) (Figure 2). This phagocytosis cannot, therefore, be
attributed to a α-glycoside specific lectin-like receptor on the
phagocyte. In contrast, all of the ß-galactosidic sugars tested, methyl-
ß-D-galactopyranoside (ß-gal), N-acetyl-galactose-amine (N-gal) and
lactose (lac),inhibited the phagocytosis of high age-density erythrocytes
from young individuals and the phagocytosis of low age-density and high
age-density erythrocytes from old individuals to the same extent (Figure 2).
None of the sugars tested inhibited the phagocytosis of IgG anti-Rh-D coated
erythrocytes. It thus appears that the phagocytosis of old erythrocytes
from young individuals and young and old erythrocytes from old individuals
cannot be attributed to phagocyte recognition of the Fc of the IgG bound to
the erythrocyte membrane but may well be attributed to a lectin-like
receptor on the phagocyte capable of recognizing ß-galactoside residues.

Table 3. Inhibition of Phagocytosis by Protein-G

ERYTHROCYTE	% ±SD
Old RBC - Old Donor	18± 9
Old RBC - Young Donor	19±11
Young RBC - Old Donor	13± 9
Anti-Rh-D Coated	94±11

Erythrocytes from young and old donors were separated on Stractan
gradients into young and old fractions. These erythrocytes and
unseparated RBC from Rh-D$^+$ donors coated with 2400 IgG per RBC
were exposed to 25 µg/ml Protein-G. The RBCwere washed and
incubated with monocytes in a phagocytosis assay. Results
are the average ±SD of 8 experiments.

It is important to note that blocking of the Fc of the IgG on these erythrocytes with Protein-G and subsequent exposure of the phagocytic cultures to any of the ß-galactosidic sugars did not increase the inhibition of phagocytosis above that seen with the ß-galactosidic sugars alone (25).

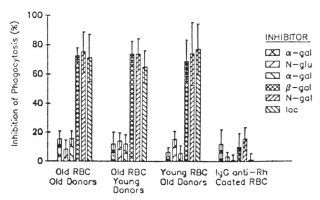

Fig. 2. The effect of various sugars on the phagocytosis of young and old erythrocytes from old donors and old erythrocytes of young donors and on IgG Anti-Rh-D coated erythrocytes.
The results are the average ±SD of 8 experiments.

In vivo sequestration of erythrocytes in CB-17-SCID mice

We have studied CB-17-SCID and control Balb/c mice in order to determine whether animals which are genetically incapable of eliciting an immune response, and thus produce no antibody (26-30), are capable of normal erythrocyte sequestration. CB-17-SCID mice suffer from a Severe Combined Immunodeficiency and are virtually agammaglobulinemic. The levels of IgG isotypes in the serum of the CB-17 SCID mice used in our experiments were 250 µl/ml (IgG1), 810 µl/ml (IgG2a), and 90 µl/ml (IgG2b), which were respectively 2/1000, 1/1000 and 3/1000 the levels observed in the control Balb/c mice. The SCID mutation has no effect on the erythrocyte compartment and these mice have hematocrits comparable to those seen in the Balb/c mouse (34) and equivalent numbers of erythrocytes per volume of blood (1.2×10^7 cells/µl). To evaluate the in vivo sequestration of a labelled cohort of nascent erythrocytes, three month old mice were given a single I.P. injection of Fe-59. To prevent possible reutilization of the Fe-59, mice were provided with non-radioactive FeSO in their drinking water starting at 24 hours after the Fe-59 injection. At weekly intervals thereafter the mice were bled (25 µl from the tail vein), the amount of residual Fe-59 in the red cell compartment was determined and corrected for decay of radioactivity. As can be seen in Figure 3, the rate of sequestration of the labelled cohort of CB-17 SCID erythrocytes was found to be identical to that of the Balb/c control and this despite the almost total lack of immunoglobulin in the serum of the CB-17-SCID mice.

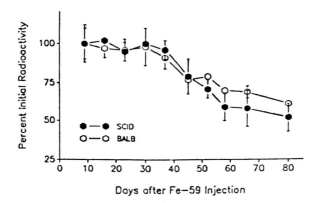

Fig. 3. Sequestration of a cohort of Fe-59 labelled erythrocytes form the
circulation of CB-17 SCID and Balb/c mice.
The results are the average ±SD of 4 mice from each strain.

DISCUSSION

     The erythrocyte throughout its sojourn in the circulation is repeatedly
exposed to environmental extremes which may lead to membrane damage which
the cell is incapable of repairing. Such membrane alterations, the nature
of which are not yet defined, are most probably those which are eventually
responsible for the sequestration of the effete erythrocyte by spenic
macrophages.

     Studies of the sequestration of the aged erythrocyte in young hosts
have led to several proposals for the mechanism(s) by which these effete
cells are recognized and removed from the circulation. Amongst these are
theories which deal with critical changes in membrane proteins (14-16),
membrane carbohydrates (11-13), and membrane lipids (17) as the red cell
ages in the circulation. It has been proposed that the phagocytic
reticuloendothelial cell either recognizes these purported membrane changes
directly (11,12,17,19) or that recognition and subsequent phagocytosis are
mediated by physiologic autoantibodies which specifically recognize the age
dependent alterations on the red cell membrane (13,16,20).

     Studies on the turnover of the erythrocyte in senescent individuals,
albeit few, have been consistent in demostrating a shortened half life of
the red cell in the circulation (6-9). Increased turnover and shortened half
life of the erythrocyte have been effectively demonstrated in the senescent
rat (6), mouse (7,8) and rabbit (9). Circumstantial evidence for a similar
shortened half life has been reported for the human in that the elderly
human has a higher reticulocyte count and increased numbers of lower density
erythrocytes (10 and Table 1).While the nature of the membrane alteration(s)
which brings about this premature sequestration of the red cell in the
elderly has not been determined, it has been demonstrated that even the
youngest of red cells of senescent individuals are depleted in the levels of
enzymes necessary to defend against oxidative damage (7,35,36). It has yet
to be determined whether the membrane alterations that bring about

sequestration of the erythrocyte in old individuals are the same as those in the young. In the light of the enzymatic defects demonstrated in the youngest erythrocytes from senescent donors, a cardinal question about the turnover of erythrocytes in the senescent individuals is whether the sequestration is time dependent. Are the oldest cells in the circulation the first to be removed, or are younger cells also marked for destruction?

We have shown that erythocytes from elderly donors are more readily phagocytosed than are those of young donors (25) and Figure 1. While high age-density (old) erythrocytes from all donors are readily phagocytosed, and young erythrocytes (low age-density fraction) from young donors are not marked for phagocytosis, those from old donors which are almost as susceptible to phagocytosis as are old erythrocytes (high age-density). These findings suggest that the sequestration of erythrocytes in senescent individuals, unlike that in young donors, may not be totally dependent upon the time the erythrocyte spends in the circulation since both low age-density and high age-density erythrocytes are marked for destruction.

As the erythrocyte ages in the circulation it becomes more susceptible to phagocytosis and carriers more membrane bound IgG. We have demonstrated that the erythrocytes of senescent individuals bear more IgG on their membranes than erythrocytes of comparable age-density fractions from young individuals (6,10 and Table 2). The phagocytosis of old and young erythrocytes from old donors and old erythrocytes from young donors are all resistant to blocking with Protein-G (25 and Table 3) demontrating that the recognition of the senescent erythrocyte is not due to the mediation of a physiologic autoantibody. The phagocytosis of these same erythrocytes is resistant to blocking with α-glycosidyl sugars and effectively blocked to an equal extent by ß-galactoside sugars (26 and Figure 2). It thus appears that the membrane alteration which brings about sequestration of the effete erythrocyte during natural aging of the red cell in both young and old hosts is related to the exposure of a ß-galactoside sugar moiety or a cross reacting conformation. Thus the mechanism(s) of recognition of the effete red cell seem to be the same whether the host individual is young or old and whether the red cell is of high age-density (in young and old individuals) or low age-density (old individuals). In addition, these results suggest that, whereas, only old erythrocytes are tagged for phagocytosis in the young individual, in the old individual the strict time in circulation dependency does not seem to hold and younger cells also marked for destruction.

Our findings suggest that erythrocyte sequestration in young and senescent individuals is not dependent upon the marking of the red cell by IgG antibody. This point has been further clarified in a study of the sequestration of erythrocytes in the circulation of mice suffering from a Severe Combined Immunodeficiency (CB-17 SCID) (26-31,34). If, indeed, antibody is necessary for the recognition of the specific alterations which occur with the aging of the erythrocyte, immuno-depressed, agammaglobulinemic animals should demontrate diminished rates of erythrocyte sequestration. These CB-17 SCID mice are virtually agammaglobulinemic and yet they have been shown to sequester their erythrocytes as well as do control mice (32 and Figure 3). It therefore appears that the role played by immunoglobulin in the removal of effete erythrocytes from the circulation is negligible.

ACKNOWLEDGEMENTS

We wish to thank Dr. David Cohen and all the staff of the Gerontology Department of the Carmel Hospital, Haifa for their assistance in selecting elderly donors. This research was supported in part by U.S.-Israel Binational Science Foundation, The Fund for the Promotion of Research at the Technion and The Technion V.P.R. Fund-Gerontology Research Fund.

REFERENCES

1. L. M. Corash, S. Piomelli, H. Chia Chan, C. Seaman and E. Gross, Separation of erythrocytes according to age on a simplified density gradient, J. Clin. Med. 84:147 (1974).
2. G. D. Bennet and M. M. B. Kay, Homeostatic removal of senescent murine erythrocytes by splenic macrophages, Exp. Hematol. 9:297 (1981).
3. D. Danon and Y. Marikovsky, Determination of density distribution of red cell population, J. Lab. & Clin. Med. 64:668 (1964).
4. J. R. Murphy, Influence of temperature and method of centrifugation on the separation of erythrocytes, J. Lab. Clin. Med. 82:334 (1973).
5. M. R. Clark, Senescence of red blood cells: progress and problems, Physiol. Rev. 68:503 (1988).
6. G. A. Glass, H. Gershon, D. Gershon, The effect of donor and cell age on several characteristics of rat erythrocytes, Exp. Hematol. 11:987 (1983).
7. E. C. Abraham, J. F. Taylor and C. A. Lang, Influence of mouse age and erythrocyte age on glutathione metabolism, Bioch. J. 174:819 (1978).
8. M. Magnani, L. Rossi, V. Stocchi, L. Cucchiarini, G. Piacentini and G. Fornaini, Effect of age on some properties of mice erythrocytes, Mech. Age Dev. 42:37 (1988).
9. T. Vomel and D. Platt, Lifespan of rabbit erythrocytes and activity of the reticulohistiocyte system, Mech. Age Dev. 17:267 (1981).
10. G. A. Glass, D. Gershon and H. Gershon, Some characteristics of the human erythrocyte as a function of donor and cell age, Exp. Hematol. 13:1122 (1985).
11. D. Danon, Recognition by macrophages of alterations in the membranes of old red cells and expelled nuclei, in: "Permeability and Function of Biological Membranes", Bolis, Katchalski, Kevens, Loevenstein, Pethica, pp. 57, North-Holland (1970).
12. D. Aminoff, Senescence and sequestration of RBC from Circulation, in: "Cellular and Molecular Aspects of Aging; The Red Cell as a Model", Eaton, Konzen, White, pp. 279, Liss, New York (1985).
13. U. Galili, I. Flechner and E. A. Rachmilewitz, A naturally occuring anti- α -galactosyl IgG recognizing senenscent human red cells, in: "Cellular and Molecular Aspects of Aging; The Red Cell as a Model", Eaton, Konzen, White, pp. 263, Liss, New York (1985).
14. M. M. B. Kay, S. R. Goodman, K. Sorensen, C. F. Whitfieid, P. Wong, L. Zaki and V. Rudloff, Senescent cell antigen is immunologically

related to band3, Proc. Natl. Acad. Sci. USA 80:1631 (1983).

15. H. U. Lutz and G. Stringaro-Wipe, Senescent red cell-bound IgG is attached to band 3 protein. Biomed. Biochim. Acta 42:S117 (1983).

16. K. Schluter and D. Drenckhan, Co-clustering of denatured hemoglobin with band 3: its role in binding of autoantibodies against band 3 to abnormal and aged erythrocytes, Proc. Natl. Sci. USA 83:6137 (1986).

17. J. J.Fidler, Macrophages and metastasis - a biological approach to cancer theraphy, Cancer Res. 45:4714 (1985).

18. D. Danon, L. Goldstein, Y. Marikovsky and E. Skutelsky, Use of cationized ferritin as a label of negative charge on cell surfaces, J Ultrastruct Res 38:500 (1972).

19. D. Aminoff, M. A. Ghalambor and C. J. Henric, Gost, galactose oxidase and sialyl transferase, substrate and receptor sites in erythrocyte senescence, in: "Erythrocyte Membrane 2: Recent Clinical and Experimental Advances , Kruckergerg, Eaton, Brewer, pp. 269 Liss, New York (1981).

20. M. Kay, Role of physiologic autoantibody in the removal of senescent human red cells, J. Supramolecular Structure 9:555 (1978).

21. H. U. Lutz, F. Bussolino, R. Flepp, S. Fasler, P. Stammler, M. D. Kazatchkine and P. Arese, Naturally occurring anti-band-3 antibodies and complement together mediate phagocytosis of oxidatively stressed human erythrocytes, Proc. Natl. Acad. Sci. USA 84:7368 (1987).

22. I. O. Szymanski, P. R. Odgren, N. L. Fortier and L. M. Snyder, Red blood cell associated IgG in normal and pathologic states, Blood 55:48 (1980).

23. M. O. Jeje, M. A. Blajchman, K. Steeves, P. Horsewood and J. G. Kelton, Quantitation of red cell-associated IgG using an immunoradiometric assay, Transfusion 24:473 (1984).

24. E. Sheiban and H. Gershon, The development of an ELISA for the determination of in situ bound IgG on erythrocytes of normal donors and specific blocking of an IgG dependent erythrophagocytosis assay by Protein-G. Submitted for publication.

25. E. Sheiban and H. Gershon, Mechanism of recognition and sequestration of young and old erythrocytes from young and old donors. Submitted for publication.

26. G. C. Bosma, R. P. Custer and M. J. Bosma, A severe combined immunodeficiency in the mouse, Nature 301:527 (1983).

27. M. R. Lieber, J. E. Hesse, S. Lewis, G. C. Bosma, N. Rosenberg, K. Mizuichi, M. J. Bosma and M. G. Gilbert, The defect in murine SCID: Joining of signal sequences but not coding segments in V(D)J recombination, Cell 55:7 (1988).

28. M. G. Kim, W. Shuler, M. J. Bosma and K. B. Marcu, Abnormal recombinant of Igh D and J gene segmants in transformed pre-B cells of SCID mice. J. Immunol. 141:1341 (1988).

29. K. Okajaki, S. Nishikawa and H. Sakano, Aberrant immunoglobulin gene rearrangement in SCID mouse bone marrow cells. J. Immunol. 141:1348 (1988).

30. B. A. Malynn, T. K. Blackwell, G. M. Fulop, G. A. Rathburn , A. J. W. Furley, P. Ferrier, L. B. Heinke, R. A. Phillips, G. Yancoupolos and F. W. Alt, The SCID defect affects the final step of the immunoglobulin VDJ recombinase mechanism, Cell 54:453 (1988).

31. K. Dorshkind, G. M. Killer, R. A. Phillips, R. G. Miller, G. C. Bosma,

M. O'Toole and M. J. Bosma, Functional status of cells from lymphoid and myeloid tissues in mice with severe combined immunodeficiency disease, J. Immunol. 132:1804 (1984).

32. H. Gershon, Normal erythrocyte sequestration in an agammaglobulinemic host: Studies in the CB-17 SCID mouse. Submitted for publication.
33. E. Beutler, C. West and K. G. Blume, The removal of leukocytes and platelets from whole blood, J. Lab. Clin. Med. 88:328 (1976).
34. R. P. Custer, G. C. Bosma and M. J. Bosma, Severe combined immunodeficiency (SCID) in the mouse: Pathology, reconstitution, neoplasms, Am. J. Pathol. 120:468 (1985).
35. G. A. Glass and D. Gershon, Decreased enzymic protection and increased sensitivity to oxidative damage in erythrocytes as a function of cell and donor age, Bioch. J. 218:531 (1984).
36. L. Bjork and A. Kronvall, Purification and some properties of streptococcal Protein G, a novel IgG-binding reagent, J. Immunol. 133:969 (1984).

A NEW MONOCLONAL ANTIBODY TO AN AGE SENSITIVE

BAND 3 TRANSMEMBRANE SEGMENT

Lucietta Ferroni[1], Anna Giuliani[1],
Stefano Marini[2], Patrizia Caprari[1],
Anna Maria Salvati[1], Saverio C. Condò[2],
Maria Teresa Ramacci[3] and Bruno Giardina[2]

[1]Istituto Superiore di Sanità, Roma, Italy
[2]II Università di Roma, Roma, Italy
[3]Sigma Tau, Pomezia, Roma, Ialy

SUMMARY

The appearance of band 3 structural modifications related to aging could be evidenced by means of monoclonal antibodies against senescence antigen. Hence in the attempt to provide an immunological marker of erythrocyte aging, we raised a monoclonal antibody against native band 3 (B6 MoAb), which seems to detect differences in the band 3 molecule from erythrocytes of different ages separated by density gradient.

Densitometric evaluation of immunoblotting patterns indicates that the in vivo aging is associated with band 3 monomer degradation. The Percoll separated fractions show a significant increase of those proteolytic fragments that bind the B6 antibody.

Finally, protease digestions of unsealed membrane ghosts have been performed to test the binding site of the B6 antibody on the band 3 molecule. The data show that the B6 antibody binds a 19 KDa chymotryptic-tryptic fragment which corresponds to a segment of the looped membrane domain whose steric structure appears to be sensitive to age.

INTRODUCTION

In recent years monoclonal antibodies against human band 3 fragments have been prepared and employed for different purposes. Topological studies of band 3 molecule arrangement have been performed (1,2) on the basis of antibody binding to proteolytic fragments obtained from sealed and unsealed ghosts. Recently, monoclonal antibodies raised against cytoplasmic end of

Red Blood Cell Aging, Edited by M. Magnani and
A. De Flora, Plenum Press, New York, 1991

351

the band 3 molecule were able to detect a correlation between cellular aging and increase of 60 KDa, 40 KDa and 20 KDa membrane fragment (3,4). This observation agrees with the hypothesis that band 3 degradation is responsible for molecular changes generating senescence antigen (5,6). We postulated that monoclonal antibody against the band 3 purified in native form (7) should better evidence the structural modifications related to cell aging.

We raised a mouse hybridoma anti-band 3 monoclonal antibody, the B6, of the IgG2a class which monitored band 3 in human normal red cells and tested it in density gradient separated red cells in view of elucidating the band 3 pattern. An effort to localize the B6 binding site on band 3 was also performed. For this, normal red cells were submitted to peptidase digestion and binding of monoclonal antibody was evaluated on red cell membrane immunoblots after SDS-PAGE.

METHODS

40 ml of fresh normal blood were collected, after informed consent, on Na$_2$-EDTA (1.5 ± 0.25 mg/ml) and processed within 2-3 hours. Leukocyte and platelet free RBCs, were obtained from filtration of the blood samples in α-cellulose/microcristalline cellulose (1:1 w/w) according to Beutler et al (8).

The RBCs were washed three times with 5 mM phosphate buffer, 150 mM NaCl and 0.1 mM PMSF, pH 7.4, centrifuged for 10 min at 2000xg. All the steps were performed between 0 and 4°C. The RBCs were lysed in 5 mM phosphate buffer and 0.1 mM PMSF (pH 8) (RBC/lysis buffer 1:30 v/v).

Filtered RBCs were washed three times, 10 min at 2000g, with HEPES buffered isotonic saline at room temperature. The erythrocyte separation was performed according to Salvo et al. (9).

Leaky human erythrocyte membrane tryptic digestion and intact erythrocytes chymotrypsin digestion followed by membrane tryptic digestion were performed according to Wainwright et al. (3). The membrane ghosts were always alkali-stripped after protease digestions to improve the band 3 detection.

Mouse monoclonal antibodies were obtained after hyperimmunization with human purified band 3 (7) as previously described (10). Screening of antibody on purified band 3 and antibody binding both on intact red blood cells and inside-out vesicles as well as the antibody isotypes were determined by ELISA.

The monoclonal antibody (B6 MoAb) purification was achieved in a Protein A-Sepharose column (binding buffer 3 M NaCl and 1.5 M glycine, pH 8.9) with 0.1 M acetate buffer, pH 5.0.

The sodium dodecyl sulphate polyacrylamide gel electrophoresis (SDS-PAGE) according to Laemmli (11) was performed at 7% or 10% acrylamide concentrations. The immunoblotting was performed according to Burnette (12).

RESULTS

Five erythrocyte fractions were recovered. The B6 MoAb was able to detect the differences between the membrane ghosts of the reticulocyte enriched lightest fraction (F1) and the densest one (F5). The erythrocyte band 3 monomer became increasingly more heterogeneous in the densest erythrocytes. The B6 seemed to be a band 3 monomer degradation marker, since: a) in lightest cells (F1) the B6 bound only the band 3 monomer; b) in the densest erythrocytes it strongly binds the 60 Kda band of the 4.5 region and increasing numbers of proteolytic fragments of this region. Furthermore in the densest fractions a new band, which ranges between 90 and 80 KDa, was present.

Table 1. Percentages* of Immunoblotting Bands Detected by B6 MoAb in Density
         Gradient Separated Normal Red Blood Cells (RBC).

RBC fractions	HMWP %	band 3 %	80-90 KDa Band %	60 KDa band %	4.5 region %
F1	--	82	--	--	--
F2	11	51	--	6	9
F3	9	48	6	8	8
F4	11	43	6	8	10
F5	9	39	10	9	13
C	7	43	--	9	10

* The table shows the percentages in a single typical experiment. The data
  obtained by additional trials were consistent with the reported ones. C:
  unseparated sample; -- not present.

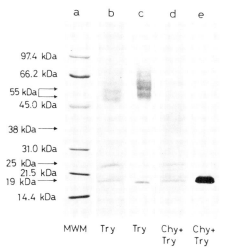

Fig. 1. SDS-PAGE (b and d) and immunoblots (c and e) of trypsin-treated
        ghosts (b and c) and trypsin-treated ghosts from chymotrypsin-
        digested red blood cells (d and e). SDS-PAGE was performed at
        10% acrylamide concentration and stained with Coomassie Blue.

The bands that are considered the most age sensitive were compared in different red blood cell fractions obtained by Percoll gradient and the densitometric percentages, detected by immunoblotting, are reported in table 1. The decreasing trend of band 3 and the increasing presence of 80-90 KDa band in the oldest fractions appear to be the most dramatic evidence; the high molecular weight polymer (HMWP) percentage seems to be unaffected, whereas the whole 4.5 region seems to increase with aging.

The SDS-PAGE of unsealed ghosts performed after trypsin digestion (Fig. 1, lane b) showed a number of band 3 fragments ranging from 55 KDa to 19 KDa detectable with Coomassie blue staining.

The 55 KDa band corresponds to the tryptic membrane-bound band 3 fragment which was heterogeneously glycosylated and hence gives diffuse bands notwithstanding the tryptic treatment. The bands with a lower Mr correspond to the cleavage products of tryptic digestion of band 3 membrane domain. Trypsin treatment of unsealed membrane ghosts, from chymotrypsin-treated erythrocytes, gave rise (Fig. 1, lane d) to a 38 KDa diffuse band corresponding to the C-terminal and of the 55 KDa membrane domain end to a number of smaller fragments whose molecular weights ranged from 30 KDa to 19 KDa after Coomassie Blue staining.

The immunoblotting pattern after trypsin digestion of unsealed ghosts showed (Fig. 1, lane c) that the epitope is located in the 55 KDa fragment. Regarding the weak band at 19 KDa it is believed to be a band 3 fragment produced in vivo by endogenous protease(s). The immunoblotting pattern of the trypsin treated membrane from chymotryptic treated erythrocytes (Fig. 1, lane e) showed that the B6 antibody binds solely the 19 KDa fragment. It must be excluded that the 19 KDa fragment corresponds to the 20 KDa fragment originated from the tryptic cleavage of the C-terminal end. In fact, the B6 antibody does not bind the 38 KDa fragment, detectable with Coomassie Blue staining, which includes the 20 KDa fragment. On the contrary the 19 KDa band closely corresponds to 17 KDa Ch-Tr N-terminal portion of 55 KDa membrane domain, which Kay et al. (13) indicate as the anion transport segment.

DISCUSSION

We raised a monoclonal antibody against the membrane erythrocyte band 3: B6 MoAb, an IgG2a class antibody, which is able to detect differences in band 3 molecule from different aged normal red blood cells separated by Percoll density gradient. The B6 MoAb does not detect the 20 KDa fragment previously observed by other authors in in vivo aged normal erythrocytes (14), but is able to bind a 80-90 KDa new band, lighter than band 3, produced by protease cleavage. This band appears in the densest erythrocyte fractions, and its increase well correlates with band 3 monomer decrease in cell aging. Furthermore the antibody binding evidences that in vivo band 3 degradation increasingly enhances the low molecular weight fragments of 4.5 region.

The reactivity of B6 MoAb with 17 KDa Ch-Tr (19 KDa in our system) proteolytic fragment of the N-terminal side of 55 KDa membrane domain suggests that the B6 MoAb binding site to band 3 differs from the epitopes located in the same membrane domain and recognized by all the previously produced antibodies. Furthermore it should be underlined that the increased binding of B6 MoAb to inside-out vesicles with respect to ghosts (data not shown), allows to postulate the epitope location on the cytoplasmic side of intramembrane 17 KDa fragment. This fragment belongs to a very looped structure which is sensitive not only to endogenous protease digestion, but also to cleavage of polysaccaride chains of different sizes that bind the band 3 membrane domain. In conclusion, the observed differences in B6 MoAb binding between the reticulocytes and the densest erythrocyte fractions suggest that both protease and glycosidase effects could be responsible for steric and structural modifications of this region of band 3 molecule in cell aging.

ACKNOWLEDGEMENTS

This work was partially supported by National Research Council (CNR) – Progetto Finalizzato Invecchiamento.

REFERENCES

1. D. M. Lieberman, and R. A. F. Reithmeier, Localization of Carboxyl terminus of band 3 to cytoplasmic side of the erythrocyte membrane using antibodies raised against a synthetic peptide, J. Biol. Chem. 263:10022 (1988).
2. P. K. K. Tai, and Carter-Su, Monoclonal antibody to the human glucose transports that differentiates between the glucose and nucleoside transports, Biochemistry 27:6062 (1988).
3. S. D. Wainwright, M. J. A. Tanner, G. E. M. Martin, J. E. Yandle, and C. Holmes, Monoclonal antibodies to the membrane domain of the human transport protein, Biochem. J. 258:211 (1989).
4. M. Czerwinski, K. Wasniowska, I. Steuden, M. Duk, A. Wiedlocha, and E. Lisowska, Degradation of the human erythrocyte membrane band 3 studied with monoclonal antibody directed againist an epitope on the cytoplasmic fragment of band 3, Eur. J. Biochem. 174:647 (1988).
5. M. M. B. Kay, K. Sorenson, P. Wong, and P. Bolton, Antigenicity, storage and aging: physiologic antibodies to cell membrane and serum proteins, Mol. Cell. Biochem. 49:65 (1982).
6. M. M. B. Kay, Band 3, the predominant transmembrane polypeptide undergoes proteolytic degradation as cells age, Monog. Dev. Biol. 17:245 (1984).
7. A. Giuliani, L. Ferroni, P. Caprari and A. M. Salvati, Band 3 purification: comparative evaluation of different extraction systems, Clin. Chem. Enzym. Comms. 3:299 (1990).
8. E. Beutler, K. G. Blume, J. C. Kaplan G. W. Lohr, B. Ramot and W. N. Valentine, International Committee for Standardization in Haematology: recommended methods for red cell enzyme analysis, Br. J. Haematol. 35:331 (1977).

9. G. Salvo, P. Caprari, P. Samoggia, G. Mariani and A. M. Salvati, Human erythrocyte separation according to age on a discontinuous "Percoll" density gradient, Clin. Chim. Acta 122:293 (1982).

10. S. Marini, G. Citro, S. di Cesare, R. Zito and B. Giardina, Production and characterization of monoclonal antibodies against DNA single ring diamine adducts, Hybridoma 7:193 (1988).

11. U. K. Laemmli, Cleavage of structural proteins by bacteriofage T4, Nature 227:680 (1979).

12. W. N. Burnette, "Western Blotting": electrophoretic transfer of proteins from sodium-dodecyl sulphate polyacrylamide gel electrophoresis to unmodified nitrocellulose and radiographic detection with antibody and radioiodinated Protein A, Anal. Biochem. 112: 195 (1981).

13. M. M. B. Kay, Localization of senescent antigen on band 3, Proc. Natl. Acad. Sci. N. Y. 81:5753 (1984).

14. M. L. Jennings, M. P. Anderson and R. Monaghan, Monoclonal antibodies against human erythrocyte band 3 protein J. Biol. Chem. 261:9002 (1986).

CHARACTERIZATION OF ANTIBODY THAT BINDS IN VIVO TO NORMAL

HUMAN RED BLOOD CELLS

Margaret R. Clark and Martin P. Sorette

Department of Laboratory Medicine
and Cancer Research Institute
University of California
San Francisco, California

INTRODUCTION

It has been well-documented that a subpopulation of normal circulating human red cells have antibody on their surfaces, and this antibody has been proposed to serve as a major mechanism for removal of senescent red cells from the circulation at the end of their lifespan (1). Antibodies of two different specificities have been reported to be enriched in high density human red cells. One is an antibody that appears to recognize an altered form of the Band 3 membrane protein (2). The other recognizes an α-galactosyl carbohydrate structure, found in small quantities in membrane glycolipids of Old World primates (3). As a basis for further understanding of the physiologic relevance of these antibodies, we have performed quantitative measurements of their contribution to in situ bound antibody on high density populations of normal human red cells. In addition, we have determined the number of binding sites of each type of antibody on different density populations that had been stripped of in situ antibody. To provide insight about the possible activity of the IgG dense cells, the IgG subclass distribution of the in situ antibody was also determined.

METHODS

Red cell preparation

Blood from normal volunteers was drawn into acid-citrate-dextrose anticoagulant. It was then passed through cellulose to remove leukocytes and platelets, and was separated into different density populations using arabinogalactan gradients as previously described (4). In particular, the gradients were designed to isolate very small cell populations (0.5-2% of total cells) at the extreme high density distribution. For experiments in which rebinding of purified antibodies was to be determined, the separated cells were first stripped of in situ bound IgG by mixing them into ice-cold pH 3 glycine-HCl buffer (5), followed by immediate centrifugation through a pH 7.4 arabinogalactan (10% w/v) cushion. Ektacytometric assay of cell deformability (6) indicated the absence of any adverse effects on cell hydration or deformability as result of this transient exposure to low pH.

Flow cytometric analysis of "in situ" bound IgG

To detect in situ bound IgG using flow cytometry, density separated red cells were first incubated with biotinylated goat anti-human IgG, washed 3 times in PBS and then incubated with fluoresceinated avidin. After additional washing they were then assayed for fluorescence using a FACSCAN flow cytometer (Becton-Dickinson, Mountain View, C.A.).

Measurement of cell boud IgG

The total amount of IgG bound to the surface of freshly separated red cells was estimated by the binding of ^{125}I-Protein A (ICN Radiochemicals, Irvine, CA), as described by Yam et al. (7). The relative amounts of the four IgG subclasses represented in in situ IgG on density separated human red cells was determined using ^{125}I-labeled monoclonal antibodies directed against each subclass. The antibodies (Calbiochem Corp., San Diego, CA and ICN Biomedicals, Inc., Lisle, IL) were labeled with ^{125}I, using Iodobeads (Pierce Chemical Co., Rockford, IL). After separation of the cells on arabinogalactan gradients and 3 washes in phosphate buffered saline (PBS: 10 mM Na phosphate, pH 7.4), aliquots of cell from a given individual were incubated in parallel with each antibody for 2 h at 37°C. The cells were washed 3 more times, and pelleted through a cushion of 10% (w/v) arabinogalactan density gradient medium to remove unbound Protein A or antibody. The residual bound label on the pellet was then quantitated using a gamma counter.

Quantitation of protein 4.1 components

Separated cell populations were used to prepare one-step ghost membranes by hemolysis in 5 mM Tris-HCl (0°C, pH 8.0) containing 1 mM EDTA and 1 mM diisopropylfluorophosphate. Membrane proteins were then separated by sodium dodecylsulfate polyacrylamide gel electrophoresis (7% acrylamide). Gels were stained with Coomassie blue, the bands corresponding to protein 4.1a and 4.1b were cut out and the dye eluted using pyridine for spectrophotometric quantitation (8).

Quantitation of anti-Gal and anti-Band 3 in eluted "in situ" IgG

Anti-Gal and anti-band 3 content of IgG eluted at pH 2.1 (5) from density separated red cells were determined using solid phase adsorption assays. Total IgG in the eluates was quantitated by a solid phase radioimmunoassay on microtitre plates using ^{125}I-Protein A (9). To determine the fraction that was anti-Gal, aliquots of the eluted IgG were incubated (1 h, room temperature) with Synsorb 115 particles, after which the samples were centrifuged and the supernatant assayed for unbound IgG by the same method used for total IgG. The anti-Gal fraction was then given by the difference in IgG content of the total eluted and that in Synsorb supernatant. For anti-band 3, aliquots of eluted IgG were incubated on microtitre plates onto which purified band 3 had been inmobilized using glutaraldehyde. After washing the plates, the antibody that bound to the immobilized antigen was quantitated using biotinylated goat anti-human IgG and ^{125}I-streptavidin.

Comparison with the total eluted IgG then gave the fraction specific for band 3.

Rebinding of purified serum anti-Gal and anti-Band 3 to stripped red cells

Anti-Gal was isolated from heat-inactivated human type AB or O plasma, using synthetic carbohydrate affinity columns (Synsorb 115, Chembiomed. Edmonton, Canada) (10). Anti-band 3 was isolated from human serum IgG using Affi-Gel affinity columns (11) (Bio Rad Laboratories, Richmond CA), to which purified band 3 protein (12) had been coupled. Density separated red cells, from which in situ antibody had been stripped by exposure to low pH (5) (glycine-HCl, pH 3) were incubated with the isolated antibodies, in autologous serum depleted of the corresponding antibody by repeated passage through an affinity column. After washing in PBS, the amount of each antibody that had rebound to the cells was detected using ^{125}I-Protein A (7). Subclass composition of affinity purified serum anti-Gal was determined using a commercial end point radial immunodiffusion assay (ICN, Lisle, IL).

RESULTS

As previously shown by several groups, in situ bound antibody is concentrated in high density red cell populations. By using density gradients that subdivided the high density cells into several small populations, we found that the antibody was particularly enriched in the most dense 1-2% of the cells (Fig. 1a). There was also enrichment in the same cells in the a component of protein 4.1, which has been proposed to be an indicator of older mature cells (13) (Fig. 1b). This supports the hypothesis that antibody binding is associated with increased red cell age.

When the most dense 1-2% of cells were stained with biotinylated anti-human IgG and fluorescent streptavidin, the increased antibody was clearly detectable by flow cytometry (Fig. 2, Table 1). In one subject the fluorescence histogram of the entire high density population was shifted to higher fluorescence, indicating that a large fraction of the cells had increased bound antibody. In another subject, the increased fluorescence appeared as a shoulder, indicating that a smaller portion of the high density population had substantially increased amounts of bound antibody. The proportion of cells with increased fluorescence in the dense cell populations was much greater than reported in previous studies, probably because of the restriction to only the most dense percent or so.

An important piece of information is available from the flow cytometric analysis that cannot be obtained from bulk assay of cell-bound IgG. Namely, it is evident that in these highly selected dense populations, there are large proportions of cells that do not carry substantially greater amounts of bound IgG than the average lower density cell. This means that the number of IgG molecules per cell calculated in the bulk assay substantially underestimates the level on the cells on which the antibody is concentrated.

For example, in Subject 1, a relatively modest proportion of dense cells (15%) had a relative fluorescence greater than 99% of the middle density cells (Table 1). In this case the average antibody level for the dense cell population would greatly underestimate the amount of antibody on the individual cells on which it was concentrated. For Subject 2 a larger proportion of the dense cells (44%) had a relative fluorescence greater that 98% of the middle density cells, and the average antibody level would be closer to that on the antibody-bearing cells. However, it is evident that a large proportion of cells still have much higher levels of bound antibody than the average. Another point of interest is that comparison of the fluorescence histogram for high density cells from Subjects 1 and 2 suggests that the antibody distribution on the antibody-bearing cells in these two samples is very similar, even though the average number of antibody molecules per cell is rather different.
When the antibody was eluted from these most dense cell populations and analyzed for the amounts that recognized Band 3 and anti-Gal, we found that together, these two antibodies constitued a substantial proportion of the total <u>in situ</u> bound antibody, from 20–60% in seven different individuals (Table 2). Of this, anti-Gal always made the greater contribution. On average, the percentage of anti-Gal in the eluted <u>in situ</u> IgG was twice that of anti-Band 3.

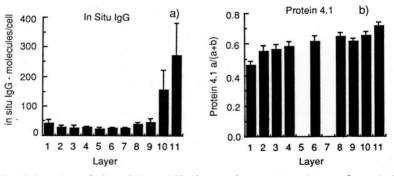

Fig. 1. Enrichment of <u>in situ</u> antibody and component a of protein 4.1 in the very highest density red cells. a) Given is the average number of molecules of IgG per cell for each layer of 11-step gradients. Error bars show the SEM. b) The values on the ordinate represent the proportion of protein 4.1 in the a component, associated with increased cell age. Layers 1 and 11 were significantly different from Layer 4 ($p < 0.05$) in their content of component 4.1 a, as judged using analysis of variance. Layers 9–11 together make up only 1% of the total cell population. Data are from 3 normal subjects.

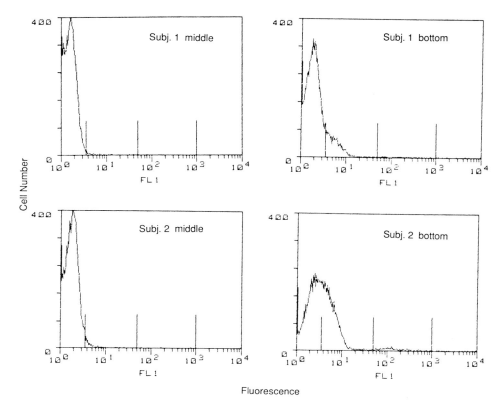

Fluorescence

Fig. 2. Flow cytometric fluorescence histograms for binding of fluorescently labeled anti-human IgG to density separated human red cells from 2 normal subjects. The middle population were taken from a density gradient layer corresponding to the median density for the total blood sample, and the bottom samples comprised the 0.5-1% most dense populations. After separation, the cells were stained with fluorescent goat-anti-human IgG to detect in situ IgG. Values along the ordinate give cell numbers, and those on the abscissa the fluorescence intensity.

We also studied the contributions of anti-Gal and anti-band 3 by determining the extent of rebinding of the purified antibodies to density-separated red cells that were then stripped of in situ antibody using low pH. These experiments also showed increased binding on anti-Gal over that of anti-Band 3 (Fig. 3).

Rebinding of purified anti-Gal was somewhat greater than that of anti-Band 3 for low and intermediate density cells as well as for high density cells. However, the summed amount of the two antibodies that could rebind to stripped cells was greater then the amount of in situ total IgG found on freshly isolated cells (Note that the amount of in situ IgG in these experiments was lower than in those reported in Table 2. This was because of less stringent selection of the very highest density red cells).

Table 1. Mean Fluorescence of Anti-IgG Labeled, Density Separated
Red Cells and Percent of Cells with Fluorescence Above
the 98-99% Range for Intermediate Density Cells.

	Mean Rel. Fluorescence	% Cells with Fl > 3.4
Subject 1:		
Top	1.48	1.4
Middle	1.54	0.7
Bottom	1.79	15.2
Subject 2		
Top	1.54	1.2
Middle	1.79	2.2
Bottom	2.18	43.9

Data correspond to flow cytometric histograms in Fig. 2.

Table 2. Anti-Gal and Anti-Band 3 Content of _in situ_ Bound IgG
on the Most Dense 1-2% of Cells from 7 Normal Individuals.

Total IgG Molecules/cell	Anti-Gal % Total IgG	Anti-Band 3 % Total IgG
90-490	9-39	5-18

Total IgG was measured on well washed cells using ^{125}I-Protein A,
and anti-Gal and anti-Band 3 content of IgG eluted from dense cells
were determined using absorbtion onto solid phase antigen and
radioimmunoassay of IgG before and after adsorbtion.

The higher levels of rebound IgG may possibily reflect exposure of
new binding sites as a result of the antibody stripping treatment.
Alternatively, it may reflect enrichment of high affinity antibodies in the
purified serum antibody preparation used for rebinding.
Analysis of IgG subclasses for total IgG bound to the most dense 0.5-1% of
cells showed that IgG_1 and IgG_3 were present on the cells from all 6
individuals studied (Fig. 4).

IgG_2 was detected on the dense cells of only 3/6 subjects, and IgG_4 was
not significantly present on any. Interestingly, the subjects whose cells
had IgG were all of blood Type O, whereas those without IgG_2 were blood
Types A and B. Cells of intermediate density showed essentially the same
pattern of IgG subclass distribution. However, the overall quantity of IgG
in each subclass was much lower than on dense cells, which was consistent
with the lower levels of total _in situ_ bound IgG In the lower density cell
populations.

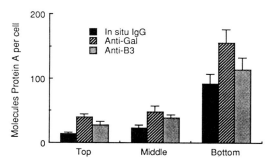

Fig. 3. Rebinding of purified anti-Gal and anti-Band 3 to the most dense
1-2% of red cells from 5 normal individuals. Cells were first
stripped of <u>in situ</u> IgG by brief exposure to pH 3.0, at 0°C.
Rebinding took place in the absence of serum and was quantitated
using ^{125}I-Protein A. Shown are the mean numbers of molecules of
Protein A binding per cell (error bars show SEM).
For reference, the first column gives the level of <u>in situ</u> IgG on
unstripped cells.

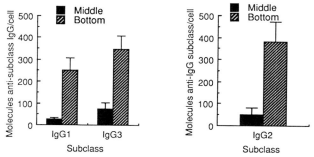

Fig. 4. IgG subclass distribution on density separated normal cells. Values
on the ordinate represent the average number of molecules/cell
of radiolabeled monoclonal antibodies directed againist each
subclass that bound to the most dense 0.5-1% population (solid
bars) and an intermediate density population (hatched bars). Values
for IgG_1 and IgG_3 are means for 6 individuals; the dense cell
value for IgG_2 is an average for the 3 subjects whose cells bound
detectable anti-IgG_2. Anti-IgG_4 did not show detectable binding.
Because the stoichiometry of binding of the probe antibodies has
not been determined, it is not know whether the values
quantitatively reflect the relative amounts of IgG subclasses
actually present on the cells.

DISCUSSION

These measurements of anti-Gal and anti-Band 3 on the most dense 1-2% of human red cells demonstrate that the in situ IgG on these cells is made up of several different specificies. Thus, no single type of binding site is responsible for the binding of autologous serum IgG to these cells. On the other hand, both anti-Gal and particularly anti-Band 3 contribute to the bound IgG out of proportion to their abundance in serum, indicating a selective exposure of specific binding sites. Because anti-Gal is present only in the serum of Old World primates, the significance of its relative abundance on human red cells is not clear. If binding of autologous antibody is a generally crucial step in the programmed turnover of mammalian or even vertebrate red cells, the spectrum of antibodies involved in such a process is likely to show some species variation.

In addition to selective binding of these two specific antibodies, the high density red cells also show selectivity in the IgG subtypes that make up this bound antibody. Although the absolute amounts of each subclass have not yet been determined, preliminary results indicate that IgG_3 is found on dense cells out of proportion to its abundance in total serum IgG. Because this subclass is the most effective in promoting phagocytosis of antibody coated red cells, its presence on dense cells may indeed have physiologic significance.

Many concerns have been raised as to whether density gradient separation provides an effective means of separating red cells on the basis of age (14). Growing evidence suggests that, in fact, the previously accepted age/density correlation is not as tight as has been previously assumed. In our own laboratory, we have obtained evidence that large proportions of relatively senescent cells are found in intermediate density populations after high resolution density gradient separation (15). Despite these concerns, it is clear that the small proportion of very highest density, normal red cells is a unique population. In this population, particularly if it is restricted to less than 1% of the total cells, there is substantial enrichment of cells that carry up to several hundred molecules of IgG on their surfaces. The concomitant enrichment in the a component of protein 4.1 in highly selected dense cell populations is consistent with the hypothesis that the antibody- bearing cells may be old cells. However, this remains to be directly demonstrated for human blood samples. To further our understanding of the significance of autologous antibody on circulating red cells, it will be important to utilize methods independent of cell density to isolate the entire senescent population and others to isolate the entire antibody-bearing cell population, in order to determine whether age and antibody binding are closely correlated.

ACKNOWLEDGMENTS

The authors gratefully acknowledge support from US PHS grant Nos. DK 32094 and HL 20985. This is publication No. 103 of the MacMillian-Cargill Hematology Research Laboratory at the University of California, San Francisco, California.

REFERENCES

1. M. M. B. Kay, Mechanism of removal of senescent cells by human macrophages "in situ", Proc. Natl. Acad. Sci. (USA) 72:3521 (1975).
2. M. M. B. Kay, Localization of senescent antigen on band 3, Proc. Natl. Acad. Sci. (USA) 81:5753 (1984).
3. U. Galilli, The natural anti-Gal antibody, the B-like antigen, and human red cell aging, Blood Cells 14:205 (1988).
4. M. R. Clark, Calcium extrusion by high density human red blood cells, Blood Cells 14:119 (1988).
5. O. P. Rekving and K. Hannestad, Acid elution of blood group antibodies from intact erythrocytes, Vox Sang. 33:286 (1977).
6. M. R. Clark, N. Mohands and S. B. Shohet, Osmotic gradient ektacytometry: comprehensive characterization of red cell volume and surface maintenance, Blood 61:899 (1983).
7. P. Yam, L. D.Petz and P. Spath, Detection of IgG sensitization of red cells with ^{125}I Staphylococcal Protein A, Am. J. Hematol. 12:337 (1982).
8. C. Fenner, R. R. Trout, D. T. Mason, N. Wickman and J. Coffelt, Quantitation of Coomassie Blue stained proteins in polyacrylamide gels based on analysis of the eluted dye, Anal. Biochem. 63:595 (1975).
9. A. P. Gee and J. J. Langone, Immunoassay using ^{125}I or enzyme labeled protein A and antigen coated tubes, Anal. Biochem. 116:524 (1981).
10. U. Galili, R. E. Mandrell, R. M. Hamedeh, S. B. Shohet and J. M. Griffiss, Interaction between human natural anti- -galactosyl immunoglobulin G and bacteria of human flora, Infect Immun. 56:1730 (1988).
11. H. U. Lutz, R. Flepp and G. Stringaro-Wipf, Naturally occurring autoantibodies to exoplasmic and cryptic regions of band 3 protein, the major integral membrane protein of human red blood cells, J Immunol. 133:2610 (1984)
12. M. F. Likacovic, M. B. Feinstein, R. I. Shaafi and S. Perrie, Purification of stabilized band 3 protein of the human erythrocyte membrane and its reconstitution into liposomes, Biochem. USA 20:3145 (1981).
13. T. J. Mueller, C. W. Jackson, M. E. Dockter and M. Morrison, Membrane skeletal alterations during in vivo mouse red cell aging. Increase in the band 4.1a:4.1b ratio, J. Clin. Invest. 79:492 (1987).
14. E. Beutler, Isolation of the aged, Blood Cells 14:1 (1988).
15. M. R.Clark, L. Corash and R. H. Jensen, Density distribution of aging, transfused human red cells, Blood 74 (Suppl. 1):217a (1989).

OPSONIC POTENTIAL OF C3b-ANTI-BAND 3 COMPLEXES WHEN GENERATED ON

SENESCENT AND OXIDATIVELY STRESSED RED CELLS OR IN FLUID PHASE

Hans U. Lutz[1], Pia Stammler[1], Daniel Kock[1]
and Ronald P. Taylor[2]

[1]Laboratory for Biochemistry
 Swiss Federal Institute of Technology
 ETH-Zentrum, Zurich, Switzerland
[2]Dept. of Biochemistry
 University of Virginia
 Charlottesville, Virginia, USA

INTRODUCTION

Covalently linked complexes of complement component C3b and IgG
(C3b-IgG complexes) were found predominantly on senescent rather than young
red cells (1,2). They contained anti-band 3 as verified after cleavage of
the hydroxylamine sensitive bond between C3b and IgG by studying the
specificity of the liberated and purified IgG molecules (3,4). Two aspects
of these C3b-IgG complexes are striking: a) their apparently exclusive
formation with anti-band 3 and b) their presumptive opsonic potency as
judged from comparison with artificially made C3b-IgG complexes (5-7).

a) Preferred generation of C3b-IgG complexes with certain IgG molecules
is kinetically possible, since activation of C3 by either the classical or
the alternative C3 convertase breaks the intramolecular thioester bond on
the chain of C3 and generates an intermediate with a short half life of
less than 60 μs (8). The intermediate reacts instantaneously with nearby
nucleophiles. Though the predominant partner is water, some nascent C3b
reacts with proteins by forming ester bonds with OH-tyrosines and amide
bonds with amino groups (9). Among proteins, immunoglobulins that form an
immune complex are the most likely candidates to become covalently bound
to nascent C3b, since the convertases that generate nascent C3b are in
immediate vicinity (9,10). In addition, immunoglobulins are preferred
partners per se, since inactivation of C3 by an immune complex results in
C3b deposition not only to the immune complex, but also to serum/plasma IgG
(11), presumably because IgG is the second most prevalent component in these
fluids and has an affinity for intact C3 (12). This affinity is, however,
not equally present on all IgG molecules, it varies among naturally

occurring antibodies and was 7–10 times higher for anti–band 3 than anti-spectrin (1,13). The binding site for C3 on anti–band 3 must differ from the antigen binding site as was concluded from one step ELISA assays using covalently immobilized C3 or band 3 protein (14): binding of anti–band 3 measured by band–3–phosphatase conjugate gave a ten time higher signal when C3 instead of band 3 protein was immobilized (15). This affinity to C3 increases the probability for a C3b deposition to anti–band 3 (4,13) and explains the preferred formation of C3b–anti–band 3 complexes once C3b deposition is induced by aggregated anti–band 3 on oligomerized band 3 protein.

b) Once a C3b–IgG complex is bound to target cells via the Fab portion of its IgG moiety, it has a high opsonic potency, since it triggers concomitantly FcR and CR1 on phagocytes (6). Moreover, phagocytic processes mediated by C3b–IgG complexes are not inhibited by serum IgG (5). In addition, C3b–IgG complexes are powerful activators of alternative pathway C3b deposition, when bound to target cells via IgG to cellular antigens rather than to C3b receptors (CR1). These complexes are up to 30 times less sensitive to inactivation by factor I and H, instead they can nucleate a C3 convertase as easily as C3b on its own (16). While nucleation of a new convertase is largely prevented on red cells by decay accelerating factor (17), formation of a potent opsonin is undoubtedly responsible for the anti–band 3 dependent stimulation of C3b deposition and phagocytosis of oxidatively stressed (1,18) and senescent red cells (4).

One aspect has so far remained elusive, the role of fluid phase C3b–anti–band 3 complexes. Based on their affinity for C3 and theoretical considerations, anti–band 3 antibodies are among fluid phase IgGs that are hit preferentially by nascent C3b. Since such complexes can be generated by any type of IC–induced complement activation (11), it appears relevant to ask whether fluid phase C3b–anti–band 3 complexes can bind to red cells or oxidatively stressed red cells and whether the bound complexes are capable of stimulating C3b deposition and phagocytosis. If this were true, IC–induced complement activation might result in accelerated clearance of red cells via generation of fluid phase C3b–anti–band 3 complexes. A process of this kind seems likely, since artificially generated C3b–IgG complexes from induced antibodies to bacterial surface antigens, when added to the fluid phase, bound to the target cell, stimulated C3b deposition, and were potent opsonins (7). Thus, the strategy has been to generate C3b–IgG complexes with anti–band 3 antibodies and study their properties. In vitro formation of C3b–IgG complexes in good yields is, however, not trivial (16). It requires a huge excess and high concentrations of IgG which was not available in case of naturally occurring anti–band 3 antibodies (19). We therefor generated artificial complexes by coupling the two proteins with a heterobifunctional cross-linker, N–succinimidyl–3– (2–pyridyldithio) propionate (SPDP), according to a published procedure that has generated C3b–x–IgG complexes with properties comparable to those of ester–linked C3b–IgG complexes (20). We studied the functional properties of C3b–x–IgG complexes generated from anti–band 3 antibodies and several control antibodies.

MATERIALS AND METHODS

Antibodies and C3b-x-IgG complexes

Naturally occuring anti-band 3 and anti-spectrin antibodies were purified from human IgG SRK (Sandoglobulin[R]) as described earlier (19,21). Dialyzed Sandoglobulin served as human IgG (IgG) and the fraction of IgG which did not bind to the spectrin and band 3 affinity columns represented IgG depleted of these antibodies (IgG⁻) C3 was purified (22) and either used for C3b-x-IgG complex formation or for binding assays. In the latter case it was ^{125}I-iodinated to a specific activity of 5-10 x 10^6 cpm/µ₃ by using chloramine T as oxidant for 45 s at room temperature (18). C3b was prepared by trypsin and separated from C3 by gel filtration on ACA 34. Artificial C3b-IgG complexes were generated by cross-linking C3b to purified IgG, naturally occuring anti-band 3, anti-spectrin antibodies, and IgG-with SPDP. We used the procedure from Reiter and Fishelson (20), except that the molar excess of SPDP was 8.5 instead of 5.5. C3b-IgG complexes obtained by SPDP cross-linking are abbreviated as C3b-x-IgG. Monoclonal anti-IgG (HB43) (23), anti-C3b antibodies (7C12) (24) were prepared as outlined, and ^{125}I-iodinated by using the IODO GEN method to a specific activity of 2 x 10^5 to 2 x 10^6 cpm/µg.

Human red blood cells were freed from white cells, washed, and treated with 200 µM diamide where indicated (18). In some experiments complement receptor 1 (CR1) was blocked by incubating red cells for 30 min at room temperature with saturating concentrations of either monoclonal anti-CR1 antibody J3D3 (a gift from Prof. M. Kazatchkine, Paris) (25) or 1B4 (26) in phosphate buffered saline containing 1% bovine serum albumin (PBS-BSA). After pelleting cells were washed once with PBS-BSA and once with PBS or PBS containing 0.1% gelatin (PBS-gel).

Binding studies with C3b-x-IgG complexes

Samples (50 or 75 µl) of washed red cells were made 40% hct in PBS and mixed with equal volumes of a solution containing C3b-x-IgG complexes and free anti-band 3, where indicated and PBS-gel. Suspensions were incubated and agitated for 30 min at room teperature, diluted with 1 ml of gelatin containing, isotonic veronal buffer (GVB⁺⁺ or EGTA-GVB buffer) (23), and centrifuged at 4°C. Pelleted red cells were either washed 3 times with PBS-BSA to determine bound IgG and C3b or were further opsonized for 30 min at 37°C with whole serum or EGTA-serum.The reaction was stopped by 3 washes with PBS-BSA and bound IgG and C3b were determined. Controls were similarly incubated with GVB⁺⁺ or EGTA-GVB respectively.

Measurement of bound IgG and C3b

Washed red cells were incubated for 1 h at room temperature with gentle agitation with saturating concentrations of mouse monoclonal ^{125}I-iodinated anti-IgG and anti-C3b in PBS-BSA. Bound label was determined by overlaying the suspension on 200 µl of phtalate oils, pelleting the cells, and measuring radioactivity in the pellet (23).

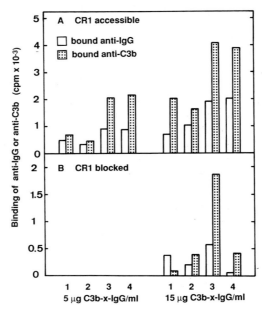

Fig. 1. Binding of C3b-x-IgG complexes to red cells that were pretreated
with 200 μM diamide in (A) and in (B) also with anti-CR1. Red cells
pretreated as indicated were incubated with 5 and 15 μg/ml of
C3b-x-IgG complexes generated with whole IgG (1), IgG⁻ (2), anti-
band 3 (3), and anti-spectrin antibodies (4). Binding of complexes
was determined by measuring the amount of bound, ^{125}I-iodinated
anti-IgG and anti-C3b monoclonal antibodies. Controls with buffer
alone were subtracted. Data in A represent the average of two
determinations from one out of 5 experiments yielding similar
results, those in B show the average of two independent experiments.

Opsonization with ^{125}I-iodinated C3

Red cells preincubated with C3b-x-IgG complexes were either washed once
or directly incubated for 30 min at 37°C in EGTA-serum or whole serum (final
concentration 15-20%) to which ^{125}I-iodinated C3 was added before starting
the reaction (1-2 x 10⁶ cpm per assay). Reactions were stopped by diluting
the reaction mixture 4-fold with cold PBS containing 10 mM EDTA. Bound
C3/C3b was determined either by centrifuging red cells through a cushion of
Ficoll in EDTA containing PBS (18) or phtalate esters (23). The two assays
yielded comparable results.

RESULTS

Binding of C3b-x-IgG complexes to oxidatively stressed red cells was
high for complexes containing anti-band 3 anti-spectrin antibodies and less
than half for those containing IgG or IgG⁻ (Fig. 1A). Bound material in
excess of that acquired _in vivo_ (control subtracted) indeed contained C3b
and IgG, since bound radioactivity increased for both types of probes
correspondingly. Binding was dose dependent and further increased beyond

that shown at 15 µg/ml. Since C3b–x–IgG complexes can interact by various ways with red cells, the apparent differences in their binding are not indicative of the type of binding. More information was obtained from incubating these complexes with red cells that were pretreated with monoclonal anti–CR1 antibodies which block binding of C3b to CR1 (25) (Fig. 1B). This greatly reduced binding for three of the complexes as judged from bound anti–C3b and lowered binding for that containing anti–band 3. The result strongly suggests that a portion of C3B–x–anti–band 3 complexes bound via the antibody moiety to CR1-blocked red cells. This was confirmed by experiments in which free anti–band 3 antibodies were added simultaneously with complexes. Anti–band 3 inhibited complex binding completely at less than half the amount of that added in form of the complex (not shown).

Though C3b–x–anti–band 3 complexes bound to red cells via their Fab portion, their binding was weak, since incubation of complex-containing red cells with EGTA-serum dissociated the acquired complexes almost completely (Fig. 2).

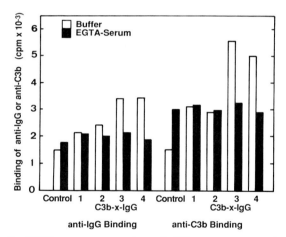

Fig. 2. Effect of EGTA-serum on bound C3b–x–IgG complexes. C3b–x–IgG complexes were bound to red cells pretreated with 200 µM diamide. Controls were incubated with PBS instead of C3b–x–IgG complexes. Cells were washed once and either incubated with GVB–EGTA buffer or EGTA serum. Bound C3b and IgG were determined. The resuls shown in Fig. 1 A and Fig. 2 are from the same experiment. Similar results were obtained in 4 other experiments. Samples are numbered as in Fig. 1.

Thus, it was not surprising that red cells tagged with C3b–x–IgG complexes in buffer were not capable of stimulating alternative pathway C3b deposition significantly (Fig. 3). In fact, C3b deposition was slightly increased at the lowest concentration of complexes, but dropped back to or even below the values obtained with control cells kept in buffer. Concentrations of C3b–x–anti–band 3 complexes lower than 1 µg/ml also stimulated C3b deposition to a similar extent (not shown). Thus, C3b-x-anti-band 3 complexes bound weakly to red cells and therefore were not capable of facilitating C3b-deposition as was expected from C3b-IgG complexes from induced antibodies.

DISCUSSION

Unlike C3b-IgG complexes made from induced antibodies to bacterial surface antigen (6,5), C3b-x-anti-band 3 complexes did not form a potent opsonin on oxidatively stressed red cells, when added to the fluid phase. Though they bound to oxidatively stressed red cells via their Fab portion in buffer, they were displaced by EGTA serum, were not able to stimulate C3b-deposition significantly and thus bound weakly. This finding was unexpected, since in vivo aged red cells contained elevated amounts of C3b-anti-band 3 complexes despite their continuous contact with plasma. The complexes of senescent red cells bound strongly, since they were not eluted by 4 isotonic washings and 3 hypotonic washings to generate membranes (3,4). The most obvious explanation is the following: anti-band 3 antibodies have a low affinity (19) and bind strongly to oligomerized band 3 protein (18,19,27). Once bound bivalently and in clusters, they withstand washing and serum addition and they acquire C3b via both complement pathways (Fig. 4A).

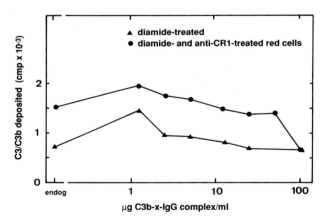

Fig. 3. C3b-deposition to red cells that acquired C3b-x-IgG complexes during preincubation in buffer. Red cells treated with 200 μM diamide (filled triangles) and also with anti-CR1 (filled circles) were preincubated with increasing concentrations C3b-x-IgG complexes as given. EGTA-serum (final concentration 15%) containing ^{125}I-iodinated C3 was then added without removing unbound complexes. Samples were osonized, diluted and bound material counted as given. Results are the average of two determinations of one out of 3 experiments yielding similar results.

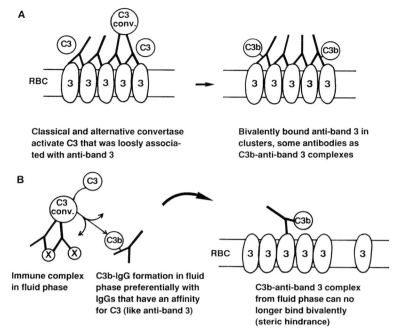

A

C3 conv.

C3   C3

RBC

Classical and alternative convertase
activate C3 that was loosly associated with anti-band 3

C3b   C3b

Bivalently bound anti-band 3 in
clusters, some antibodies as
C3b-anti-band 3 complexes

B

C3

C3 conv.

C3b

X   X

Immune complex   C3b-IgG formation in fluid
in fluid phase   phase preferentially with
                   IgGs that have an affinity
                   for C3 (like anti-band 3)

C3b

RBC

C3b-anti-band 3 complex
from fluid phase can no
longer bind bivalently
(steric hindrance)

Fig. 4. Hypothetical scheme of how C3b-anti-band 3 complexes are generated
and how their opsonic potency depends on where they were formed.
This figure summarizes the present understanding of how anti-band 3
antibodies and C3b interact with red cells when C3b-anti-band 3
complexes are generated on red cell bound anti-band 3 clusters (A)
or on fluid phase anti-band 3, e.g. by artificial generation or in
the course of immune complex-induced C3b deposition to serum IgG
(B). The scheme is based on the assumption that real C3b-anti-band 3
complexes have similar properties as C3b-x-anti-band 3 complexes.
It differs from that published earlier (4) which did not specify
between a monovalently and bivalently bound C3b-anti-band 3 complex.
Except for "C3 conv." which means C3 convertase, the figure is self
explanatory.

In contrast to this, bivalent interaction with oligomerized band 3
protein might be sterically impaired for preformed C3b-x-anti-band 3
complexes. Thus, C3b-x-anti-band 3 complexes added to the fluid phase may
bind exclusively with only one Fab portion and are therefore easily
displaced by serum concentrations of free anti-band 3 antibodies (Fig. 4B).
This seems likely, since competitive binding experiments in buffer showed
that free anti-band 3 inhibited binding at concentrations which were lower
than half that present in C3b-anti-band 3 complexes. It can not, however, be
excluded that the inability of C3b-x-anti-band 3 complexes to bind
bivalently was due to the fact that the antibodies were cross-linked to C3b
by SPDP. C3b and IgG might be bound randomly in artifical complexes rather
than at the site mediated by the opened thioester bond of the α chain of
C3b in real C3b-IgG complexes. This argument calls for studies with the real
C3b-anti-band 3 complexes which are under way. On the other hand, the

argument is weak particularly in case of anti-band 3 antibodies that have an affinity for C3 (13,15). Thus, it is conceivable that this affinity directs a proper interaction of C3b with anti-band 3 antibodies also in case of artificial complex formation which is a slow process allowing mutual interactions to occur.

In case these findings can be reproduced and verified with real C3b-anti-band 3 complexes they would be of utmost relevance to our understanding of red cell clearance. They imply that IC-induced C3b-deposition to serum IgG and in particular to anti-band 3 will not generate species that are vulnerable to red cells. Experiments aimed at testing this prediction confirmed it: "innocent bystander opsonization" of red cells by C3b was not increased by elevated concentrations of exogenous anti-band 3 in serum (Lutz, Ernest and Stammler, unpublished). The reason for the inability of monovalently bound C3b-x-band 3 complexes to act as strong opsonins is their poor affinity to band 3 protein. Thus the findings further imply that any mutation which increases the affinity of anti-band 3 will render them more vulnerable. A slight increase in affinity may suffice to elict an IC-dependent opsonozation and accelerated clearance, while an even higher affinity may give rise to autoimmune hemolysis. This speculation is not at odds with recent findings suggesting that certain immune hemolytic anemia are indeed due to antibodies against band 3 protein (28). Moreover, Masouredis's group even found high antiband 3 reactivity immunoprecipitation in eluates from red cells of some apparently normal donors (1 in 2-5000 donors) (29). It is not known whether these antibodies shared similarities with naturally occurring anti-band 3 antibodies and whether they were signs of subsequent autoimmune hemolysis or predisposed for IC-induced hemolytic anemia.

ACKNOWLEDGEMENTS

This work was supported by grants to HUL from the Swiss National Science Foundation (31-25175.88), the Swiss Federal Institute of Technology (0.330.089.37/6), and the Central Laboratory of the Swiss Red Cross, Berne. We also acknowledge support by a NIH grant (AR 24083) to RPT which helped initiate this collaboration. Human IgG was a gift from the Central Laboratory of the Swiss Red Cross, Berne. We thank Dr. M. Kazatchkline for monoclonal anti-CR1 antibodies and Dr. S. P. Masouredis and coworkers for providing a preprint of their paper (reference 29).

REFERENCES

1. H. U. Lutz, S. Fasler, P. Stammler, F. Bussolino and P. Arese, Naturally occurring anti-band 3 antibodies and complement in phagocytosis of oxidatively-stressed and in clearance of senescent red cells, Blood Cells 14:175 (1988).
2. H. U. Lutz, P. Stammler, C. Further and S. Fasler, "Anti-Bande 3"-Antikorper aktivieren Komplement über den alternative Weg, Schweiz. med. Wschr. 117:1821 (1987).

3. S. Flaser, PhD Thesis No. 8777 ETH, Charakterisierung erythrozyten-assoziarter IgG-und Komplementfaktoren, ETH-Zurich:, 1989.

4. H. U. Lutz, Erythrocyte Clearence, in "Blood Cell Biochemistry, 1 Erythroid Cells", J. R. Harris, Plenum Press, New York and London (1990) p. 81.

5. L. F. Fries, S. A. Siwik, A. Malbran and M. M. Frank, Phagocytosis of target particles bearing C3b-IgG covalent complexes by human monocytes and polymorphonuclear leucocytes, Immunology 62:45 (1987).

6. A. Malbran, M. M. Frank and L. F. Fries, Interactions of monomeric IgG bearing covalently bound C3b with polymorphonuclear leucocytes, Immunology 61:15 (1987).

7. K. A. Joiner, L. F. Fries, M. A. Schmetz and M. M. Frank, IgG bearing covalently bound C3b has enhanced bactericidal activity for Escherichia coli O111, J. Exp. Med. 162:877 (1985).

8. R. B. Sim, T. M. Twose, D. S. Paterson and E. Sim, The covalent-binding reaction of complement component C3, Biochem. J. 193:115 (1981).

9. Y. Takata, N. Tamura and T. Fujita, Interaction of C3 with antigen-antibody complexes in the process of solubilization of immune precipitates, J. Immunol. 132:2531 (1984).

10. K. J.Gadd and K. B. M. Reid, The binding of complement component C3 to antibody-antigen aggregates after activation of the alternative pathway in human serum, Biochem. J. 195:471 (1981).

11. R. J. Jacobs and M. Reichlin, Generation of low M. W., C3-bearing immunoglobulin in human serum, J. Immunol. 130:2775 (1983).

12. J. Kulics, E. Rajnavölgyi, G. Füst and J. Gergely, Interaction of C3 and C3b with immunoglobulin G, Mol. Immunol. 20:805 (1983).

13. H. U. Lutz, S. Fasler and P. Stammler, An affinity for complement C3 as a possible reason for the potency of naturally occurring antibodies in mediating tissue homeostasis, in: "Beiträge zur Infusionstherapie, "K. H. Bässler, A. Grünert, H. Reissigl and K. Widlam, Karger, Basel (1989).

14. H. U. Lutz, P. Stammler and E. A. Fischer, Covalent binding of detergent-solubilized membrane glycoproteins to 'Chemobond" plates for ELISA, J. Immunol. Meth. 129:211 (1990).

15. H. U. Lutz, P. Stammler and S. Fasler, How naturally occurring anti-band 3-antibodies stimulate C3b deposition to senescent and oxidatively stressed red blood cells, Biomed. Biochim. Acta 49:224 (1990).

16. L. F. Fries, T. A. Gaither, C. H. Hammer and M. M. Frank, C3b covalently bound to IgG demonstrates a reduced rate of inactivation by factors H and I, J. Exp. Med. 160:1640 (1984).

17. A. Nicholson-Weller, J. Burge, D. T. Fearon, P. F. Weller and K. F. Austen, Isolation of a human erythrocyte membrane glycoprotein with decay-accelerating activity for C3 convertases of the complement system, J. Immunol. 129:184 (1982).

18. H. U. Lutz, F. Bussolino, R. Flepp, S. Fasler, P. Stammler, M. D. Kazatchkine and P. Arese, Naturally occurring anti-band 3 antibodies and complement together mediate phagocytosis of oxidatively stressed human red blood cells, Proc. Natl. Acad. Sci. USA 84:7368 (1987).

19. H. U. Lutz, R. Flepp and G. Stringaro-Wipf, Naturally occurring autoantibodies to exoplasmic and cryptic regions of band 3 protein of human red blood cells, J. Immunol. 133:2610 (1984).

20. Y. Reiter and Z. Fishelson, Targeting of complement to tumor cells by heteroconjugates composed of antibodies and of the complement component C3b, J. Immunol. 142:2771 (1989).

21. H. U. Lutz and G. Wipf, Naturally occurring autoantibodies to skeletal proteins from human red blood cells, J. Immunol. 128:1695 (1982).

22. C. H. Hammer, G. H. Wirtz, L. Renfer, H. D. Gresham and B. F. Tack, Large scale isolation of functionally active components of the human complement system, J. Biol. Chem. 256:3995 (1980).

23. J. C. Edberg, L. Tosic, E. L. Wright, W. M. Sutherland and R. P. Taylor, Quantitative analyses of the relationship between C3 consumption, C3b capture, and immune adherence of complement-fixing antibody/DNA immune complexes, J. Immunol. 141:4258 (1988).

24. L. Tosic, W. M. Sutherland, J. Kurek, J. C. Edberg and R. P. Taylor, Preparation of monoclonal antibodies to C3b immunization with c3b(i)-Sepharose, J. Immunol. Meth. 120:241 (1989).

25. J. Cook, E. Fischer, C. Boucheix, M. Mirsrahi, M.-H. Jouvin, L. Weiss, R. M. Jack and M. D. Kazatchkine, Mouse monoclonal antibodies to the human C3b receptor, Mol. Immunol. 22:531 (1985).

26. J. C. Edberg, E. Wright and R. T. Taylor, Quantitative analyses of the binding of soluble complement-fixing antibody/dsDNA immune complexes to CR1 on human red blood cells, J. Immunol. 139:3739 (1987).

27. M. Beppu, A. Mizukami, M. Nagoya and K. Kikugawa, Binding of anti-band 3 autoantibody to oxidatively damaged erythrocytes, J. Biol. Chem. 265: 3226 (1990).

28. E. J. Victoria, S. W. Pierce, M. J. Branks and S. P. Masouredis, IgG red blood cell autoantibodies in autoimmune hemolytic anemia bind to epitopes on red blood cell membrane band 3 glycoprotein, J. Lab. Clin. Med. 115:74 (1990).

29. S. W. Pierce, E. J. Victoria and S. P. Masouredis, Red cell autoantibodies characterized by competitive inhibition of [125]I Rh alloantibody binding and by immunoprecipitation of membrane proteins, J. Lab. Clin. Med. in press (1990).

CONTRIBUTORS

Prof. Gino Amiconi
Dipartimento di Scienze Biochimiche
Università di Roma "La Sapienza"
Piazzale Aldo Moro, 5
00185 - ROMA

Prof. Paolo Arese
Dipartimento di Genetica,
Biologia e Chimica Medica
Università di Torino
Via Santena 5 bis
10126 - TORINO

Prof. Ernest Beutler
Department of Molecular and
Experimental Medicine
Scripps Clinic
10666 North Torrey Pines Road
92037 LA JOLLA, CA
U.S.A.

Prof. Augusta Brovelli
Dipartimento di Biochimica
Università di Pavia
Via Bassi, 21
27100 - PAVIA

Prof. Maria Domenica Cappellini
Istituto di Medicina Interna
Università di Milano
Via Bassi, 21
20122 - MILANO

Prof. Margaret Clark
School of Medicine
Cancer Research Institute
University of California
San Francisco - CA
U.S.A.

Prof. George L. Dale
Research Institute of Scripps Clinic
10666 Nothe Torrey Pines Road
92037 - LA JOLLA, CA

Prof. Clive Ellory
Department of Physiology
University Oxford
Parks Road
OXI 3PT OXFORD ENGLAND

Dr. Lucietta Ferroni
Istituto Superiore di Sanità
V.le Regina Elena, 299
ROMA

Prof. Antonio De Flora
Istituto di Chimica Biologica
Università degli Studi di Genova
V.le Benedetto XV, 1
16132 - GENOVA

Prof. Patrizia Galletti
Istituto di Biochimica
delle Macromolecole
I Facoltà di Medicina e Chirurgia
Via Costantinopoli, 16
80138 - NAPOLI

Prof. Liliane Gattegno
School of Medicine
74, Rue Marcel Cachin
93012 Bobigny
FRANCE

Prof. Harriet Gershon
Technion Faculty of Medicine
P.O.B. 9649
31096 - HAIFA ISRAEL

Prof. Bruno Giardina
Dipartimento di Medicina Sperimentale
Università Tor Vergata
Via O. Raimondo, 1
00173 - ROMA

Prof. Arthur Haas
Department of Biochemistry
Medical College of Wisconsin
8701 Watertown Plank Road
53226 Milwaukee, Wisconsin U.S.A.
Dr. Sara Horn
Ben-Gurion University
of the Negev
BEER-SHEVA   ISRAEL

Dott. Diego Ingrosso
Istituto di Biochimica
delle Macromolecole
Università di Napoli
Via Costantinopoli, 16
80138 - NAPOLI

Prof. Rose M. Johnstone
Dept. of Biochemistry
McGill University
3655 Drummond
H3GIY6 - MONTREAL   CANADA

Prof. Marguerite M.B. Kay
Department of Internal Medicine
College of Medicine
Texas A&M University
1901 S. First Street
76504 - TEMPLE - TEXAS U.S.A.

Prof. Philip S. Low
Herbert C. Brown
Laboratory of Chemistry
Pardue University
47097-369 - WEST LAFAYETTE - IN U.S.A.

Prof. Hans Lutz
Laboratorium f. Biochemie
ETH-Zeutrum
CH 8092 - ZURICH SWITZERLAND

Prof. Mauro Magnani
Istituto di Chimica Biologica
Università di Urbino
Via A. Saffi, 2
61029 - URBINO (PS) ITALY

Prof. Laura Mazzanti
Istituto di Biochimica
Facoltà di Medicina e Chirurgia
Università di Ancona
Via Ranieri
60131 - ANCONA   ITALY

Dott. Vanna Micheli
Istituto di Chimica Biologica
Università di Siena
Pian dei Mantellini, 44
53100 - SIENA   ITALY

Dr. Andrea Mosca
Dipartimento Scienze e
Tecnologie Biomediche
Via Olgettina, 60
20123 - MILANO   ITALY

Dr. Koko Murakami
College of Physicians of Surgeous
Columbia University
Division of Pediatric
Hematology Oncology
10032 - NEW YORK   U.S.A.

Prof. Paolo Antonio Nassi
Dipartimento di Scienze Biochimiche
Università degli Studi di Firenze
V.le Morgagni, 50
FIRENZE   ITALY

Prof. Sergio Piomelli
College of Physicians of Surgeous
Columbia University
Division of Pediatric
Hematology Oncology
10032 - NEW YORK   U.S.A.

Prof. Michele Samaja
Dip. Scienze e Tecnologie Biomediche
Ist. Scient. San Raffaele
Via Olgettina, 60
20132 - MILANO   ITALY

Prof. Gerald E.J. Staal
Department of Haematology
University Hospital Ultrecht
P.O. Box 85500
3508 GA UTRECHT   THE NETHERLANDS

Prof. Vilberto Stocchi
Istituto di Chimica Biologica
Università di Urbino
Via A. Saffi, 2
61029 – URBINO (PS) ITALY

Prof. Emanuel E. Strehler
ETH Laboratorium fur Biochemie
CH–8092 ZURICH SWITHZERLAND

Prof. Wiliam Valentine
Department of Medicine
Ucla School of Medicine
Center for the Health Sciences
10833 Le Conte Avenue
80024 173 – LOS ANGELES – CA U.S.A.